CONTENTS

Primatology in Regards to Reported Global Primate Megafauna

Dakota Frandsen

First Printing, 2026

1

The Allure of the Unknown Primate

The human fascination with the unknown primate, with its echoes of our own evolutionary past and its shadowy presence in the fringes of our world, is far from a modern phenomenon. For millennia, tales of large, ape-like beings have woven through the tapestry of human culture, whispered around campfires, etched into ancient lore, and passed down through generations. These "whispers from the wilderness" offer a fascinating, albeit often fragmented, historical record, hinting at a persistent belief in creatures that defy easy categorization, creatures that occupy the liminal space between the known and the speculative. As we embark on a scientific inquiry into these persistent narratives, it is crucial to acknowledge the deep roots of this fascination, to understand how these stories have evolved, and to identify the recurring themes that suggest something more than mere fantasy.

One of the earliest documented accounts that might resonate with our modern understanding of large, unknown primates comes from ancient Mesopotamia. The Epic of Gilgamesh, a sweeping epic poem dating back to the third millennium BCE, describes a wild man named Enkidu. Initially, Enkidu lives amongst the beasts of the field, his body covered in hair, his diet comprising grass and the milk of wild animals. He is depicted as a powerful, hirsute being, initially untamed and uncivilized, who eventually encounters humanity. While Enkidu is eventually integrated into human society, his primal origins and wild nature bear a striking resemblance to the archetypal wild man or ape-like creature that recurs in folklore worldwide. This early literary depiction suggests that even in antiquity, humanity pondered the existence of beings that blurred the lines between human and animal, between civilization and the wild.

Moving further East, ancient Greek mythology offers figures like the satyrs and, perhaps more relevantly, the Cyclops. While not strictly ape-like, the Cyclops, particularly Polyphemus, is described as a giant, brutish, and solitary creature inhabiting a remote cave, a being of immense physical power that preys on travelers. The emphasis on size, isolation, and a fearsome presence in a wild environment aligns with some of the characteristics attributed to later ape-man cryptids. Furthermore, the concept of the "wild man" was prevalent across various Indo-European cultures. These figures, often depicted as powerfully built, hairy individuals living in forests or mountains and sometimes exhibiting brutish or bestial behavior, can be seen as precursors to later descriptions of ape-like beings. They represent a primal, untamed aspect of nature, a reflection of humanity's own wilder origins and the potent mystery of the untamed landscape.

The rich tapestry of indigenous cultures around the globe provides a particularly fertile ground for exploring historical accounts of large, unknown primates. These oral traditions, passed down through countless generations, often contain detailed descriptions of creatures that bear a striking resemblance to modern cryptids. In the dense forests and vast wildernesses of North America, numerous Native American tribes have long spoken of large, hairy bipeds inhabiting the deep woods. These beings are often described with remarkable consistency: tall, powerfully built, covered in dark hair, and possessing immense strength. They are sometimes portrayed as shy and elusive, avoiding human contact, while at other times as territorial or even aggressive.

For example, the Pacific Northwest, a region now synonymous with Bigfoot sightings, has a long history of indigenous legends about such creatures. The Coast Salish people, for instance, have tales of "Sasquatch," a term that has become globally recognized for the phenomenon. These accounts often describe a creature that walks upright, is covered in hair, and inhabits remote, forested areas. Similarly, indigenous peoples across North America, from the Cree of Canada to the Cherokee of the southeastern United States, have their own variations of these ape-like beings, often incorporating them into their creation stories, spiritual beliefs, and cautionary tales. These traditions are not merely fanciful stories; they often serve as a reflection of the deep ecological knowledge held by these communities, understanding the subtle signs of life in their environment and the potential for its hidden inhabitants.

In the high Himalayas, the legend of the Yeti, or Abominable Snowman, has captured the global imagination for centuries. This creature is typically described as a large, ape-like being with thick, shaggy fur, inhabiting the treacherous, snow-covered mountains. Local folklore, particularly among the Sherpa people, speaks of "metoh-kangmi," or "snowman," a creature that is both feared and respected. These accounts, often intertwined with spiritual beliefs about mountain deities and spirits, describe a powerful, elusive being adapted to the harsh alpine environment. Early Western explorers encountering these legends in the late 19th and early 20th centuries documented these tales, bringing them to international attention and igniting a sustained curiosity. The consistency in descriptions from diverse sources within the region—different ethnic groups and villages sharing remarkably similar accounts over extended periods—lends a degree of historical weight to these narratives.

Further afield, in the dense rainforests of Central Africa, pygmy populations have long spoken of creatures that align with our concept of large, unknown primates. Stories of the "Mokélé-mbembé," a creature often described as a sauropod-like reptile, are well-known, but alongside these are accounts of large, ape-like beings. These creatures, sometimes referred to by various local names, are often described as more powerful and less arboreal than known apes, inhabiting remote swamps and forests. While these accounts are often filtered through the lens of scientific inquiry focused on dinosaurs, the underlying descriptions of large, powerful, and elusive forest dwellers resonate with the broader theme of unknown megafauna primates.

Australia, too, has its own rich tradition of cryptid lore, with the Yowie being the most prominent example. Aboriginal Dreamtime stories, predating European settlement by tens of thousands of years, contain references to large, hairy, bipedal beings inhabiting the continent's wilder regions. These creatures are often depicted as powerful, territorial, and sometimes fearsome, occupying a niche within the ancient Australian landscape. European settlers and subsequent generations have continued to report sightings of the Yowie, describing a creature similar in build and appearance to North American Bigfoot, adapted to the diverse Australian environment, from dense rainforests to rugged bushland. The longevity and cultural pervasiveness of these stories among indigenous communities suggest a deep-seated understanding of, or at least a strong belief in, such beings within their ancestral lands.

The recurring themes across these diverse cultures and geographical locations are striking and form the bedrock of our investigation. The consistent description of a large, bipedal, hairy, ape-like creature is perhaps the most prominent. While the specific details might vary—the color of the fur, the exact height, the temperament of the creature—the fundamental archetype remains remarkably similar. This consistency, spanning continents, cultures, and millennia, challenges the notion that all such reports can be dismissed as mere fantasy or misidentification. It suggests that there may be an underlying phenomenon that has been perceived and interpreted in various ways by different societies.

Another recurring theme is the creature's elusiveness and its association with remote, undisturbed wilderness areas. Whether it is the deep forests of the Pacific Northwest, the towering peaks of the Himalayas, the dense jungles of Central Africa, or the remote bushlands of Australia, these reports invariably emerge from regions where human presence is minimal, and the natural environment remains largely intact. This ecological correlation is significant. Large animals, especially those with substantial dietary and spatial needs, require extensive, unfragmented habitats to survive and maintain viable populations. The consistent association of these creatures with such environments lends an ecological plausibility to their potential existence, suggesting that if they do exist, they are likely masters of evasion in landscapes where they can remain hidden.

The intelligence and sometimes semi-humanoid characteristics attributed to these beings also form a consistent thread. They are rarely depicted as simple, unthinking beasts. Instead, many accounts suggest a degree of awareness, cunning, and even a capacity for deliberate avoidance of humans. This anthropomorphic quality, while often a source of sensationalism, also speaks to a deeper human psychological connection. We project onto these unknown entities qualities that resonate with our own sense of self, our own evolutionary trajectory, and our place in the natural world. The ape-like form itself is a potent symbol, a mirror reflecting our own origins and raising questions about the potential for other branches of our evolutionary tree to have persisted, hidden from our view.

Furthermore, the persistent presence of these legends in indigenous cultures, often integrated into spiritual or cosmological frameworks, suggests that for these communities, these beings are not simply curiosities but integral parts of their ancestral lands and their understanding of the natural world. These oral traditions represent a form of embodied knowledge, passed down through practice and lived experience, and their longevity and consistency cannot be easily dismissed. While scientific methodology demands empirical evidence, it would be remiss to ignore the depth and breadth of these historical narratives, which have often predated formal scientific investigation by centuries, if not millennia.

The evolution of these tales also offers insight. Early accounts often described a more primal "wild man" figure, perhaps a reflection of early human encounters with the untamed aspects of nature or with early hominid species. As human understanding of biology and primatology advanced, and as scientific expeditions penetrated remote regions, the descriptions began to coalesce more specifically around the form of a large ape. This evolution in description, while potentially influenced by emerging scientific knowledge, also suggests a refinement of observations and interpretations, moving from general concepts of the "wild" to more specific biological possibilities.

It is crucial, however, to approach these historical accounts with a critical yet open mind. We must acknowledge the potential for misidentification of known animals, the influence of folklore and myth, the impact of cultural biases, and, in more recent times, the possibility of hoaxes. Yet, within the vast corpus of these ancient narratives and anecdotal reports, there exists a remarkable degree of consistency. This consistency, spanning vast geographical distances and temporal scales, forms the historical backdrop against which modern scientific inquiry into the possibility of unknown primates must be framed. These whispers from the wilderness, these echoes of a hidden primate, have persisted for so long because they tap into something fundamental in the human psyche—a curiosity about the unknown, a respect for the wild, and a deep-seated wonder about the diversity of life that may still exist beyond the edges of our explored world. They set the stage not for immediate acceptance, but for rigorous investigation, for a scientific quest to discern the verifiable from the fanciful, and to explore the possibility that these ancient tales might indeed hold a kernel of biological truth. By examining these historical whispers, we begin to understand the enduring allure of the unknown primate and the profound human fascination that drives our quest to find it.

The enduring human narrative of the unknown primate, a thread woven through history and across cultures, naturally leads us to a crucial question: what exactly are we referring to when we speak of these elusive creatures? Within the context of this exploration, we are not broadly discussing any and all mysterious animals. Instead, we are focusing our attention on a specific category that, for the sake of clarity and scientific inquiry, we shall define as **cryptid megafauna primates**. This designation serves to delineate our subject matter from a multitude of other cryptids, such as lake monsters, winged beasts, or reptilian entities, which fall outside the purview of our primatological and ecological focus.

Our definition of cryptid megafauna primates centers on a distinct set of characteristics that consistently emerge from a vast array of anecdotal reports, historical accounts, and indigenous traditions. At its core, this definition encompasses large, bipedal, ape-like beings. The "megafauna" aspect signifies their considerable size, distinguishing them from known primate species such as chimpanzees, gorillas, or orangutans, and placing them in a size category comparable to or exceeding the largest extant primates. The "primate" designation is derived from their predominantly ape-like morphology, as reported, and the implication that they share a common ancestry with other primates, even if their evolutionary path has diverged significantly and remained undiscovered by science.

The most pervasive and defining characteristic across numerous descriptions is their

bipedal locomotion. While some anecdotal evidence might suggest quadrupedal movement, the overwhelming majority of credible accounts, particularly those from regions like the Pacific Northwest of North America and the Himalayas, describe creatures that habitually walk upright on two legs. This bipedalism is often described as powerful and deliberate, sometimes with a gait that differs subtly from human walking, perhaps appearing more lumbering or knuckle-dragging when observed in certain contexts, though fundamentally upright. This upright stance is a key feature that distinguishes them from known quadrupedal apes and suggests a unique evolutionary adaptation.

Complementing their bipedalism is their

hirsute nature. Reports consistently describe these beings as being covered in thick, dense hair or fur, which varies in color from dark brown, black, and reddish-brown to, in the case of the Yeti, a paler, grayish-white or sandy hue. The length and texture of this hair are also frequently mentioned, with descriptions ranging from short, coarse fur to long, shaggy locks. This dense pelage is interpreted as an adaptation to cold or temperate climates, providing insulation against harsh environmental conditions, a trait that aligns with the ecological niches these creatures are reported to inhabit.

The

ape-like morphology is another critical component of our definition. While precise anatomical details are often scarce and vary, common descriptions include a broad, powerful frame, muscular build, a relatively short neck, a large head, and a pronounced brow ridge. The facial features are often described as ape-like, with a prognathous jaw and a less developed cranial capacity compared to humans, though often perceived as possessing a degree of intelligence. The hands and feet are typically depicted as large and robust, adapted for terrestrial locomotion and potentially for grasping. The absence of a tail is another consistently reported feature, aligning them with the Hominoidea superfamily (apes), rather than the Cercopithecoidea (monkeys).

Size is a fundamental aspect of the "megafauna" designation. Reports commonly place the height of these beings between six and ten feet (approximately 1.8 to 3 meters), with some accounts even suggesting larger individuals. Their weight is correspondingly immense, often estimated to be several hundred kilograms. This sheer scale sets them apart from any known primate species and implies substantial dietary and habitat requirements, which are crucial considerations for ecological plausibility.

It is vital to differentiate these cryptid megafauna primates from other types of reported phenomena. For instance, while Bigfoot or Yeti narratives sometimes incorporate elements of the supernatural or shamanic beliefs, our focus remains on the

biological entity. We are seeking to identify a potentially undiscovered species of primate, not a spirit or a mythical creature in the purely supernatural sense, although acknowledging the cultural significance of such beliefs is important for understanding the historical context.

Furthermore, we must distinguish them from cases of misidentification of known animals. While a bear walking upright or a lone primate escaping from captivity can generate compelling eyewitness accounts, the persistent descriptions of unique morphology, gait, and behavior across vast geographical areas and timeframes suggest something beyond simple misidentification. The consistency in features like the supraorbital ridge, the bipedal gait, and the sheer size, often reported by experienced outdoorsmen, trackers, and indigenous peoples with intimate knowledge of their environment, lends weight to the hypothesis of an unknown species.

The term "ape-like" is used deliberately. While they share superficial similarities with known apes, the specific combination of traits – particularly the consistent bipedalism and the reported cranial features – suggests that if they are indeed primates, they represent a lineage distinct from the great apes we are familiar with. Some researchers hypothesize they could be a surviving member of an archaic hominin lineage, or a highly specialized branch of the ape family tree that evolved unique adaptations for terrestrial life in specific environments.

The geographical distribution of these reports is also a key definitional parameter. While similar creatures are reported worldwide, our primary focus within this book will be on the most widely documented and consistently described examples, such as Sasquatch/Bigfoot in North America, the Yeti in the Himalayas, and potentially the Yowie in Australia. These regions share characteristics of being vast, largely undeveloped wilderness areas, which are ecologically suitable for supporting large, elusive mammals.

To summarize, for the purposes of this book, a

cryptid megafauna primate is defined as a large, bipedal, ape-like creature, typically covered in hair, exhibiting a size and physical build significantly exceeding that of known primates. It is a being predominantly associated with remote, wild terrestrial environments and characterized by its elusiveness. This definition is not an assertion of existence, but rather a framework for scientific investigation, guiding our search for evidence and distinguishing our subject from other cryptid phenomena or misidentified known animals. It is this specific type of creature, persistently appearing in lore and anecdotal reports, that bridges the gap between ancient myth and the potential for modern scientific discovery, setting the stage for a rigorous examination of the evidence.

The transition from anecdotal accounts and folklore to a tangible, scientifically verifiable entity hinges on the ability to define and categorize these reported phenomena with enough precision to enable targeted research. The prevailing descriptions of what we are terming the cryptid megafauna primate—Sasquatch, Yeti, and their global counterparts—paint a remarkably consistent picture, despite the vast distances and diverse cultures from which these reports originate. This consistency is not merely coincidental; it forms the very foundation upon which a scientific inquiry must be built. Without a clear understanding of what we are looking for, the search becomes unfocused and prone to misinterpretation.

The cornerstone of this definition is undoubtedly the

bipedal locomotion. Unlike known great apes such as gorillas, orangutans, and chimpanzees, which are primarily quadrupedal or exhibit facultative bipedalism, the cryptid megafauna primate is almost universally depicted as habitually walking on two legs. Eyewitness accounts, particularly those from individuals with extensive experience in wilderness tracking and observation, often detail a gait that is distinct from human locomotion. It is frequently described as more lumbering, perhaps with a wider stance, or with a noticeable sway from side to side, suggesting a different spinal structure and pelvic anatomy compared to humans. This persistent observation of upright movement is a critical factor that distinguishes these creatures from any known ape species and places them within a hypothetical evolutionary niche occupied by creatures like extinct hominins, or perhaps a unique evolutionary divergence among apes. The very act of walking upright requires significant anatomical adaptations, including a specialized pelvis, a stable lumbar spine, and a well-balanced head, all of which would be indicative of a distinct biological entity. The robusticity of the reported leg bones and the powerful musculature associated with bipedalism in these descriptions further reinforce this defining characteristic.

Equally prominent in the descriptive lexicon is the

hirsute nature of these beings. Reports consistently emphasize a thick, dense covering of hair or fur, which serves as a primary visual identifier. The color spectrum is diverse, ranging from the dark, almost black or deep brown observed in North American Sasquatch reports, to the reddish-brown hues sometimes mentioned, and the pale, grayish-white or sandy coloration associated with the Himalayan Yeti. The texture of this pelage is also frequently commented upon, described as shaggy, coarse, or matted, suggesting adaptations for insulation and protection against the elements. This dense fur coat is a logical adaptation for survival in the often cold and rugged environments where these creatures are most frequently reported, such as dense forests, mountainous terrain, and snow-covered landscapes. The thickness and distribution of this hair would also be a significant differentiating factor from known primates, many of which have relatively sparse body hair compared to their overall size. The potential for seasonal shedding or variations in pelage density based on climate could also be considered within a biological framework.

The

ape-like morphology provides the primate designation. While not identical to extant apes, the general form is consistent with this classification. Key features often mentioned include a large, broad torso, powerful musculature, a relatively short neck that appears to merge directly into the head, and a large, sometimes somewhat flattened, facial structure. The cranial features are particularly intriguing, with repeated descriptions of a pronounced brow ridge (supraorbital torus) and a sloping forehead, reminiscent of fossil hominins or certain ape species. The jawline is often described as prognathous, meaning the jaw protrudes forward, and the teeth, though rarely observed directly, are presumed to be large and adapted for a diet that likely includes significant plant matter, but possibly also occasional animal protein. The hands and feet are consistently reported as large and powerfully built, with long digits, capable of grasping and providing significant traction. The absence of a tail is a universal characteristic in these descriptions, firmly placing them within the primate infraorder of Hominoidea, which includes humans, chimpanzees, gorillas, orangutans, and gibbons. This morphological consistency, spanning diverse geographical locations, suggests a common underlying biological template.

The sheer scale of these creatures, defining their "megafauna" status, is another crucial element. Reports consistently place their height well beyond that of any known living primate, typically ranging from 1.8 to over 3 meters (6 to 10 feet). This considerable stature implies a substantial body mass, with estimates often exceeding 200 kilograms (440 pounds) and sometimes reaching much higher. Such a size necessitates a significant territory, a consistent food supply, and an ecological niche that can sustain a large predator or omnivore. This aspect alone raises significant questions about their ability to remain undetected by modern human activities, such as logging, mining, and widespread recreation in wilderness areas. The ecological implications of such large-bodied animals are profound and are a key area for scientific consideration.

When distinguishing these cryptid megafauna primates from other phenomena, precision is paramount. We are not focusing on creatures that exhibit overtly reptilian traits, such as the popularly reported Mokélé-mbembé, which is often described as a sauropod-like dinosaur remnant. Nor are we concerned with aerial cryptids like the Mothman, or aquatic beings like lake monsters. Our definition strictly limits the scope to terrestrial, ape-like primates. This deliberate narrowing of focus allows for the application of existing primatological knowledge, ecological principles, and forensic analysis techniques.

The distinction from known animals is also critical. While many sightings can be attributed to misidentifications of bears, particularly when standing on hind legs, or other familiar wildlife, the persistent and detailed descriptions of unique morphological features—such as the brow ridge, cranial shape, and proportions of the limbs—suggest that these reports often go beyond simple errors in perception. The consistent reporting of a creature that is undeniably primate-like, yet distinctly not human or any known ape, by individuals with deep familiarity with wildlife, underscores the potential for a genuine unknown. Reports from indigenous communities, often imbued with spiritual or cultural significance, have described these beings for centuries, predating widespread Western exploration and scientific observation. These oral traditions represent a deep, ingrained understanding of the local fauna, and their consistent descriptions of such creatures warrant serious consideration.

The question of evolutionary lineage is central to understanding these beings. Are they a surviving species of extinct hominin, such as

Gigantopithecus, the giant ape that once roamed Asia? Or do they represent an entirely novel evolutionary branch within the primate family tree, possessing unique adaptations for their specific environments? The term "ape-like" allows for both possibilities. It acknowledges the superficial resemblance to known apes while leaving open the possibility of a more ancient or divergent lineage. This ambiguity is not a weakness but a reflection of the current state of knowledge; until definitive physical evidence is obtained, their precise taxonomic placement remains speculative.

The ecological context of these sightings is also a crucial part of our definition. These creatures are almost invariably reported from vast, remote, and relatively undisturbed wilderness areas. From the dense, temperate rainforests of the Pacific Northwest and the rugged, mountainous regions of the Himalayas, to the ancient forests of Australia and the dense jungles of Central Africa, these environments offer the necessary cover, food resources, and isolation to potentially support large, elusive mammals. The very "megafauna" designation implies a significant ecological footprint, requiring extensive habitats that are increasingly scarce in the modern world. The geographical distribution of these reports, therefore, is not random but strongly correlated with areas of substantial, contiguous wilderness.

In essence, the cryptid megafauna primate is not a vague notion of a "hairy man" or a generic "wild beast." It is a specific type of biological entity, defined by a consistent suite of physical and behavioral characteristics that align it with the primate order, albeit as an unknown and significantly larger form. This definition provides a concrete target for scientific investigation, allowing researchers to move beyond generalized speculation and towards hypothesis-driven research. It is this clearly defined, yet still mysterious, entity that captures the imagination and fuels the scientific quest to understand the hidden biodiversity of our planet. By establishing this working definition, we create a bridge between the historical whispers of folklore and the rigorous demands of scientific inquiry, preparing the ground for the critical analysis of evidence that lies ahead.

The allure of the unknown primate, as we have begun to define it, beckons us into realms where folklore and biology intertwine. Yet, for any exploration to transcend mere storytelling and enter the sphere of genuine scientific understanding, it must be anchored in a robust methodology. This is where the **scientific method** becomes our indispensable compass, guiding us through the often murky waters of anecdotal evidence and speculative claims. It is not a rigid dogma, but a dynamic process, a systematic approach to acquiring knowledge about the natural world, demanding evidence, logic, and a willingness to be proven wrong.

At its heart, the scientific method is a cyclical process of observation, hypothesis formation, prediction, experimentation, and analysis. We begin by observing phenomena that defy current explanation. In our context, these are the recurring patterns in sightings of large, bipedal, ape-like creatures, often in remote wilderness areas. These observations, gathered from a multitude of sources—from indigenous lore to modern eyewitness accounts—spark questions. Could these reports represent a biological reality, an as-yet-undiscovered species? This leads to the formation of a

hypothesis: a testable explanation for the observed phenomena. In this case, the hypothesis might be that a population of large, undiscovered hominoids, or a highly divergent ape lineage, exists in certain geographical regions.

However, a hypothesis, no matter how compelling, remains mere conjecture until it can be rigorously tested. This is where the principle of

falsifiability becomes paramount. A scientific hypothesis must be capable of being proven false. If a hypothesis cannot, in principle, be disproven, it falls outside the realm of science and into that of dogma or faith. For example, a hypothesis that states "invisible, undetectable unicorns live on the moon" is unfalsifiable because there is no conceivable observation or experiment that could definitively prove its falsehood. In contrast, a hypothesis such as "Sasquatch is a distinct biological species inhabiting the Pacific Northwest forests" is falsifiable. If, through extensive surveys, genetic analysis of environmental DNA, and thorough investigation of all reported sightings, no evidence supporting the existence of such a creature emerges, then the hypothesis can be considered falsified, or at least, not supported by current data.

The scientific method demands empirical evidence – data that can be observed, measured, and verified. This is where the study of cryptid megafauna primates faces its most significant challenges. The evidence typically presented consists of eyewitness testimonies, blurry photographs, ambiguous footprints, and occasionally, anecdotal accounts of vocalizations or scat. While these can be valuable starting points, they are inherently subjective and prone to misinterpretation, hallucination, or fabrication. The scientific approach necessitates moving beyond these initial reports to seek concrete, verifiable evidence. This might include:

Physical specimens: A carcass, bones, or even live capture would provide irrefutable biological material for study. This would allow for detailed anatomical analysis, genetic sequencing, and taxonomic classification, placing the creature within the tree of life.

Biological traces: Hair samples, feces (scat), or blood found in situ that can be analyzed to reveal DNA distinct from any known species. This is the domain of environmental DNA (eDNA) analysis, a powerful tool increasingly used in biodiversity surveys. If eDNA consistent with a large, unknown primate were repeatedly found in specific locations, it would be compelling evidence.

Clear, unambiguous photographic or video evidence: While the vast majority of photographic evidence for cryptids is of poor quality, theoretically, a clear, verifiable image or video capturing the creature in detail, free from any signs of manipulation or misidentification, could serve as strong evidence.

Verifiable fossil evidence: Discoveries of fossils belonging to a previously unknown large primate lineage in regions where contemporary sightings are reported would lend significant weight to the hypothesis of a surviving species.

The process of scientific investigation is also characterized by its

rigor and skepticism. Skepticism in science is not cynicism; it is a healthy, critical attitude that demands evidence before accepting a claim. It means exploring all plausible conventional explanations for a phenomenon before resorting to extraordinary ones. For instance, an unusual footprint is first examined for signs of human alteration, misidentification of a known animal's track, or even geological anomalies. Only after these conventional explanations have been thoroughly ruled out would scientists entertain the possibility of an unknown animal. This principle is encapsulated in Occam's Razor, which suggests that the simplest explanation that fits the evidence is usually the correct one. In the context of cryptids, this often means concluding that a sighting is a misidentification of a bear, a known ape in unusual circumstances, or a deliberate hoax, rather than immediately postulating the existence of a new species.

Furthermore, the scientific method emphasizes

reproducibility. If a phenomenon is real and can be studied scientifically, then the observations and experiments leading to a particular conclusion should, in principle, be reproducible by other independent researchers. If a single scientist claims to have found evidence of a cryptid, the scientific community will await confirmation from other laboratories or researchers. The lack of reproducibility is a significant red flag for claims of anomalous phenomena.

The challenge with cryptid research, particularly concerning cryptid megafauna primates, lies in the fact that the very qualities that make them elusive—their intelligence, their remote habitats, and their apparent avoidance of human detection—also make them incredibly difficult to study using conventional scientific methods. This does not invalidate the scientific method, but it means that the burden of proof is exceptionally high. Extraordinary claims require extraordinary evidence, as the saying goes.

Consider the case of the Giant Panda. For centuries, tales of a large, bear-like animal in the mountainous regions of China circulated among local populations. However, it wasn't until the late 19th and early 20th centuries that Western scientists began to obtain concrete evidence—pelts, skulls, and eventually live specimens. The scientific method was applied in stages: initial reports (observations), formation of hypotheses about the animal's existence and classification, expeditions to gather specimens (experimentation), and subsequent analysis of the physical evidence (data interpretation). The eventual classification of the Giant Panda as a unique species,

Ailuropoda melanoleuca, was the result of this rigorous, evidence-based process.

Applying this to our subject, the process must be driven by a structured research approach, rather than by wishful thinking or sensationalism. This involves:

1.

Systematic Data Collection: Instead of relying on sporadic reports, establishing protocols for collecting eyewitness accounts, recording detailed environmental data at sighting locations, and mapping recurring patterns. This might involve structured interviews with witnesses, geological and botanical surveys of reported habitats, and acoustic monitoring for unusual vocalizations.

2. **Targeted Field Research:** Based on data analysis, identifying specific areas with higher probabilities of encounters and conducting systematic field surveys. This could include setting up camera traps in strategic locations, deploying scent lures, or conducting extensive track surveys during periods of high reported activity.

3. **Interdisciplinary Collaboration:** Bringing together experts from various fields—primatology, zoology, genetics, anthropology, ecology, forensics, and even psychology—to analyze the evidence from multiple perspectives. A primatologist can assess the plausibility of reported morphology and behavior, a geneticist can analyze any biological samples, an ecologist can evaluate habitat suitability, and a forensic expert can assess the authenticity of footprints or other physical traces.

4. **Peer Review and Replication:** Any findings, whether they support or refute the existence of a cryptid megafauna primate, must be subjected to rigorous peer review. This means presenting the methodology and results to other qualified scientists for critique and validation. If promising evidence is found, the research must be replicable by independent scientific teams.

The scientific method, when applied diligently, serves as a powerful filter, separating the plausible from the fantastical. It demands that we acknowledge the limits of our current knowledge and approach the unknown with both an open mind and a critical eye. It does not dismiss the possibility of undiscovered species out of hand, but it insists that the evidence supporting such extraordinary claims must be equally extraordinary. For the cryptid megafauna primate, this means a sustained, methodical, and evidence-based investigation, one that is prepared to embrace the reality of its existence if the data warrant it, but also equipped to explain away its elusive nature through conventional means if that is where the evidence leads. It is this disciplined approach that elevates the pursuit of the unknown primate from the realm of myth to the potential domain of scientific discovery.

The contemporary surge of interest in the possibility of undiscovered megafauna primates is not a sudden anomaly but rather a convergence of several powerful factors. While tales of elusive, ape-like beings have echoed through human history, the last few decades have witnessed a distinct re-energization of this pursuit, propelled by technological advancements and a refined scientific understanding of ecological principles. This modern quest is characterized by a more systematic approach, leveraging tools and knowledge that were simply unavailable to earlier generations of explorers and naturalists. It is a period where the whispers of folklore are being met with the sophisticated hum of modern investigative techniques, creating an environment ripe for potential discovery, or at least, for a more thorough and compelling exploration of the unknown.

One of the most significant drivers of this renewed interest is the dramatic evolution of tracking and remote sensing technologies. Gone are the days when cryptozoologists relied solely on foot patrols, binoculars, and patient observation in the hopes of a fleeting glimpse. Today, sophisticated tools allow for a much broader and deeper understanding of remote environments. Satellite imagery, for instance, can provide high-resolution topographical maps that highlight potential habitats previously inaccessible or poorly understood. Thermal imaging cameras, capable of detecting body heat in complete darkness, can be deployed in known sighting areas to identify large, warm-blooded creatures moving through dense foliage, even when visual confirmation is impossible. These devices transform nights in the wilderness from a period of relative inactivity into prime opportunities for detection.

Furthermore, the proliferation of motion-activated trail cameras, often employed by wildlife biologists for studying elusive fauna, has also played a crucial role. These cameras, scattered across vast tracts of forest and mountains, can capture images and video footage of animals that pass by, day or night, without human intervention. While many such cameras capture images of known species, the potential for them to record an anomaly—a large, unidentifiable primate—is ever-present. The sheer volume of data collected by these networks of cameras, analyzed both by dedicated researchers and citizen scientists, offers an unprecedented opportunity to document the presence of creatures that might otherwise remain hidden. This passive data collection method significantly increases the chances of encountering evidence of a rare or elusive animal.

Acoustic monitoring devices, once prohibitively expensive and complex, are now more accessible and can be deployed to record sounds over extended periods. The potential for large, unknown primates to possess unique vocalizations—howls, calls, or roars—is a recurring theme in anecdotal reports. By deploying arrays of microphones in remote areas with a history of sightings, researchers can analyze soundscapes for unusual patterns, frequencies, or amplitudes that do not correspond to known fauna. Advanced algorithms can then sift through hundreds of hours of audio data, flagging potential anomalies for human review. This acoustic surveillance complements visual methods, providing another layer of sensory input to the search.

Beyond technological advancements, a deeper understanding of ecological niches and habitat modeling has also fueled modern investigations. Scientists now possess a far more sophisticated grasp of how species interact with their environment, the dietary requirements, territorial needs, and population densities necessary for a species to survive and reproduce. This ecological knowledge allows for more informed predictions about where a large, undiscovered primate might exist. Rather than searching randomly, researchers can now utilize data on climate, vegetation, water sources, and prey availability to identify areas that would be most conducive to supporting a population of such creatures.

The concept of "refugia"—areas that have remained relatively stable environmentally, allowing species to persist even as surrounding landscapes change—is particularly relevant. Scientists can analyze historical climate data and geological records to identify regions that may have provided sanctuary for large primates through past ice ages or other significant environmental shifts. These stable, often remote, regions, such as dense rainforests, extensive mountain ranges, and isolated archipelagos, become prime targets for focused investigation. The identification of such ecologically viable areas significantly narrows the search parameters, making the quest more scientifically tractable.

Moreover, the study of evolutionary biology and genetics has provided new avenues for investigation. While direct evidence remains elusive, the understanding of primate evolution and diversification allows scientists to consider plausible evolutionary pathways for an unknown species. For example, if a new primate were to exist, where would it fit on the evolutionary tree? Would it be a surviving lineage of a known extinct hominoid, a highly divergent branch of an existing ape family, or something entirely unexpected? While speculative, these considerations guide the search for potential biological markers and inform the interpretation of any anomalous findings.

The sheer volume and consistency of anecdotal reports, when viewed through this modern lens, also contribute to the renewed scientific curiosity. While individual eyewitness accounts are insufficient for scientific proof, a pattern emerging from numerous reports, often from credible sources and across different geographical locations within a defined region, cannot be entirely dismissed. These reports, when cross-referenced with known ecological data and potentially corroborated by technological surveillance, begin to form a more compelling narrative. The modern investigator is more equipped to critically assess these reports, filtering out misidentifications and hoaxes, and identifying recurring elements that might point to a genuine phenomenon.

The development of citizen science initiatives has also become a critical component of modern cryptozoological research. Platforms and organizations now exist that allow individuals to report sightings, submit photographic or video evidence, and participate in data analysis. This democratizes the investigative process, tapping into a vast network of observers who are often intimately familiar with their local environments. While the quality of submissions can vary widely, these platforms act as powerful data-gathering hubs, potentially identifying areas of intense interest or recurring anomalies that warrant further professional investigation. The aggregation of such distributed information allows for a more comprehensive overview of reported phenomena than any single researcher could achieve alone.

Furthermore, the growing understanding of animal behavior and intelligence has shifted perceptions. We are increasingly aware of the remarkable capacity of many animals to evade detection, especially those that are highly intelligent, solitary, or inhabit extremely dense or remote terrains. The traditional image of a clumsy, easily observable beast is being replaced by a more nuanced understanding of how intelligent, adaptable creatures might successfully avoid human contact for decades, even centuries. This realization makes the absence of definitive proof less of a disqualifier and more of a testament to the potential sophistication of such elusive beings.

The increasing interconnectivity facilitated by the internet and social media has also played a significant role. Information, both credible and otherwise, now travels at an unprecedented speed. This has allowed for the rapid dissemination of sighting reports, research findings (both professional and amateur), and discussions within communities of interest. While this can amplify misinformation, it also facilitates collaboration, allows for the rapid pooling of resources for expeditions, and enables a global conversation about the possibilities of undiscovered megafauna primates. It brings together diverse individuals—biologists, anthropologists, technologists, and enthusiasts—who might otherwise never have crossed paths, fostering a collective drive to explore the unknown.

Finally, the very nature of scientific inquiry is iterative. Discoveries are rarely made in a single, dramatic revelation. Instead, they are built upon layers of previous work, incorporating new data, new technologies, and new theoretical frameworks. The current resurgence of interest in large, unknown primates represents just such an iterative process. It builds upon the legacy of earlier investigations, acknowledges their limitations, and leverages the powerful tools and refined understanding of the present to pursue answers with renewed rigor and optimism. This confluence of advanced technology, ecological insight, vast amounts of collected data, and a global network of interested individuals has created a unique moment, transforming the modern quest for giant apes from a fringe pursuit into a compelling area of scientific and public fascination. It is this convergence that makes "now" the opportune moment for such an exploration.

The study of unconfirmed species, particularly large primates like the Yeti or Sasquatch, inevitably traverses a landscape punctuated by deeply entrenched skepticism and fervent belief. These two poles, often perceived as diametrically opposed, are in fact intertwined threads in the tapestry of human inquiry into the unknown. Understanding this dynamic is crucial for any serious exploration of the subject, moving beyond simplistic dichotomies to appreciate the complex interplay of evidence, perception, and psychology.

At one end of the spectrum lies unwavering skepticism. This viewpoint is rooted in a rigorous application of scientific methodology, demanding empirical proof that meets established standards for biological classification. Skeptics emphasize the vast amount of research already conducted on primate populations globally, the pervasive nature of human exploration in most terrestrial environments, and the logistical challenges of concealing a large, breeding population of animals. They point to the consistent lack of definitive physical evidence – no authenticated carcasses, no bones, no verifiable DNA samples that have withstood independent scrutiny, and no clear, unambiguous photographic or video evidence that cannot be attributed to known animals, hoaxes, or misidentifications. From this perspective, the absence of such evidence, despite decades of anecdotal reports and some high-profile expeditions, strongly suggests that these creatures do not exist. The burden of proof, they argue, lies squarely with those who propose the existence of an unknown species, and until that burden is met with irrefutable, repeatable, and verifiable data, the most rational conclusion is non-existence. This stance is not necessarily dismissive of the people who report sightings;

rather, it is a commitment to a scientific framework that prioritizes evidence over assertion. It acknowledges that human perception can be fallible, influenced by a myriad of factors, from suggestion and expectation to optical illusions and genuine misinterpretations of familiar phenomena.

Conversely, at the other extreme, we find fervent belief. This conviction often stems from a deep personal experience of witnessing something inexplicable, a profound respect for folklore and indigenous knowledge, or a philosophical stance that embraces the possibility of undiscovered wonders in the natural world. For those who hold this view, the sheer volume and consistency of eyewitness accounts, often from individuals who appear credible and have no apparent motive to deceive, become compelling evidence in themselves. They may interpret the lack of definitive proof not as evidence of absence, but as a testament to the creature's intelligence, elusiveness, and ability to thrive in environments that are difficult for humans to access or survey. Some proponents of this view may also feel that the scientific establishment is inherently biased against such discoveries, or that it is too conservative to accept evidence that challenges established paradigms. They might draw parallels to historical discoveries of previously unknown species that were initially met with disbelief. The rich tapestry of oral traditions and cultural narratives surrounding these beings across different continents and centuries can also form a significant part of their belief system, suggesting a shared human intuition or observation that transcends mere fabrication.

However, the space between these two poles is where the most productive scientific inquiry resides. This middle ground is characterized by a stance of critical open-mindedness, a willingness to consider the possibility of undiscovered primate species without succumbing to unfounded claims, and a commitment to employing rigorous scientific methods to investigate the phenomenon. It is a position that acknowledges the inherent limitations of our current knowledge and the vastness of unexplored ecosystems, while simultaneously demanding that any claims be supported by robust evidence.

Psychologically, several factors contribute to the persistence of these belief systems. Confirmation bias is a powerful influence, whereby individuals tend to seek out, interpret, and remember information that confirms their pre-existing beliefs. For someone predisposed to believe in Bigfoot, a blurry photograph or a distant, unidentified sound in the woods can easily be interpreted as evidence, while similar phenomena from known animals might be overlooked or dismissed. Conversely, a skeptic might readily find alternative explanations for any piece of purported evidence, fitting it into their existing worldview. The desire for mystery and the allure of the unknown also play a significant role. In an increasingly mapped and understood world, the idea of a large, undiscovered creature offers a tantalizing glimpse into a primal past and a realm of untamed nature. This can be particularly appealing to individuals who feel disconnected from the natural world or who are disillusioned with the perceived predictability of modern life.

Sociologically, the dissemination of information, particularly through the internet and social media, has amplified both the claims and the skepticism surrounding cryptid primates. Online forums, dedicated websites, and social media groups provide platforms for believers to share their stories and evidence, fostering a sense of community and reinforcing their convictions. These platforms also provide fertile ground for the spread of misinformation, hoaxes, and sensationalized accounts, which can unfortunately muddy the waters for genuine investigation. Simultaneously, these same platforms allow for a rapid and widespread dissemination of skeptical analyses, debunking efforts, and scientific critiques, contributing to a robust public debate. The framing of these discussions often involves deeply held cultural narratives and archetypes, where the unknown primate can represent the wild, the ancient, or even a threat to the known order.

A balanced scientific approach, therefore, seeks to navigate these psychological and sociological currents without being swayed by emotional appeals or prejudiced dismissals. It begins with an assumption of neutrality, treating each report or piece of potential evidence with an open but critical mind. The first step in such an approach is

evidence assessment. This involves meticulously examining any submitted data – photographs, videos, audio recordings, hair samples, footprints, and eyewitness testimonies – for authenticity, consistency, and plausibility. Skeptical analysis is vital here, applying techniques to detect manipulation, identify known subjects, and analyze environmental factors that might explain the observation. For instance, a footprint might be analyzed for dermal ridges, mid-tarsal break, or other characteristics that differentiate it from known primate anatomy. Hair samples would be subjected to microscopic and genetic analysis to determine their origin. Photographic and video evidence would be scrutinized for signs of digital alteration, motion blur that suggests a known animal at speed, or inconsistencies in scale and lighting.

Eyewitness testimony, while valuable, is treated with caution. While individuals may genuinely believe what they saw, memory is a reconstructive process, prone to distortion, confabulation, and suggestion. A skilled investigator would conduct interviews that are structured to elicit factual details without leading the witness, exploring factors such as the witness's distance from the subject, lighting conditions, duration of observation, and any prior knowledge or expectations the witness may have had. Cross-referencing multiple testimonies from independent witnesses in the same vicinity can sometimes reveal patterns, but even then, shared perceptual biases or environmental factors cannot be ruled out without further corroboration.

The scientific community's inherent demand for **repeatability and verifiability** presents a significant hurdle for the study of cryptids. A single sighting, even by a seemingly credible individual, is insufficient to establish the existence of a new species. What is required are reproducible findings, such as consistent tracks leading to a reliable nesting site, or repeated sightings that can be corroborated by independent observers or technological means. The ultimate goal is the collection of unambiguous biological evidence – a specimen, a clear DNA profile, or definitive fossil remains – that can be subjected to peer review and integrated into the established framework of zoology.

Furthermore, a scientific approach necessitates **ecological plausibility**. The existence of a large primate species requires a viable population capable of sustaining itself over generations. This involves considerations of habitat availability, food resources, reproductive rates, and genetic diversity. Researchers must ask: Does the proposed habitat contain sufficient resources to support a population of this size? What are the potential food sources? What is the estimated breeding population size required to avoid genetic bottlenecks? Are there any existing species that fill a similar ecological niche? If a new species were to exist, its presence would have ecological implications that might be detectable through broader environmental surveys.

The challenge lies in finding a way to bridge the gap between anecdotal reports and rigorous scientific proof without dismissing the former entirely or demanding the impossible. This is where the concept of the **scientific frontier** becomes relevant. The frontiers of science are not always about sensational discoveries; they are often about pushing the boundaries of our understanding, developing new methodologies, and exploring the fringes of what is known. In the context of cryptid primates, the scientific frontier involves:

1.

Advanced Technology Deployment: As discussed previously, leveraging technologies like thermal imaging, acoustic monitoring, and extensive camera trapping in areas with persistent reports can move beyond subjective accounts to objective data collection. The analysis of this data, even if it yields negative results, is scientifically valuable, helping to refine hypotheses about habitat and behavior.

2.

Targeted Field Research: Instead of broad, undirected searches, scientific expeditions would focus on specific regions identified through ecological modeling, analysis of consistent sighting patterns, or the potential presence of isolated refugia. Such expeditions would be staffed by trained biologists and equipped with appropriate scientific tools for sample collection and environmental assessment.

3.

Citizen Science Collaboration: While skeptics may be wary of amateur input, well-structured citizen science initiatives can be invaluable. By providing standardized protocols for reporting, sample collection (e.g., for DNA analysis of scat or hair), and image submission, researchers can tap into a vast network of observers while maintaining a degree of data quality control. This can help identify potential areas of interest that warrant more intensive professional investigation.

4.

Interdisciplinary Approaches: The study of cryptid primates is not solely the domain of zoologists. Collaboration with anthropologists, ecologists, geneticists, forensic scientists, and even folklorists can provide diverse perspectives and analytical tools. For instance, understanding the cultural context of indigenous legends might offer clues about ancient primate distributions or environmental changes.

5.

Focus on Indirect Evidence: While direct evidence is the ultimate goal, the scientific community can and should rigorously investigate credible indirect evidence. This includes properly analyzing footprint casts for anatomical details, systematically collecting and analyzing hair or scat samples found in high-probability areas, and correlating reported sightings with environmental anomalies or geological data.

The scientific pursuit of unknown species, including large primates, is a delicate balancing act. It requires the intellectual honesty to acknowledge the limitations of current knowledge and the courage to explore possibilities that lie beyond our established understanding. At the same time, it demands an unwavering commitment to the principles of scientific inquiry: evidence-based reasoning, rigorous methodology, peer review, and falsifiability. To dismiss all claims outright is to close the door on potential discovery, clinging to a certainty that the natural world often proves to be mistaken. Conversely, to embrace every claim without critical assessment is to abandon the very rigor that defines scientific progress, succumbing to wishful thinking and unsubstantiated narratives. The true frontier lies in the disciplined, open-minded, and evidence-driven exploration of the unknown, acknowledging that the absence of evidence is not always evidence of absence, but it is always a call for more, and better, evidence. The ongoing fascination with large, elusive primates serves as a potent reminder that, despite our advances, the planet still holds mysteries, and the edge of our knowledge remains an invitation to explore.

2

Primate Anatomy and the Case for Gigantism

The skeletal structure of any organism is the fundamental scaffold upon which its entire physiology and ecological niche are built. For primates, this framework is a testament to millions of years of evolutionary refinement, optimizing for arboreal locomotion, manual dexterity, and a complex sensory apparatus. When considering the theoretical possibility of gigantism within this order, the skeletal system becomes a primary focal point, presenting both constraints and, potentially, pathways for extraordinary size increases. The bones of extant primates are a marvel of adaptation, and understanding their inherent properties – density, composition, structural integrity, and growth patterns – is crucial to assessing the biomechanical feasibility of a primate significantly larger than any currently known.

Primates, in general, possess a skeletal architecture that balances strength with relative lightness, a necessity for many arboreal species who must navigate a three-dimensional environment with agility. This is particularly evident in the limb bones. The long bones, such as the femur and humerus, are robust, but their internal structure is often characterized by a well-developed medullary cavity, which reduces overall weight. The cortical bone, the dense outer layer, is thickest in areas subjected to the greatest mechanical stress, reflecting the principle of Wolff's Law, which posits that bone adapts to the loads under which it is placed.

However, this very density and structure, while providing strength, also represents a significant weight penalty. For a primate to achieve truly gigantiform proportions, this inherent compromise between strength and weight would need to be dramatically altered or overcome.

Bone density is a critical factor. The bones of larger mammals, in general, tend to be denser and more robust than those of their smaller counterparts. This increased density provides the necessary compressive strength to withstand the greater gravitational forces and impact loads associated with a larger body mass. Consider the difference between a squirrel and a horse; the horse's femurs are vastly thicker and denser, capable of supporting its multi-ton frame. In primates, we see a spectrum. The skeletal elements of a gorilla, the largest living primate, are considerably more robust than those of a capuchin monkey. The gorilla's skull is massive, its vertebral column is thick, and its limb bones are exceptionally dense and muscularly anchored. This suggests that, within the primate lineage, there is a predisposition for skeletal reinforcement as body size increases. However, the scaling of bone strength with body mass is not linear. A simple linear increase in bone dimensions would not suffice to support a proportionally larger body; rather, bone diameter typically scales to the power of 2/3 of body mass, while bone strength scales to the power of 3/2. This means that as body mass increases, bone strength must increase at a disproportionately faster rate to maintain structural integrity.

The internal architecture of primate bones is also a subject of interest. Trabecular bone, the spongy, porous network within the bone, plays a significant role in shock absorption and distributing stress. In larger primates, this trabecular network is likely to be more extensively developed and interconnected, providing greater resistance to fracture under load. The arrangement of these trabeculae often aligns with the principal lines of stress within the bone, a remarkable example of biological engineering that minimizes material usage while maximizing strength. For a giant primate, we might hypothesize a more intricate and robust trabecular system, capable of dissipating the immense forces generated during locomotion.

The vertebral column presents another fascinating area of consideration. In most primates, the vertebral column is designed to support the torso while allowing for significant flexibility, crucial for maneuvering in arboreal environments. However, in terrestrial, large-bodied primates like gorillas and orangutans, the lumbar region shows adaptations for greater stability. The lumbar vertebrae are larger and more rigidly articulated, reflecting the increased load they bear. A hypothetical giant primate, especially one that spends a significant amount of time on the ground, would likely possess an exceptionally robust vertebral column. The intervertebral discs, the shock absorbers between vertebrae, would need to be exceptionally thick and resilient to prevent compression fractures. The spinous processes, which serve as attachment points for powerful back muscles, would also need to be greatly elongated and thickened to provide the necessary leverage and strength to support a massive head and torso. The sheer mass of such a vertebral column would represent a significant portion of the creature's overall skeletal weight.

The skull of a primate is a complex mosaic of bones that houses the brain, sensory organs, and provides attachment for the powerful muscles of mastication. In larger primates, the skull is invariably more robust. The cranial vault is thicker, and the facial bones are more substantial, particularly in species that consume tough, fibrous vegetation. The sagittal crest, a bony ridge running along the midline of the skull in some primates (like gorillas), serves as an anchor for the temporalis muscles, which are responsible for closing the jaw. The development of such a crest is directly proportional to the forces required for chewing. A hypothetical gigantopithecine primate, with a diet that might include extremely tough plant material or even carrion, could potentially develop an even more pronounced sagittal crest or other cranial buttressing mechanisms to support colossal jaw musculature. The sheer volume of the cranial cavity itself would also be a significant factor, albeit one that doesn't directly contribute to structural support in the same way as the rest of the skeleton.

The appendicular skeleton – the limbs – would require substantial modification to support a greatly increased body mass. The long bones would need to be significantly thicker and more densely mineralized. The joints, such as the hip and knee, would bear immense loads. The articular cartilage, which cushions the ends of bones in a joint, would need to be exceptionally thick and robust to prevent wear and tear. The ligaments and tendons that stabilize these joints would also have to be proportionally stronger and more resilient. For terrestrial locomotion, the hind limbs would likely be more developed than the forelimbs, as is typical in large terrestrial mammals, to provide the primary propulsive force. However, if the hypothetical giant primate retained some arboreal tendencies, the forelimbs might still be substantial, perhaps for brachiation (swinging through trees) or for supporting the body's weight during climbing, albeit at a much grander scale. This dual requirement could lead to a unique and perhaps less efficient limb morphology compared to purely terrestrial or purely arboreal specialists.

The hands and feet would also reflect adaptations to size and potential locomotion. While primates are characterized by grasping hands and feet, a massive primate would likely have reduced dexterity in its digits, with thicker, more robust bones and less independent movement of individual digits, focusing more on weight-bearing and powerful gripping rather than fine manipulation. The metacarpals and metatarsals would be significantly broadened and strengthened. In some terrestrial primates, the foot has evolved to be more plantigrade (walking on the sole of the foot), which offers greater stability. A gigantiform primate might exhibit a more pronounced plantigrade stance, with a reinforced tarsus and metatarsus to efficiently distribute its immense weight.

The scaling of bone dimensions is a critical constraint. If we imagine a primate with the skeletal proportions of a chimpanzee but scaled up to the size of a rhinoceros, the bone-crushing forces would be catastrophic. The strength of a bone is roughly proportional to the square of its cross-sectional area, while its weight is proportional to the cube of its length. As an animal increases in size, its weight increases cubically, while the strength of its bones only increases quadratically. This fundamental biomechanical principle is why we don't see elephants with the delicate limb structure of a gazelle. To overcome this, a giant primate would likely exhibit a different body plan, with much thicker, more columnar limbs, a broader pelvis, and a more compact torso, prioritizing stability and load-bearing over agility. Think of the stout, pillar-like legs of an elephant compared to the gracile limbs of a monkey.

Furthermore, the process of ossification and bone remodeling throughout a primate's life would need to be incredibly robust. The growth plates, responsible for long bone elongation, would need to remain active for a longer period or be of a larger size to achieve truly immense stature. The continuous remodeling of bone tissue, driven by mechanical stress and hormonal signals, would need to efficiently repair microfractures and adapt to the constantly escalating loads. The metabolic demands of maintaining such a colossal skeletal structure, including the continuous synthesis and mineralization of bone tissue, would be immense, requiring a highly efficient digestive system and a substantial caloric intake.

The possibility of gigantism in primates is intrinsically linked to the evolutionary pressures that might favor such a development. In environments with abundant food resources, reduced predation pressure on adults, and a need for sheer physical dominance, selection might favor larger body size. However, the biomechanical challenges posed by the skeletal system are formidable. It is not simply a matter of "scaling up" existing primate anatomy. Instead, it would necessitate a radical re-engineering of the primate skeleton, likely involving trade-offs in agility, dexterity, and perhaps even cognitive complexity, as resources are diverted to maintaining a massive physical form. The skeletal structure would need to evolve to become more robust, denser, and potentially adopt a more columnar morphology, moving away from the gracile and agile frame typical of many arboreal primates. The sheer mass and structural requirements of such a skeleton are among the most significant hurdles to overcome for a gigantiform primate to exist within the known evolutionary possibilities of the order. Understanding these limitations, as well as the inherent capacities for adaptation within primate skeletal evolution, forms the bedrock for any scientifically grounded discussion about the feasibility of exceptionally large, undiscovered primates.

The immense skeletal framework described in the preceding discussion provides the foundation, but it is the musculature that imbues this structure with power and enables locomotion. For a primate to attain gigantiform proportions, its muscular system would need to be profoundly augmented, transcending the capabilities of even the largest extant species. The fundamental principles governing muscle function – contraction, force generation, and energetic expenditure – remain constant, but the sheer scale of a giant primate would necessitate dramatic evolutionary modifications.

At the core of this discussion lies the concept of muscle-to-bone ratio. In smaller, agile primates, muscles are often lean and efficiently integrated with the skeleton, facilitating rapid, precise movements. However, as body mass increases, the forces exerted on bones during locomotion also increase exponentially. To counteract these forces and generate sufficient propulsive power, muscle mass must increase disproportionately. This doesn't simply mean larger versions of existing muscles; it implies a fundamental shift in their architecture and metabolic capacity. We can hypothesize that a gigantiform primate would possess a significantly higher muscle-to-bone ratio compared to its smaller relatives. This would manifest as thicker, more voluminous muscles, capable of generating the colossal forces required to move a body weighing potentially hundreds of kilograms, or even tons. The muscle fibers themselves might also be adapted, perhaps with a greater proportion of slow-twitch fibers for sustained effort during long-distance terrestrial travel, or a greater reliance on anaerobic metabolism for bursts of power, though the latter would be energetically unsustainable for prolonged activity in such a large animal.

The attachment points of these hypothetical colossal muscles are of paramount importance. Evolution would have favored the development of larger and more pronounced bony landmarks – tuberosities, crests, and epicondyles – on the bones of a giant primate. These serve as the anchor points for tendons, the connective tissues that transmit muscular force to the skeleton. Imagine the deltoid tuberosity on the humerus, or the tibial tuberosity on the tibia, but scaled up dramatically. These would need to be broad, rugged surfaces to accommodate the immense cross-sectional area of the tendons connecting to truly massive pectoral and leg muscles. The supraspinous and infraspinous fossae of the scapula, for example, which anchor the muscles of the rotator cuff, would likely be considerably expanded in a gigantiform primate to provide stability and power to the shoulder joint under extreme loads. Similarly, the linea aspera on the femur, the roughened ridge on the posterior aspect of the femur that serves as the origin for several thigh muscles, would likely be more pronounced and extensive, reflecting the immense power required to extend the hip and propel the body forward.

The energetic demands of moving such a massive frame are staggering. Muscle contraction is an energy-intensive process, and the larger and more powerful the muscles, the greater their metabolic requirements. A giant primate would necessitate a highly efficient metabolism, capable of processing vast quantities of food to fuel its locomotory needs. This implies a digestive system that can extract maximum energy from its diet, potentially favoring foods that are calorie-dense or easily digestible, despite the common primate tendency towards herbivory or omnivory. The respiratory and circulatory systems would also need to be significantly enhanced to deliver oxygen and nutrients to the enormous muscle masses and to efficiently remove metabolic waste products. High cardiac output, a large lung capacity, and a dense network of capillaries within the muscles would be essential adaptations.

Considering locomotion, the shift from quadrupedalism or arborealism to a more terrestrial existence, particularly bipedalism, presents unique challenges and requires specific muscular adaptations. While many primates exhibit some degree of bipedalism, it is often facultative and energetically costly for all but the largest great apes. For a gigantiform primate, efficient and sustained bipedalism would necessitate a profound reshaping of its muscular architecture. The gluteal muscles, particularly the gluteus maximus, would need to be exceptionally well-developed to extend the hip and prevent the torso from falling forward. In humans, the gluteus maximus is a primary driver of bipedal locomotion; in a giant primate, it would likely be the largest and most powerful muscle in the body, anchoring to a significantly enlarged ilium and femur. The quadriceps femoris group, responsible for extending the knee, would also be crucial, providing stability and power for each stride. The vastus medialis, lateralis, and intermedius, along with the rectus femoris, would need to be robust to support the immense weight of the body and to absorb the impact of each footfall.

The plantar flexors of the ankle, primarily the gastrocnemius and soleus, would be equally vital for generating propulsive force. Their origins on the posterior aspect of the tibia and fibula would need to be extensive, and their insertions on the calcaneus (heel bone) via the Achilles tendon would have to be exceptionally strong to push the body forward. The entire lower limb musculature would likely evolve to be more robust and columnar, resembling that of large terrestrial mammals rather than the more gracile limbs of smaller primates. This would involve a simplification of some muscle groups, potentially sacrificing fine motor control in the feet for increased power and stability.

The stabilizing muscles of the core and trunk would also play an indispensable role in bipedal locomotion. The erector spinae muscles, running along the vertebral column, would need to be exceptionally developed to maintain an upright posture and to counterbalance the weight of the head and upper body. The abdominal muscles, including the rectus abdominis and the obliques, would provide crucial support for the viscera and contribute to postural stability. The forces generated by the limbs during walking and running would be transmitted through the pelvis and spine, requiring a highly integrated and powerful muscular network to prevent instability and injury.

The evolution of brachiation, the mode of locomotion involving swinging by the arms, is a hallmark of many arboreal primates. While the skeletal adaptations for brachiation in large primates are well-documented, the muscular requirements for a gigantiform brachiator would be immense. The shoulder girdle would need to be exceptionally strong, with massive deltoids, pectorals, and latissimus dorsi muscles to power the swing. The biceps and triceps would need to be powerfully developed to flex and extend the elbow, and the forearm muscles would require extraordinary strength for gripping branches. However, the energetic cost of brachiation increases significantly with body mass, and it is improbable that even a massive primate could sustain this mode of locomotion for extended periods or over great distances. Thus, while remnants of arboreal musculature might persist, the primary locomotory muscles would likely be geared towards terrestrial movement.

The potential for a gigantiform primate to exhibit quadrupedal locomotion, either terrestrial or semi-arboreal, would also involve significant muscular adaptations. If it retained a quadrupedal stance, the forelimbs would need to be substantially muscularized to bear a significant portion of the body's weight. The pectoral muscles would be massive, anchoring to a broad sternum and clavicle (if present and robust). The muscles of the shoulder, elbow, and wrist would need to be powerful enough to support and propel the animal. The spinal musculature would be crucial for maintaining the integrity of the back during locomotion, especially if the animal had a more elongated body form. The development of the hind limbs would still be critical for propulsion, but the distribution of forces between the fore and hind limbs would differ from that of a bipedal primate.

The interplay between muscle and tendon strength is another critical consideration. As muscles increase in size and force output, the tendons that connect them to bone must also scale accordingly. Tendons are dense connective tissues, but they have limits to the tensile forces they can withstand. A disproportionately weak tendon could rupture under the immense forces generated by a giant primate's musculature, leading to severe injury or death. Therefore, evolutionary pressures would undoubtedly favor the thickening and strengthening of tendons alongside muscle and bone development. The Achilles tendon, for instance, would need to be exceptionally robust to transmit the propulsive forces of the gastrocnemius and soleus muscles. Similarly, the rotator cuff tendons and the patellar tendon would require significant reinforcement.

The very biomechanics of muscle contraction at a larger scale present interesting avenues for speculation. While the fundamental mechanisms of actin-myosin cross-bridge cycling remain the same, the sheer volume of muscle tissue involved could influence the efficiency and speed of contraction. Heat generation is also a significant factor; larger muscles produce more metabolic heat, which would need to be dissipated effectively to prevent hyperthermia. This reinforces the need for a well-developed circulatory system and potentially adaptations for radiating heat, such as larger surface areas or specialized vascularization in certain body regions.

Furthermore, the neurological control of such a massive muscular system would be a remarkable feat of biological engineering. The motor neurons innervating these colossal muscles would need to be larger and capable of transmitting strong signals. Coordinating the complex interplay of hundreds of muscles to achieve smooth and efficient locomotion, whether bipedal or quadrupedal, would require a sophisticated nervous system. The cerebellum, responsible for coordinating movement and balance, would likely be highly developed.

In conclusion, the musculature of a gigantiform primate would represent a profound departure from that of its extant relatives. It would be characterized by immense volume and power, with muscles scaled to generate the forces necessary to move a body of extraordinary size. This would necessitate significant adaptations in muscle fiber type, attachment points, energetic efficiency, and neurological control. The move towards sustained bipedalism, in particular, would demand a radical redesign of the lower limb and core musculature, prioritizing power, stability, and load-bearing capacity over the agility and dexterity typical of many primate species. The energetic cost would be immense, requiring a highly efficient metabolic and physiological system. Ultimately, the musculature is the engine that would drive the locomotion of a giant primate, and its evolution would be intrinsically linked to the skeletal modifications and the ecological pressures that might favor such an astonishingly large primate.

The immense skeletal framework and powerful musculature discussed previously form the corporeal basis of a hypothetical gigantiform primate. However, to fully comprehend the potential of such a creature, we must delve into the realm of its most defining organ: the brain, and its direct correlate, intelligence. The cranial capacity of a primate is not merely a measure of skull volume; it is a proxy for the size of the central processing unit that governs behavior, cognition, and adaptation. In the context of gigantism, the question arises: would an increase in body size correlate with an increase in brain size, and what implications would such an augmentation hold for the primate's cognitive capabilities?

The relationship between body size and brain size in mammals, including primates, is a complex interplay of evolutionary pressures and metabolic constraints. While it is a tempting oversimplification to assume that larger bodies automatically equate to larger brains, the primate order exhibits a distinct trend: a disproportionately larger brain relative to body mass compared to many other mammalian groups. This phenomenon, often referred to as encephalization, is a key factor in the remarkable cognitive abilities observed across primate species, from the intricate social dynamics of baboons to the sophisticated tool use of chimpanzees. For a gigantiform primate, this trend of encephalization, if it were to manifest, could lead to a brain of truly unprecedented scale.

Consider the concept of cranial capacity. In extant primates, this varies dramatically. The diminutive pygmy marmoset boasts a cranial volume of mere milliliters, while the robust gorilla possesses a cranial capacity around 500-600 cubic centimeters (cc). Humans, on average, have a cranial capacity of approximately 1350 cc, a figure that, relative to our body mass, is exceptionally high. If a gigantiform primate were to follow this encephalization trajectory, its cranial capacity could potentially dwarf that of even the largest known hominids. Imagine a primate with a body mass exceeding that of a rhinoceros, possessing a cranial vault capable of housing a brain several times larger than our own. This would not merely be a quantitative increase; it could represent a qualitative leap in cognitive potential.

The implications of such a substantial brain are profound, particularly when considering the potential for enhanced problem-solving skills. Primate intelligence is often characterized by its adaptability and flexibility. They are not rigidly programmed automatons; their behavior is shaped by learning, experience, and the ability to devise novel solutions to challenges. A larger brain, with its increased neuronal density and more complex neural networks, could facilitate more sophisticated forms of reasoning, planning, and memory. This might manifest in advanced foraging strategies, where the primate could remember vast territories, identify seasonal food sources, and even coordinate group efforts to exploit resources. The ability to anticipate the actions of others, a hallmark of primate social intelligence, could be amplified, leading to more intricate social hierarchies and alliances.

Furthermore, the capacity for tool use, a trait observed in various primate lineages, could be significantly advanced. While many primates use simple tools, a giant primate with enhanced cognitive abilities might develop more complex tool manufacturing and application. This could range from modifying natural objects with greater precision to creating multi-step tools for specific tasks, such as extracting inaccessible food items or defending its territory. The foresight required to fashion and utilize tools effectively, especially in a dynamic environment, would be a direct product of advanced cognitive processing.

The potential for complex communication is another area where a larger brain could have a significant impact. While vocalizations and gestures are common, a primate with heightened cognitive capacity might develop more nuanced forms of communication. This could involve a richer lexicon of calls, more elaborate gestural sequences, or even symbolic representation, allowing for the transmission of complex information about danger, food, or social relationships. The capacity for abstract thought, however rudimentary, could emerge, enabling the primate to understand concepts beyond immediate sensory input.

Consider the concept of self-awareness and theory of mind. While the evidence for these in non-human primates is debated, a primate with a significantly larger and more complex brain might exhibit a more pronounced understanding of itself as an individual and of others as distinct entities with their own intentions and perspectives. This could lead to sophisticated social maneuvering, empathy, and even altruistic behavior, further complicating its ecological and social interactions.

The energetic cost of a large brain is substantial. Brain tissue is metabolically expensive, requiring a constant supply of glucose and oxygen. For a gigantiform primate, supporting such a large brain would necessitate an equally robust metabolic system. This aligns with our previous discussions on musculature and locomotion; the energy demands for moving a colossal body are already immense, and the addition of a hyper-encephalized brain would compound these requirements. This suggests that such a creature would likely be an opportunistic feeder, capable of processing a wide range of foodstuffs, or inhabit an environment particularly rich in calorie-dense resources.

The very structure of the primate brain offers clues to potential cognitive enhancements. The cerebral cortex, responsible for higher-level cognitive functions such as sensory perception, conscious thought, and voluntary movement, is particularly enlarged in primates. Within the cortex, specific regions, such as the prefrontal cortex, are associated with executive functions like planning, decision-making, and working memory. An expansion and increased convolution of these areas in a gigantiform primate could translate into superior cognitive abilities. The neocortex, in particular, has undergone significant expansion throughout primate evolution, and this trend could be amplified in a hypothetical giant primate. This expansion is not simply about more neurons; it is also about the intricate connectivity between them, forming the basis of complex thought processes.

Furthermore, the development of the cerebellum, crucial for motor control, coordination, and balance, would likely be significant, especially if the primate exhibits sophisticated locomotion, be it bipedal or arboreal. However, the cerebellum also plays a role in cognitive functions, including language and attention, suggesting that even these traditionally motor-oriented brain regions could contribute to enhanced intelligence.

The comparative neurology of extant primates provides a valuable framework. Chimpanzees, our closest living relatives, exhibit remarkable cognitive skills, including problem-solving, rudimentary tool use, and complex social interactions. Their brain-to-body mass ratio is significantly higher than that of many other mammals. Extrapolating from this, a primate that is orders of magnitude larger, but still adheres to the evolutionary trajectory of primate encephalization, would possess a brain that could unlock cognitive potentials far beyond our current understanding.

It is also important to consider that intelligence is not a monolithic entity. Different primates excel in different cognitive domains. Some are masters of spatial memory, navigating complex environments with ease. Others are adept at social learning, acquiring skills by observing and imitating others. A gigantiform primate could potentially possess a more balanced and multifaceted intelligence, excelling across a broader range of cognitive tasks. This would enhance its adaptability and survival prospects in a wide array of ecological niches.

The challenges of living in a complex environment, whether it be navigating dense forests, exploiting fluctuating food resources, or managing intricate social relationships, all exert selective pressures favoring increased cognitive abilities. For a creature of immense size, these pressures could be amplified. For instance, coordinating the movements of a massive body, especially in a treacherous environment, would require sophisticated motor control and spatial awareness. The need to find sufficient food to sustain both the body and the brain would necessitate advanced foraging strategies and potentially the ability to exploit resources that are inaccessible to smaller animals.

The study of fossil primates, while limited in terms of potential giants, has revealed trends in cranial development that could inform our speculation. Changes in cranial shape and the presence of specific cranial features can sometimes indicate relative brain size and complexity. While direct evidence of a truly gigantic primate with a correspondingly large brain is absent from the fossil record, the evolutionary trajectory observed in primates, with its consistent trend towards increased encephalization, provides a compelling basis for considering the potential cognitive capacities of such a hypothetical being.

In essence, the cranial capacity and potential intelligence of a gigantiform primate represent a fascinating speculative frontier. If such a creature were to evolve, driven by the same evolutionary principles that have shaped extant primates, it is plausible that it would possess a brain of extraordinary size and complexity. This would, in turn, unlock cognitive abilities far exceeding those of its smaller relatives, enabling advanced problem-solving, complex communication, sophisticated social behavior, and a remarkable capacity for adaptation. The energetic costs would be considerable, but the evolutionary advantages conferred by such intelligence could provide the crucial edge for survival and dominance in its ecological niche, making it not just a marvel of physical scale, but also a titan of intellect.

The contemplation of a gigantiform primate, while rooted in anatomical and cognitive extrapolation, inevitably leads to a crucial, tangible question: how would we know if such a creature existed? Beyond the hypothetical skeletal structures and oversized brains, the physical world would bear its indelible mark. This is where the principles of forensic anthropology and the meticulous analysis of trace evidence become paramount. If a primate of such colossal proportions were to inhabit our planet, it would, by necessity, leave behind a trail of biological clues, detectable by scientific methods, that could serve as irrefutable proof of its existence. The challenge lies not only in recognizing these traces but in interpreting them correctly, distinguishing them from the known, and understanding the unique signatures they would carry.

Forensic anthropology, the study of human remains for legal purposes, provides a powerful framework for approaching the potential evidence left by an unknown, large primate. While its primary domain is human skeletal material, the underlying methodologies – the examination of bone morphology, the analysis of taphonomic processes (the changes to remains after death), and the reconstruction of biological profiles – are directly applicable to any large vertebrate. For a gigantiform primate, the sheer scale of its remains would be the most immediate distinguishing characteristic. Imagine footprints, not of a deer or a bear, but of a bipedal or quadrupedal creature whose foot was the size of a small automobile. These impressions, preserved in mud, sand, or volcanic ash, would be the first, most dramatic pieces of evidence. The depth of the impression, the stride length, and the distinct morphology of a primate foot – the presence of opposable digits, for instance, even if massive – would be key identifiers. The geological context of these prints would be vital, indicating their age and the environment in which they were formed. Analysis of the surrounding sediment could reveal details about the creature's gait, speed, and even its weight distribution, providing further clues about its locomotion and mass.

Beyond footprints, other significant trace evidence would include hair and scat. The sheer size of the primate would necessitate correspondingly large hair shafts, potentially of a texture and diameter unlike any known mammal. Microscopic analysis of hair, a cornerstone of forensic science, would be crucial. This would involve examining the cuticle (the outer layer), the cortex (the middle layer containing pigment), and the medulla (the innermost core). For a hypothetical giant primate, these characteristics might be magnified, offering unique microscopic signatures. Pigmentation patterns, banding, and the overall structure of the hair could provide clues to its taxonomic placement within the primate order, even if the specific species were unknown. The density of hair growth would also be a factor; a creature of such size would likely have a substantial coat for insulation, and shedding would be a continuous process, leading to more frequent discovery of shed hairs in its habitat.

Fecal matter, or scat, would offer a treasure trove of information. The volume alone would be staggering, indicative of a digestive system processing immense quantities of food. Analysis of scat would go beyond mere identification of size and shape. Microscopic examination of undigested food particles, such as plant fibers, seeds, or chitinous insect exoskeletons, would reveal dietary habits, providing crucial insights into the primate's ecological niche. DNA analysis, even from degraded scat samples, could potentially be revolutionary. While challenges with degradation in the environment are significant, advancements in ancient DNA retrieval are constantly pushing the boundaries of what is possible. If viable DNA could be extracted, it could be compared against existing genomic databases to confirm its primate origin and potentially to identify its phylogenetic relationship to known primate lineages. Furthermore, the presence of specific gut flora within the scat could offer further clues about its digestive physiology and diet.

Perhaps the most direct, albeit challenging, evidence would be bone fragments. Even a creature of immense size does not exist without mortality. Fractured bones, weathered fragments, or even complete skeletal elements discovered in situ would provide undeniable proof. Forensic anthropologists would be tasked with analyzing these remains. They would assess the degree of post-mortem decomposition, identify signs of scavenging by other animals, and examine any evidence of trauma, such as fractures indicative of falls, fights, or environmental hazards. The sheer size and density of these bones would be immediately apparent, differentiating them from any known fauna. Radiographic analysis, even of fragmented pieces, could reveal internal structures, bone density, and potential signs of pathology or stress fractures, offering insights into the life and health of the individual. Isotopic analysis of bone collagen could reveal details about the primate's diet and geographic origin, tracing its movements and feeding grounds. The growth rings within bone, similar to tree rings, could even provide information about its age at death.

The interpretation of such evidence presents a unique set of challenges. Differentiating between the remains of a hypothetical gigantiform primate and other large fauna, or even geological formations that might mimic organic structures, would require rigorous scientific scrutiny. For instance, unusual geological concretions or fossilized wood might, at first glance, be mistaken for bone fragments by an untrained observer. The expertise of paleontologists and comparative anatomists would be crucial in this initial phase of identification. When dealing with potential hair, distinguishing it from that of known large mammals like elephants or rhinos would be a primary step, relying on microscopic and, if possible, genetic analysis. Similarly, large footprints could be mistaken for geological anomalies or the tracks of known megafauna if not carefully examined for primate-specific characteristics.

The scientific methods employed would draw heavily from established forensic and biological disciplines. For footprint analysis, techniques such as photogrammetry and 3D scanning could be used to create detailed models of impressions, allowing for precise measurement and comparison. Sediment analysis would involve palynology (the study of pollen and spores) to reconstruct the environment, and geochronology to date the deposits. For hair and fiber analysis, microscopy, spectrophotometry (to analyze color), and chemical analysis to identify unique protein structures would be employed. In the realm of DNA, advancements in next-generation sequencing would be critical, even if initial samples are degraded or present in low quantities. The process of contaminant removal and the use of specialized extraction protocols would be essential to obtain reliable genetic data.

The sheer scale of the evidence would necessitate adaptations in existing methodologies. Imagine needing specialized equipment to excavate, transport, and analyze a bone fragment the size of a small car, or to process scat that requires industrial-scale sieving. The development of new analytical tools or the modification of existing ones would likely be required. Furthermore, the rarity of such discoveries would mean that any evidence found would be of immense scientific importance, demanding the highest levels of meticulousness and verification. Collaboration between disparate scientific fields – forensic science, paleontology, primatology, genetics, and geology – would be essential. Each discipline would bring its unique expertise to bear on the fragmented puzzle pieces, slowly assembling a picture of this elusive creature.

One of the most exciting potential breakthroughs would come from molecular paleontology and the analysis of ancient proteins. Even if DNA is too degraded for sequencing, proteins can persist in organic remains for much longer periods. Techniques like mass spectrometry can identify unique protein sequences, providing a molecular fingerprint that could confirm primate ancestry and potentially even identify specific ancestral lineages. This could be applied to hair, bone, and even, in exceptionally preserved circumstances, to scat.

The concept of bio-signatures would also be critical. A large primate would not only leave physical traces but also alter its environment in subtle yet detectable ways. For instance, changes in vegetation patterns around a primate's foraging areas, distinct breakage patterns on branches it uses for locomotion or feeding, or even the presence of unusual microbial communities associated with its waste products could constitute indirect evidence. These bio-signatures would require ecological surveys and the application of remote sensing technologies, such as satellite imagery or drone-based hyperspectral imaging, to detect anomalies in the landscape that might correlate with the presence of an unknown, large primate.

The challenge of substantiating the existence of an unknown species through trace evidence is immense, but not insurmountable. History is replete with examples of scientific discoveries that began with seemingly isolated, enigmatic clues. The discovery of the Okapi, for instance, began with fragmented accounts from indigenous peoples and elusive physical traces, eventually leading to its formal scientific description. For a gigantiform primate, the initial evidence might be anecdotal reports, strange footprints, or unusual scat. However, applying the rigorous methodologies of forensic anthropology and trace evidence analysis, coupled with the power of modern scientific instrumentation and interdisciplinary collaboration, would provide the tools to transform these tantalizing clues into concrete, undeniable proof. The quest for such evidence demands patience, keen observation, and an unwavering commitment to scientific inquiry, pushing the boundaries of our understanding of the natural world and the potential for life within it. The silent forests and remote wildernesses of our planet may still hold secrets, and the meticulous examination of their physical traces is our best hope of uncovering them.

The journey to conceptualize a gigantiform primate is not one solely of imagination or speculative fiction; it is deeply rooted in the rigorous science of comparative anatomy. By dissecting the skeletal structures, muscular arrangements, and functional morphology of our closest living relatives, as well as those that have long since vanished into the fossil record, we can begin to sketch a plausible anatomical blueprint for such an extraordinary creature. This comparative approach allows us to extrapolate, with a degree of scientific grounding, the potential form, locomotion, and overall capabilities of a megafauna primate. It is through understanding the variations and commonalities within the primate order that we can construct a framework for what a truly colossal ape might entail.

Our point of departure must be the hominoid superfamily – the great apes and humans. Within this group, we find a remarkable range of sizes and adaptations, offering valuable insights. Consider the gorilla,

Gorilla gorilla, the largest living primate, with males reaching up to 1.8 meters (5.9 feet) in standing height and weighing as much as 200 kilograms (440 pounds). Their robust skeletal structure, particularly the dense bones of the limbs and torso, is a testament to the demands of their terrestrial lifestyle and powerful musculature, essential for quadrupedal locomotion and knuckle-walking. The sheer thickness of a gorilla's femur, for instance, provides a foundational understanding of how significant mass is supported.

In parallel, the orangutan (

Pongo genus) presents a different arboreal model, with males of *Pongo abelii* and *Pongo pygmaeus* reaching up to 1.4 meters (4.6 feet) and weighing around 90 kilograms (200 pounds). Their elongated arms, specialized shoulder joints, and prehensile feet are adaptations for a life spent primarily in the trees. While smaller than gorillas, their skeletal proportions, particularly the relative length of their upper limbs, highlight the diversity of form within the great apes and suggest alternative locomotor strategies that a larger primate might or might not employ.

Then there are ourselves,

Homo sapiens. Our bipedal locomotion, a defining characteristic, is supported by a suite of specialized anatomical features: a broad pelvis that cradles the abdominal organs, a splayed foot with a robust heel bone, and an S-shaped vertebral column that distributes weight efficiently. The precise angle of the femur, articulating with the knee joint, and the forward-facing foramen magnum at the base of the skull are all crucial for upright posture. Extrapolating from our bipedalism to a significantly larger primate introduces immediate questions about structural integrity and the energetic costs of locomotion. A bipedal primate of enormous size would require proportionally thicker and stronger bones, larger joint surfaces to distribute stress, and a significantly expanded musculature, particularly in the gluteal and hamstring regions, to facilitate movement and maintain balance. The biomechanics of such a gait would be profoundly different from our own, likely necessitating a slower, more deliberate pace to avoid excessive strain on the skeletal system.

However, to truly grasp the potential for gigantism within the primate lineage, we must turn our gaze to the fossil record, a repository of extinct forms that push the boundaries of what we consider typical. The most compelling example is

Gigantopithecus blacki, an extinct ape that roamed Southeast Asia from perhaps 9 million years to as recently as 200,000 years ago. This is not merely a slightly larger ape; *Gigantopithecus* was truly colossal. Based on fossilized teeth and a few jaw fragments, paleontologists estimate that *Gigantopithecus* stood at an imposing height of around 3 meters (10 feet) and weighed between 200 and 550 kilograms (440 to 1200 pounds), making it the largest primate known to have ever lived.

The teeth of

Gigantopithecus are particularly informative. They are remarkably large and robust, with thick enamel, suggesting a diet consisting of tough, fibrous plant matter, such as bamboo, fruits, and bark. This dietary adaptation for processing coarse vegetation is a key feature that distinguishes it from other great apes, which tend to have more varied diets that include insects and animal matter. The sheer size of the molars and premolars, along with the broad, flat surfaces, indicates immense grinding power. This would have necessitated a powerful jaw musculature, anchored to a proportionally larger and more robust skull and mandible. While no complete skull has been found, reconstructions based on jaw fragments suggest a broad, prognathic face, distinct from the flatter faces of modern humans and chimpanzees. The mandible itself would have been exceptionally deep and thick to support these forces.

The limited fossil evidence suggests that

Gigantopithecus was likely a quadruped, or at least capable of efficient quadrupedal locomotion, perhaps with occasional bipedal stances. Its limb bones, if any are ever definitively identified and attributed to the genus, would likely be massively proportioned, far exceeding those of modern gorillas. Imagine femurs and tibias as thick as tree trunks, capable of supporting immense weight. The joints, such as the hip and knee, would have been correspondingly large and robust, designed to withstand pressures far beyond those experienced by any living primate. The structure of the pelvis would also be a critical point of comparison. For efficient bipedalism, a broad, basin-shaped pelvis is essential. While *Gigantopithecus* may not have been habitually bipedal in the human sense, any capacity for upright posture would have required modifications to its pelvic structure, though likely less pronounced than those seen in *Homo sapiens*.

When comparing

Gigantopithecus to modern great apes, the differences in overall body plan become stark. While gorillas are powerfully built, *Gigantopithecus* likely represented an evolutionary pathway towards extreme mass. This gigantism would have had profound implications for its locomotion. Knuckle-walking, as seen in gorillas and chimpanzees, might have been a primary mode of terrestrial movement, but the sheer scale would have altered the mechanics. The distance covered with each stride would be significantly greater, and the forces transmitted through the limbs would be immense. The center of gravity would also be higher, requiring adaptations in musculature and skeletal bracing for stability.

The potential for a giant, bipedal ape beyond

Gigantopithecus necessitates a careful consideration of the anatomical trade-offs involved in extreme bipedalism. While humans have evolved a highly efficient bipedal gait, it comes with its own vulnerabilities, such as increased susceptibility to back pain and knee problems. For a primate vastly larger than us, these challenges would be amplified exponentially. A bipedal gait for a creature weighing half a ton or more would require an exceptionally strong and stable vertebral column, capable of withstanding immense compressive forces. The lumbar vertebrae would likely be significantly larger and more robust than those of humans, with broader articular facets to prevent slippage. The sacrum, the fused vertebrae that form the posterior part of the pelvis, would also need to be exceptionally strong to support the weight of the upper body.

The pelvis itself would be a critical area of investigation. For a bipedal ape, the pelvis would need to be broad and bowl-shaped to support the abdominal viscera and provide broad attachment points for powerful gluteal muscles, which are essential for extending the hip and propelling the body forward. The iliac blades would likely be expanded laterally, and the pubic symphysis would need to be reinforced to withstand the stresses of bipedal locomotion. The angle of the femoral neck and shaft, as well as the structure of the hip joint, would also be key indicators of bipedal adaptation. A more vertical femoral shaft and a well-developed femoral head articulating with a deep acetabulum would be expected.

Furthermore, the foot anatomy would be crucial. While modern humans have a non-opposable hallux (big toe) and a pronounced arch for shock absorption and propulsion, a giant bipedal ape might exhibit a slightly different foot structure. While opposability of the digits is a defining primate characteristic, it might be reduced or modified in a large, terrestrial species to provide a more stable platform for walking. The calcaneus (heel bone) would likely be significantly enlarged to absorb impact, and the metatarsals and phalanges would be robust, designed for weight-bearing rather than grasping.

Considering the potential for such a creature to inhabit remote, forested environments, arboreal capabilities might still be a factor, even for a large primate. While a 3-meter tall, 500-kilogram ape is unlikely to be a regular canopy dweller in the same way as an orangutan, some degree of climbing or brachiating ability cannot be entirely ruled out, particularly in younger individuals or for specific foraging strategies. This would imply adaptations in the shoulder girdle, such as a broad scapula positioned dorsally on the rib cage, and a highly mobile shoulder joint with a deep glenoid fossa, allowing for a wide range of arm movement. The clavicle would need to be strong to transmit forces from the arm to the trunk.

The cranial anatomy of a hypothetical giant primate is another area where comparative analysis proves invaluable. While the teeth of

Gigantopithecus suggest a herbivorous or omnivorous diet with a heavy reliance on tough plant materials, a primate with more generalized primate feeding habits but on a larger scale would likely possess different cranial features. A larger brain size, as discussed in earlier chapters, would necessitate a larger cranial vault. However, primate brain size does not always scale linearly with body size. The encephalization quotient (EQ) – a measure of relative brain size compared to what would be expected for an animal of its size – would be a critical parameter. If our hypothetical giant primate possessed cognitive abilities comparable to or exceeding those of modern great apes, its brain would indeed be substantial, requiring a proportionally larger cranial capacity.

The foramen magnum, the opening at the base of the skull through which the spinal cord passes, would likely be positioned more anteriorly and inferiorly in a habitually bipedal creature, reflecting a more balanced head posture. In contrast, a more quadrupedal primate would typically have a more posteriorly placed foramen magnum. The muscle attachment sites on the occipital bone would also provide clues about head and neck posture. Strong nuchal musculature would be required to support a large head, especially if it were to become more prognathic, extending forward from the neck.

Comparative anatomy also allows us to consider the potential for sexual dimorphism. In many primate species, males are significantly larger than females, a trait often associated with increased competition for mates. If a giant primate species exhibited significant sexual dimorphism, the anatomical differences between males and females could be extreme. Males might possess even more robust skeletal structures, larger canines (though perhaps less pronounced if the diet was primarily herbivorous), and more pronounced cranial crests for muscle attachment.

Beyond the skeletal system, comparative anatomy extends to the musculature. The sheer mass of a giant primate would necessitate a proportionately massive musculature to facilitate movement, maintain posture, and perform daily activities. The gluteal muscles, quadriceps, and hamstrings would be particularly well-developed for bipedal locomotion. For quadrupedalism, the pectoral muscles, deltoids, and back muscles would be heavily employed. The distribution of muscle mass would also be critical for balance and efficiency.

When we consider the possibility of a primate that bridges the gap between known hominoids and the colossal

Gigantopithecus, perhaps a creature that retained some degree of arboreal agility while being significantly larger than a gorilla and habitually bipedal, the anatomical synthesis becomes even more intricate. Such a being would represent a remarkable evolutionary balancing act. Its skeletal structure would need to be robust enough to support immense weight during terrestrial locomotion, yet flexible enough to allow for movement through complex forest environments. The proportions might lean towards a more stocky build than that of a modern human, with shorter, thicker limbs relative to torso length, optimizing for stability over speed.

The study of primate paleontology, including the enigmatic

Australopithecus and *Homo* genera, provides further grist for this anatomical mill. While these hominids are generally within the size range of modern humans or slightly larger, their diverse locomotion strategies – from arboreal adaptations in early australopithecines to the fully committed bipedalism of *Homo erectus* – offer models for how gait and skeletal morphology evolve. For instance, the curved finger and toe bones of *Australopithecus afarensis* suggest climbing abilities, while the robusticity of their limb bones indicates significant terrestrial activity. Extrapolating these trends to a much larger scale, while not directly applicable, highlights the continuum of adaptation within the hominoid lineage.

Ultimately, comparative anatomy provides the essential framework for envisioning a gigantiform primate not as a mere monstrous fantasy, but as a biologically plausible entity. By scrutinizing the skeletal structures, muscle attachments, and functional adaptations of existing and extinct primates, we can construct a detailed hypothetical blueprint. This blueprint, grounded in the observable realities of primate evolution, allows us to ask specific questions about the potential anatomy of such a creature: the required bone density, the necessary muscle mass, the probable limb proportions, the pelvic and vertebral adaptations for locomotion, and the cranial features that would support its physiology and diet. It is this scientific foundation that allows us to move from the realm of pure speculation to that of informed hypothesis, preparing us to recognize the distinctive anatomical signatures that a truly colossal primate would, by necessity, leave behind.

3

Ecological Niches and Habitat Requirements

The preceding chapter delved into the anatomical plausibility of a gigantiform primate, drawing parallels with existing and extinct ape species to construct a hypothetical blueprint. Now, we turn our attention to the ecological stage upon which such a creature might perform. The existence of any species, particularly one of significant size and presumably elusive nature, is inextricably linked to the availability of suitable habitats. Identifying these potential strongholds requires a methodical survey of the globe, pinpointing regions that offer not only the necessary resources but also the relative sanctuary from anthropogenic pressures that a large, rare animal would demand.

Our search begins with an appreciation for the fundamental ecological requirements of large primates. These are animals with substantial energetic needs, necessitating access to abundant food sources. For a creature hypothetically akin to

Gigantopithecus, with its inferred reliance on tough plant matter, this implies extensive areas of dense vegetation, encompassing diverse floras capable of sustaining a large herbivore or omnivore throughout the year. Forests, in their myriad forms, emerge as the prime candidates. Tropical and subtropical rainforests, with their unparalleled biodiversity and year-round growing seasons, represent the most productive ecosystems on Earth. Similarly, temperate rainforests, though less diverse, can support substantial biomass. Even montane forests, with their unique microclimates and specialized flora, could offer refuge and sustenance.

Beyond the sheer availability of food, the structure of the habitat plays a crucial role. Large primates, especially those with a significant terrestrial component to their locomotion, require landscapes that facilitate movement while also providing cover. Dense undergrowth, a complex canopy structure, and varied terrain – including valleys, slopes, and plateaus – all contribute to a habitat that can conceal a large animal and offer a mosaic of foraging opportunities. Steep canyons, rugged mountain ranges, and extensive river systems can act as natural barriers, fragmenting human access and potentially isolating populations of elusive fauna.

Furthermore, water availability is a non-negotiable requirement. Large animals have high hydration needs, and proximity to reliable water sources – rivers, lakes, and even consistent rainfall – is essential. These water bodies also often dictate the distribution of vegetation, creating riparian corridors that can serve as vital travel routes and foraging grounds.

The critical factor that elevates a potentially suitable ecosystem to a viable habitat for a large, undiscovered primate is the degree of human encroachment. Large animals, with their extensive home ranges and dietary needs, are particularly vulnerable to habitat fragmentation and direct human conflict. Areas that have experienced significant deforestation, agriculture, urbanization, or industrial development are far less likely to support a population of such a creature. Therefore, our focus must be on the planet's most remote and least disturbed regions.

Using Geographic Information Systems (GIS) and ecological data, we can begin to delineate these potential zones. These sophisticated tools allow us to overlay various datasets, including land cover, elevation, precipitation, human population density, and infrastructure development, to identify areas that meet our criteria. We can identify vast tracts of unbroken forest, mountainous regions with limited road access, and large river basins that have remained largely untouched by human activity.

Let us consider the implications of this approach for specific biogeographic regions. Southeast Asia, the historical home of

Gigantopithecus, remains a compelling candidate. The vast, contiguous rainforests of Borneo and Sumatra, for instance, are characterized by their immense scale, rugged interior topography, and historically low human population densities in many areas. The dense vegetation provides ample cover and a rich diversity of plant life, including bamboo species which may have formed a significant part of the diet for its extinct relative. The mountainous spine of Borneo, in particular, with its challenging terrain and limited accessibility, presents an ideal scenario for an elusive, large primate. Similarly, the remote rainforests of mainland Southeast Asia, encompassing parts of Myanmar, Laos, Vietnam, and Cambodia, offer expansive, often politically unstable, regions where large tracts of forest remain unsurvivable to extensive human settlement. The biodiversity of these regions is well-documented, and the presence of numerous large mammals suggests a capacity to support substantial fauna.

Moving eastward, the islands of Melanesia, particularly New Guinea and its surrounding islands, represent another significant frontier. New Guinea, the world's largest tropical island, is characterized by an exceptionally rugged interior dominated by towering mountain ranges, deep valleys, and vast expanses of pristine rainforest. The island's relatively small human population, concentrated along the coasts and major river valleys, has left the interior largely unexplored and ecologically intact. The sheer scale of its wilderness, coupled with its isolation, makes it a prime candidate for harboring undiscovered large fauna. The altitudinal zonation of New Guinea's forests also creates a remarkable diversity of habitats, from lowland swamps to alpine meadows, each potentially supporting specialized populations.

The Amazon Basin, the largest rainforest on Earth, is another obvious contender. Its immense size and the relatively low human population density in its deepest interior provide a vast potential refuge. The sheer immensity of the Amazon, spanning nine countries, means that even with significant deforestation in certain areas, enormous tracts of primary forest remain. The complex network of rivers acts as both a barrier and a conduit for wildlife, and the dense vegetation offers unparalleled concealment. While many of the Amazon's larger fauna are well-documented, the sheer scale and the challenging conditions for exploration mean that the possibility of undiscovered species, particularly primates of unusual size, cannot be discounted. Regions with minimal road access, such as parts of the western Amazon in Peru or remote areas of the Brazilian Amazon, are of particular interest.

Africa, with its diverse array of primates, also holds potential. While much of Africa's primate diversity is well-known, the vast, largely inaccessible rainforests of the Congo Basin present a significant area of interest. This immense rainforest, second only to the Amazon in size, is characterized by its dense vegetation, extensive river systems, and historically low human population density in its core. Despite ongoing challenges related to resource extraction and political instability, large areas of the Congo Basin remain remarkably wild. The gorillas and chimpanzees that inhabit these forests demonstrate the capacity of the ecosystem to support large primate populations. A hypothetical larger primate might occupy a niche within these same environments, exploiting resources that are less accessible to its smaller relatives. Remote areas within the Democratic Republic of Congo, the Republic of Congo, and parts of Cameroon and the Central African Republic are particularly compelling.

Even more temperate regions, often overlooked in the search for large, tropical primates, could harbor surprises. The temperate rainforests of the Pacific Northwest of North America, for example, while lacking the primate diversity of tropical regions, possess a dense forest structure and remote mountainous areas that have historically provided refuge for large, elusive mammals like the Bigfoot of folklore. While the direct comparison to a primate is speculative, it highlights the principle that large, undiscovered animals can persist in unexpected environments, provided they offer sufficient resources and isolation. Similarly, the boreal forests and mountain ranges of the Russian Far East, particularly areas bordering China and North Korea, are characterized by vast, rugged wilderness with a sparse human population. These regions are known to support large mammals such as bears and tigers, indicating a healthy ecosystem capable of sustaining significant biomass.

When applying GIS analysis to these regions, several key indicators emerge. Firstly, we look for contiguous areas of forest cover exceeding a certain threshold, perhaps tens of thousands of square kilometers, to accommodate the potential home range of a large primate. Secondly, we assess the degree of fragmentation, prioritizing areas with minimal road networks, human settlements, and agricultural land. Thirdly, we consider the ruggedness of the terrain, favoring regions with significant elevation changes, steep slopes, and natural barriers that impede human access. Finally, we examine indices of biodiversity and primary productivity, as these are directly correlated with the availability of food resources.

The confluence of these factors – dense, diverse forests, rugged terrain, and minimal human disturbance – creates the ideal crucible for the undetected persistence of a large, elusive primate. These are not simply areas of wilderness; they are ecological refugia, pockets of the planet where evolutionary processes may have continued largely unhindered by modern human influence. These are the landscapes that invite further exploration, the places where the shadow of an unknown giant might, just might, still fall. It is within these remote, biodiverse expanses that the scientific endeavor to find evidence of such a creature must focus its efforts. The ecological niche it occupies, the resources it requires, and the very remoteness that allows it to remain hidden are all dictated by the geography and ecology of these extraordinary, under-explored regions. The challenge lies in the sheer scale of these areas and the inherent difficulty in surveying them comprehensively. Yet, it is precisely this vastness and inaccessibility that offers the most compelling possibility for the continued existence of a creature that has evaded scientific detection thus far.

The preceding chapter laid the groundwork by exploring the anatomical and evolutionary plausibility of a gigantiform primate, drawing upon existing and extinct ape species to construct a hypothetical physical profile. Now, we pivot to the ecological theatre upon which such a creature might have played its part. The very existence of any species, especially one of considerable size and presumed elusiveness, is fundamentally bound to the availability of suitable habitats. Identifying these potential strongholds necessitates a methodical global survey, pinpointing regions that offer not only the essential resources but also the relative sanctuary from human pressures that a large, rare animal would demand. Our investigation begins with an appreciation for the core ecological needs of large primates. These are animals with significant energetic requirements, demanding access to abundant food sources. For a hypothetical creature akin to *Gigantopithecus*, with its inferred reliance on fibrous plant matter, this implies extensive areas of dense vegetation, encompassing diverse flora capable of sustaining a large herbivore or omnivore year-round. Forests, in their myriad forms, emerge as the prime candidates. Tropical and subtropical rainforests, with their unparalleled biodiversity and continuous growing seasons, represent the most productive ecosystems on Earth. Similarly, temperate rainforests, though less diverse, can support substantial biomass. Even montane forests, with their unique microclimates and specialized flora, could offer refuge and sustenance. Beyond the sheer availability of food, habitat structure is crucial. Large primates, particularly those with a significant terrestrial locomotion component, require landscapes that facilitate movement while offering cover. Dense undergrowth, complex canopy structure, and varied terrain – including valleys, slopes, and plateaus – all contribute to a habitat that can conceal a large animal and provide a mosaic of foraging opportunities. Steep canyons, rugged mountain ranges, and extensive river systems can act as natural barriers, fragmenting human access and potentially isolating populations of elusive fauna. Furthermore, water availability is non-negotiable. Large animals have high hydration needs, and proximity to reliable water sources – rivers, lakes, and consistent rainfall – is essential. These water bodies often dictate vegetation distribution, creating riparian corridors vital for travel and foraging. The critical factor elevating a potentially suitable ecosystem to a viable habitat for a large, undiscovered primate is the degree of human encroachment. Large animals, with their extensive home ranges and dietary needs, are particularly vulnerable to habitat fragmentation and direct human conflict. Areas with significant deforestation, agriculture, urbanization, or industrial development are less likely to support such a creature. Therefore, our focus must be on the planet's most remote and least disturbed regions. Using Geographic Information Systems (GIS) and ecological data, we can delineate these potential zones. These tools allow us to overlay datasets like land cover, elevation, precipitation, human population density, and infrastructure development. We can identify vast tracts of unbroken forest, mountainous regions with limited road access, and large river basins largely untouched by human activity. Southeast Asia, the historical home of *Gigantopithecus*, remains a compelling candidate. The vast rainforests of Borneo and Sumatra, for instance, are characterized by their immense scale, rugged interior topography, and historically low human population densities in many areas. The dense vegetation offers ample cover and a rich diversity of plant life, including bamboo species that may have been a dietary staple for its extinct relative. The mountainous spine of Borneo, with its challenging terrain and limited accessibility, presents an ideal scenario for an elusive, large primate. Similarly, the remote rainforests of mainland Southeast Asia, encompassing parts of Myanmar, Laos, Vietnam, and Cambodia, offer expansive, often politically unstable, regions where large forest tracts remain largely free from extensive human settlement. The biodiversity of these regions is well-documented, and the presence of numerous large mammals suggests a capacity to support substantial fauna. Moving eastward, the islands of Melanesia, particularly New Guinea and its surrounding islands, represent another significant frontier. New Guinea, the world's largest tropical island, is characterized by an exceptionally rugged interior dominated by towering mountain ranges, deep valleys, and vast expanses of pristine rainforest. The island's relatively small human population, concentrated along coasts and major river valleys, has left the interior largely unexplored and ecologically intact. The sheer scale of its wilderness, coupled with its isolation, makes it a prime candidate for harboring undiscovered large fauna. The altitudinal zonation of New Guinea's forests creates remarkable habitat diversity, from lowland swamps to alpine meadows, each potentially supporting specialized populations. The Amazon Basin, the largest rainforest on Earth, is another obvious contender. Its immense size and the relatively low human population density in its deepest interior provide a vast potential refuge. The sheer immensity of the Amazon, spanning nine countries, means that enormous tracts of primary forest remain despite significant deforestation in certain areas. The complex river network acts as both a barrier and a conduit for wildlife, and the dense vegetation offers unparalleled concealment. While many of the Amazon's larger fauna are well-documented, the sheer scale and the challenging conditions for exploration mean the possibility of undiscovered species, particularly primates of unusual size, cannot be discounted. Regions with minimal road access, such as parts of the western Amazon in Peru or remote areas of the Brazilian Amazon, are of particular interest. Africa, with its diverse primate array, also holds potential. While much of Africa's primate diversity is well-known, the vast, largely inaccessible rainforests of the Congo Basin present a significant area of interest. This immense rainforest, second only to the Amazon in size, is characterized by dense vegetation, extensive river systems, and historically low human population density in its core. Despite ongoing challenges related to resource extraction and political instability, large areas of the Congo Basin remain remarkably wild. The gorillas and chimpanzees inhabiting these forests demonstrate the ecosystem's capacity to support large primate populations. A hypothetical larger primate might occupy a niche within these same environments, exploiting resources less accessible to its smaller relatives. Remote areas within the Democratic Republic of Congo, the Republic of Congo, and parts of Cameroon and the Central African Republic are particularly compelling. Even more temperate regions, often overlooked in the search for large, tropical primates, could harbor surprises. The temperate rainforests of the Pacific Northwest of North America, for example, while lacking the primate diversity of tropical regions, possess a dense forest structure and remote mountainous areas that have historically provided refuge for large, elusive mammals like the Bigfoot of folklore. While direct comparison to a primate is speculative, it highlights the principle that large, undiscovered animals can persist in unexpected environments if they offer sufficient resources and isolation. Similarly, the boreal forests and mountain ranges of the Russian Far East, particularly areas bordering China and North

Korea, are characterized by vast, rugged wilderness with a sparse human population. These regions support large mammals like bears and tigers, indicating a healthy ecosystem capable of sustaining significant biomass. Applying GIS analysis to these regions reveals key indicators. Firstly, we seek contiguous forest cover exceeding tens of thousands of square kilometers to accommodate the potential home range of a large primate. Secondly, we assess fragmentation, prioritizing areas with minimal road networks, human settlements, and agricultural land. Thirdly, we consider terrain ruggedness, favoring regions with significant elevation changes, steep slopes, and natural barriers impeding human access. Finally, we examine biodiversity and primary productivity indices, correlated with food resource availability. The confluence of these factors – dense, diverse forests, rugged terrain, and minimal human disturbance – creates an ideal crucible for the undetected persistence of a large, elusive primate. These are ecological refugia, pockets of the planet where evolutionary processes may have continued largely unhindered by modern human influence. These are the landscapes that invite further exploration, the places where the shadow of an unknown giant might still fall. It is within these remote, biodiverse expanses that the scientific endeavor to find evidence of such a creature must focus. The ecological niche it occupies, the resources it requires, and the very remoteness that allows it to remain hidden are all dictated by the geography and ecology of these extraordinary, under-explored regions. The challenge lies in the sheer scale of these areas and the difficulty of comprehensive surveying. Yet, it is precisely this vastness and inaccessibility that offers the most compelling possibility for the continued existence of a creature that has evaded scientific detection thus far.

Forest Ecosystems: The Primate Stronghold

The overwhelming majority of known primate species are intrinsically linked to forest environments. This is no accident; forests, particularly tropical and subtropical rainforests, represent the pinnacle of terrestrial biodiversity and ecological productivity. For a hypothetical large primate, these arboreal and terrestrial sanctuaries would provide not only the fundamental necessities for survival but also the critical elements for successful reproduction and the maintenance of a viable population. Understanding these forest ecosystems as potential strongholds for an unknown giant requires a detailed examination of their key characteristics: canopy structure, water availability, and the sheer diversity and abundance of vegetation.

The canopy of a mature forest is far more than just a ceiling of leaves. It is a complex, three-dimensional environment that dictates light penetration, humidity levels, and temperature regulation. For a large primate, this canopy offers invaluable benefits. Firstly, it provides crucial shelter. The dense foliage and interlocking branches offer protection from harsh weather conditions, such as torrential rains and intense solar radiation, and also serve as a vital defense against aerial predators. For a creature of significant size, the ability to move through this dense, multi-layered environment, utilizing larger branches for support and navigating through the interstitial spaces, would be essential. The structure of the canopy also influences the distribution of food resources. Many fruits, flowers, and leaves are found in the upper strata, making a primate adept at arboreal or semi-arboreal locomotion highly advantageous. Even for a primarily terrestrial primate, the canopy's edge and the forest floor beneath it are rich foraging grounds, benefiting from the nutrient fall from above. The presence of large, well-established trees, often with buttress roots and wide girths, is also significant. These provide stable platforms for resting, sleeping, and even nesting sites. Furthermore, the structural complexity of a mature forest canopy, with its diverse array of tree species, epiphyte growth, and vine networks, creates microhabitats that can support a wide range of food sources. This is particularly important for large herbivores or omnivores, which require a constant and varied intake of plant matter, fruits, and possibly insects or small invertebrates. The sheer volume of biomass contained within these forest ecosystems is staggering, offering a continuous supply of sustenance throughout the year, a critical factor for a large animal with substantial daily caloric needs.

Water is the lifeblood of any ecosystem, and for large primates, its consistent availability is non-negotiable. Forests, especially rainforests, are characterized by high precipitation levels and the presence of numerous water bodies. Rivers, streams, swamps, and even ephemeral pools created by rainfall all serve as vital sources of hydration. The proximity of these water sources is not merely for drinking; they also influence the types of vegetation that flourish in their vicinity. Riparian zones, the areas along rivers and streams, are often characterized by particularly lush and diverse plant growth, providing an enhanced foraging buffet. These watercourses can also act as natural corridors, facilitating movement through the forest for both the primate and its prey or food sources, while also acting as barriers to other, potentially competing, species or predators. The dense vegetation surrounding water sources also offers additional cover and concealment, crucial for a large, potentially shy animal. Furthermore, many primates engage in social behaviors centered around water sources, such as bathing, grooming, or communal drinking. Such behaviors, if extrapolated to a larger primate, would suggest that areas with reliable and accessible water bodies would be of significant importance to its daily routine and social structure. The ability to navigate and utilize these aquatic features, perhaps even to cross smaller streams or lakes, could also influence its dispersal patterns and range expansion. The presence of extensive river systems, like those found in the Congo Basin or the Amazon, creates vast, interconnected forest landscapes that are difficult to traverse for humans but perfectly suited for the dispersal of large arboreal or semi-terrestrial animals.

The diversity and abundance of vegetation are perhaps the most critical components of a forest ecosystem for sustaining large primates. Unlike smaller primates, which may specialize in narrow dietary niches like insects or specific types of fruit, a large primate would likely require a broad spectrum of plant matter to meet its substantial energetic demands. This includes a wide variety of leaves, shoots, fruits, seeds, and possibly roots and tubers. The sheer botanical richness of tropical rainforests is astounding, with hundreds of tree species in a single hectare, each offering potentially different nutritional profiles and palatability. For a large herbivore or omnivore, this diversity ensures that even if certain food sources are seasonally unavailable, others will be present to fill the dietary gap. Consider the ecological requirements of existing large primates, such as the lowland gorillas (

Gorilla gorilla) and the mountain gorillas (*Gorilla beringei beringei*) of Africa. Lowland gorillas, inhabiting dense rainforests and swamp forests, are predominantly herbivorous, consuming a vast array of leaves, stems, fruits, and occasionally insects. Their diet can include over 100 different plant species. Mountain gorillas, in contrast, living in montane forests, have a diet primarily composed of leaves, stems, and roots, adapted to the cooler climate and different flora. Both species demonstrate the capacity of forest ecosystems to support large primate biomass through a diverse plant diet. The hypothetical gigantiform primate, with its potentially larger body size and digestive system, would require even greater quantities and varieties of vegetation. This might favor forest types with a high proportion of nutrient-rich browse, abundant fruiting trees, and perhaps a significant understory or ground layer that can provide tubers or roots. The presence of specific plant families known for their palatability and nutritional value to primates, such as palms, figs, and legumes, would be a strong indicator of a suitable habitat. Bamboo, a staple for the extinct *Gigantopithecus blacki*, also thrives in certain forest environments, particularly in Southeast Asia and parts of the Himalayas, suggesting that forests with a strong bamboo presence could have been a crucial element of its habitat. The structure of the understory is also important, providing ground-level foraging opportunities and pathways for movement.

The Congo Basin, often referred to as the "Green Lung" of Africa, serves as a quintessential example of a forest ecosystem capable of supporting immense primate populations. This vast expanse of tropical rainforest, second only to the Amazon in size, is home to a remarkable diversity of primate species, including the world's largest living primates: the Western and Eastern Gorillas. These forests are characterized by a dense, multi-layered canopy, a rich understory, and an extensive network of rivers and swamps. The sheer scale of the Congo Basin, covering millions of square kilometers and spanning several countries, provides vast tracts of relatively undisturbed forest. The high rainfall and consistent temperatures support year-round plant growth, ensuring a continuous supply of food for gorillas and other primates. Gorillas, with their substantial body mass, require extensive foraging grounds. A troop of lowland gorillas, for instance, might cover several square kilometers in a single day while foraging. The presence of numerous fruiting trees, such as various species of

Pandanus, *Gilbertiodendron*, and *Dialium*, provides essential dietary components. They also consume significant quantities of leaves and stems from herbaceous plants found in the forest understory and clearings. The gorillas' reliance on these varied food sources highlights the importance of forest heterogeneity – the presence of different forest types, including lowland rainforest, swamp forest, and montane forest at higher elevations. This heterogeneity ensures a mosaic of food availability throughout the year and across different altitudes. For a hypothetical larger primate, the conditions found in the Congo Basin would be highly favorable. The dense vegetation provides excellent concealment, while the extensive waterways offer hydration and travel routes. The presence of large, established trees would provide resting and sleeping sites, and the sheer volume of plant matter available would support a significant energetic demand. The relatively low human population density in the deep interior of the basin, coupled with the challenging terrain and dense vegetation, further enhances its potential as a refuge for large, elusive fauna.

Similarly, the ancient rainforests of Southeast Asia, the historical territory of

Gigantopithecus, offer a compelling blueprint. Islands like Borneo and Sumatra, with their immense primary rainforests, possess the critical characteristics. These islands are characterized by a highly diverse flora, with numerous species of figs, palms, and other fruit-bearing trees that are vital food sources for primates. The rugged interior topography, with its steep mountains and deep valleys, creates natural barriers that limit human access and preserve vast, unbroken tracts of forest. This terrain also provides a complex environment, offering varied foraging opportunities across different altitudes and slopes. The presence of bamboo, a significant component of the *Gigantopithecus* diet, is also notable in many Southeast Asian forest ecosystems. These forests, particularly the dipterocarp rainforests, are renowned for their complex canopy structure, with emergent trees reaching considerable heights, creating a multi-layered environment that offers both food and shelter. The consistent rainfall supports year-round growth, and numerous rivers and streams provide essential water sources. For a large primate, these conditions would be ideal. The dense vegetation offers unparalleled concealment, while the abundance and variety of food sources would meet substantial dietary needs. The ruggedness of the terrain would further aid in maintaining isolation from human populations, allowing a large, potentially slow-reproducing animal to persist undetected. The historical evidence of *Gigantopithecus* itself within these regions strongly suggests that these forest ecosystems possess the inherent capacity to support a large primate.

The Amazon Basin represents another global hotspot for forest ecosystems that could potentially harbor a large, undiscovered primate. As the largest tropical rainforest on Earth, its sheer scale is almost incomprehensible. Spanning over 5.5 million square kilometers, it contains an unparalleled diversity of plant and animal life. The Amazon is characterized by a dense, multi-layered canopy, a humid climate, and an intricate network of rivers, the largest of which is the Amazon River itself. This vast river system acts as a natural highway, facilitating dispersal for many species, while also creating diverse aquatic and semi-aquatic habitats. The sheer abundance and variety of plant life in the Amazon are staggering, offering a rich larder for herbivores and omnivores. Numerous species of palms, figs, and other fruiting trees are found throughout the basin, providing a continuous supply of food. The understory is also dense and varied, offering opportunities for foraging on leaves, shoots, and roots. For a large primate, the Amazon would offer immense tracts of prime habitat. The dense vegetation would provide excellent concealment, and the abundance of food sources would meet significant energetic requirements. The presence of numerous rivers and streams would provide essential hydration and potential travel corridors. While many of the Amazon's larger fauna are well-documented, the sheer immensity of the region, coupled with the challenging conditions for exploration in its deepest interiors, means that the possibility of undiscovered species, particularly large primates, cannot be entirely discounted. Regions with limited human infrastructure and road access, such as parts of the western Amazon in Peru, Ecuador, or remote areas of Brazil and Colombia, represent particularly compelling potential strongholds. These areas combine vast expanses of primary forest with challenging terrain and limited human intrusion, creating ideal conditions for a large, elusive animal to thrive undetected.

Beyond the tropics, even some temperate forest ecosystems could theoretically provide a niche. The temperate rainforests of the Pacific Northwest of North America, for instance, while lacking the primate diversity of their tropical counterparts, possess a dense forest structure and rugged mountainous interiors that have historically supported large, elusive mammals. These forests are characterized by a dense canopy of conifers, a thick understory, and significant annual rainfall. While the available plant food might be less diverse and abundant than in tropical rainforests, the sheer scale of the wilderness, particularly in areas like the Olympic Peninsula or parts of British Columbia, offers significant potential for concealment and the support of a large herbivore. The presence of large mammals such as bears and elk in these regions indicates the ecosystem's capacity to sustain substantial biomass. For a primate adapted to cooler climates and a different suite of food sources, such as tough plant matter, roots, and berries, these temperate forests could offer a viable, albeit less likely, refuge. Similarly, the vast boreal forests and mountainous regions of the Russian Far East, particularly areas bordering China and North Korea, are characterized by rugged wilderness with a sparse human population. These regions are known to support large mammals, including bears and Siberian tigers, suggesting a healthy ecosystem capable of sustaining significant biomass. While these are not traditional primate habitats, the principle remains: large, undisturbed forest ecosystems, irrespective of latitude, can provide the necessary resources and isolation for large, undiscovered fauna.

In essence, forest ecosystems, from the steamy depths of tropical rainforests to the cool, damp environs of temperate woodlands, represent the ultimate primate strongholds. They offer the essential combination of abundant food, crucial shelter, reliable water sources, and, most importantly, the dense vegetation and complex terrain necessary for concealment and isolation. The continued existence of a large, undiscovered primate is most probable within the largest, most intact, and least disturbed forest ecosystems on Earth, where the intricate web of life provides the resources and the sanctuary for such a magnificent creature to remain hidden from the ever-watchful eyes of human civilization. These are the ancient cathedrals of nature, where evolution may still be crafting its largest secrets.

Mountainous Regions and Alpine Zones: The Realm of the Unseen

While the verdant depths of tropical rainforests and the sprawling expanses of temperate woodlands represent the archetypal habitats for primate life, the rugged, often inhospitable terrains of mountainous regions and alpine zones present a fascinating, albeit more speculative, set of ecological niches. These environments, characterized by dramatic elevation changes, extreme temperature fluctuations, and unique flora, are intrinsically linked in folklore and cryptozoology with the existence of large, elusive hominoids. The most prominent example, of course, is the Yeti, or Abominable Snowman, whose legendary habitat is firmly rooted in the high Himalayas. Moving beyond myth, however, it is crucial to scientifically assess the ecological feasibility of a large primate successfully inhabiting and thriving in such demanding landscapes.

The primary challenges for any large mammal in mountainous and alpine environments are manifold. Foremost among these is the availability of food. Unlike the continuous, year-round abundance found in many lowland forests, mountainous regions often experience severe seasonal limitations. The growing season can be short, and the flora at higher altitudes is typically less diverse and less calorically dense than its lowland counterparts. Alpine meadows, while picturesque, often support only a limited range of grasses and herbaceous plants. Higher still, above the tree line, vegetation becomes sparse, consisting of hardy mosses, lichens, and low-lying shrubs. For a large primate, whose caloric requirements would be substantial, this presents a significant hurdle. Its diet would likely need to be highly specialized, or it would require access to a broader altitudinal range, allowing it to descend to lower elevations during harsher periods. The hypothetical primate would need to be an adept forager, capable of exploiting ephemeral food sources, such as ripening berries in late summer, nutrient-rich roots unearthed from rocky soil, or perhaps even seasonal animal prey. The nutritional content of alpine vegetation, often high in fiber and low in readily digestible carbohydrates, would necessitate a robust digestive system, similar to that seen in extant mountain-dwelling herbivores like the yak or the mountain goat. The digestive physiology required to break down tough, fibrous plant material efficiently would be a crucial adaptation.

Climate presents another formidable obstacle. Mountainous regions are subject to drastic diurnal and seasonal temperature swings. Winters can bring heavy snowfall, freezing temperatures, and limited daylight hours. For a large primate, maintaining core body temperature would be a significant energetic drain. Adaptations such as a denser coat of fur, a thicker layer of subcutaneous fat, or specialized metabolic strategies would be essential for survival. The sheer difficulty of movement on steep, often treacherous terrain also plays a critical role. Navigating scree slopes, traversing snowfields, and climbing sheer rock faces would require exceptional physical prowess. While many mountain fauna, such as the ibex or the snow leopard, have evolved specialized adaptations for such environments – powerful limbs, specialized hooves, and incredible agility – a large primate would face unique challenges. Its presumed arboreal or semi-arboreal ancestry might not translate well to such extreme vertical environments, necessitating a significant evolutionary shift towards terrestrial locomotion adapted for steep inclines and uneven surfaces. The morphology of such a creature might therefore diverge significantly from its lowland relatives, perhaps exhibiting broader feet for better grip, a more robust skeletal structure to withstand impacts, and powerful hindquarters for prodigious leaps.

Despite these challenges, certain mountainous regions do offer unique advantages that could support a large, elusive primate. The very inaccessibility of these areas provides a degree of isolation and refuge from human encroachment, a critical factor for any large, rare animal. Vast mountain ranges, with their extensive cave systems and dense, high-altitude forests (where they exist), can offer shelter and concealment. The altitudinal zonation of mountain ecosystems creates a mosaic of habitats, each with its own distinct flora and fauna. A large primate might exploit different zones seasonally, descending to lower, more productive slopes in winter and ascending to alpine meadows in summer. The rugged terrain also creates natural barriers, potentially leading to the evolution of isolated populations with unique adaptations.

The Himalayas, the iconic home of the Yeti legend, exemplify these characteristics. This colossal mountain range boasts some of the planet's highest peaks, vast tracts of wilderness, and a dramatic altitudinal gradient that transitions from subtropical forests at the foothills to barren alpine deserts and glaciers at extreme altitudes. While direct evidence of a large primate inhabiting these upper reaches remains absent, the ecosystem does support a remarkable array of adapted fauna. The snow leopard (

Panthera uncia), for instance, thrives in the high-altitude rocky outcrops and alpine meadows, preying on bharal (blue sheep) and other mountain ungulates. Its thick, camouflaged fur, powerful build, and specialized paws are testaments to the extreme adaptations required for survival. Similarly, the Himalayan tahr (*Hemitragus jemlahicus*) and the bharal itself are herbivores that have evolved to graze on the sparse vegetation of the alpine zone. These animals demonstrate that a substantial biomass can indeed be supported by these seemingly barren landscapes, albeit through specialized feeding strategies and efficient energy utilization. If a large primate were to exist in the Himalayas, it would likely occupy a niche similar to these adapted herbivores, perhaps focusing on the more abundant vegetation found in the subalpine forests and meadows, and descending to lower altitudes during the harshest winter months. The existence of extensive rhododendron forests, known for their vibrant blooms and dense foliage at mid-altitudes, could also provide a significant food source during certain seasons. Furthermore, the numerous cave systems within the Himalayas offer potential denning sites or shelter from severe weather. The cultural narratives of the region, while steeped in folklore, often describe encounters with large, ape-like beings, suggesting a long-standing presence, however elusive.

Other mountainous regions around the world also present intriguing possibilities. The Andes Mountains in South America, with their vast extents and diverse ecosystems ranging from cloud forests to high-altitude puna grasslands, could theoretically support a large, undiscovered primate. While the known primate diversity of South America is concentrated in the Amazonian lowlands, the sheer scale and remoteness of the Andean Cordillera might offer overlooked refuges. The cloud forests, often draped in mist and teeming with epiphytic growth, offer a unique environment with high humidity and a wealth of arboreal resources. As one ascends, the vegetation transitions to puna, a treeless high-altitude grassland dominated by tussock grasses and hardy shrubs, a landscape that would present similar dietary challenges to the alpine zones of the Himalayas. The vicuña (

Vicugna vicugna) and guanaco (*Lama guanicoe*), wild relatives of the llama and alpaca, are well-adapted to these high-altitude environments, subsisting on the sparse vegetation and demonstrating remarkable physiological adaptations to thin air and cold. A hypothetical primate here would likely need to adopt similar strategies, potentially utilizing lower-altitude forests during colder periods.

In Central Asia, the Tian Shan and Pamir mountain ranges, often referred to as the "roof of the world," represent vast, rugged wildernesses with limited human populations and challenging climatic conditions. These regions are characterized by dramatic elevation changes, deep valleys, and extensive alpine meadows interspersed with high-altitude deserts. The flora, though limited in diversity at the highest elevations, can be surprisingly rich in certain valleys and on lower slopes, supporting ungulates like the argali sheep (*Ovis ammon*) and the ibex (*Capra ibex*). For a large primate, these regions would necessitate significant adaptations for cold tolerance and efficient foraging on tough plant matter. The presence of robust herbivores in these areas suggests that sufficient biomass exists to support a larger animal, provided it possesses the necessary adaptations. The remoteness and inaccessibility of these mountain systems offer excellent potential for a species to remain undetected.

Even in regions not typically associated with primate habitation, mountainous environments might offer speculative niches. The Atlas Mountains of North Africa, for example, are home to a population of Barbary macaques (*Macaca sylvanus*), the only macaque species found outside of Asia and a remarkable example of a primate adapting to a cooler, more temperate, and rugged environment. While significantly smaller than a hypothetical gigantiform primate, their presence demonstrates that primate life can indeed persist in mountainous terrains outside the tropics. The Barbary macaque inhabits cedar and oak forests at altitudes between 1,000 and 1,700 meters, demonstrating a reliance on arboreal resources and foraging on the ground for fruits, seeds, and invertebrates. A much larger primate in such a region would require a more substantial and varied food base, perhaps exploiting higher altitude grasslands and foraging for roots and tubers.

The concept of a large primate in a mountainous, alpine zone invariably leads to considerations of thermoregulation and energy expenditure. Unlike their lowland counterparts, animals in these environments must constantly battle the cold. A large body size, while beneficial for maintaining a stable core temperature, also presents a larger surface area for heat loss. Therefore, any primate inhabiting such regions would likely possess adaptations to mitigate this. A thick, shaggy pelage, similar to that of the Yeti of legend, would be a crucial protective layer against the cold and wind. Subcutaneous fat reserves would also play a vital role, providing insulation and an energy store during lean periods. Metabolic adaptations, such as a lower basal metabolic rate during colder periods or the ability to enter torpor-like states, could also be advantageous. The fur of mountain-dwelling mammals often differs in density and color from their lowland relatives. For example, the fur of the snow leopard is exceptionally dense and matches the rocky, snowy terrain, providing both insulation and camouflage. A hypothetical mountain primate would likely exhibit similar traits, with a thick, perhaps grizzled or reddish-brown coat, providing excellent camouflage against the rocky slopes and sparse vegetation.

The terrain itself would necessitate specialized locomotion. While a primate might retain some arboreal capabilities in the forested lower slopes, its primary mode of movement in the higher alpine zones would likely be terrestrial. This would require powerful limbs, broad feet for stability on uneven ground, and perhaps a more plantigrade or semi-plantigrade stance for better weight distribution and shock absorption. The ability to navigate steep inclines, rocky scree, and potentially snow-covered slopes would be paramount. Such adaptations would be akin to those seen in mountain ungulates, but adapted for a primate's more flexible skeletal structure and limb articulation. The hands, while perhaps less dexterous than those of a forest-dwelling primate, might be adapted for grasping and climbing on rock faces or digging for roots.

The food sources available in mountainous and alpine zones present the most significant ecological constraint. While extensive forests at lower altitudes could provide a substantial base of leaves, fruits, and bark, the higher alpine meadows and rocky slopes offer a more limited range of edible plants. These might include hardy grasses, sedges, herbaceous flowering plants, berries, and roots. A large primate would likely need to be highly opportunistic, consuming a wide variety of available vegetation. The presence of nutrient-rich alpine flowers and berries during the short summer months could provide crucial energy boosts. Roots and tubers, accessible through digging, could form a significant part of the diet, especially during colder periods when above-ground vegetation is scarce. Some mountain herbivores, like marmots, are known to consume insects and grubs, suggesting that a large primate might also supplement its herbivorous diet with invertebrates, particularly if they are abundant in certain microhabitats.

The legend of the Yeti, while largely unsubstantiated by concrete scientific evidence, highlights the enduring human fascination with large, mysterious creatures inhabiting the world's highest mountain ranges. These stories often depict a large, bipedal ape-like being, covered in thick, white or reddish-brown fur, dwelling in the remote valleys and snow-covered peaks of the Himalayas. While cryptozoological accounts are often anecdotal and prone to misidentification, they reflect a deep-seated belief that such environments, precisely because of their remoteness and mystery, could harbor undiscovered fauna. The ecological plausibility of such a creature hinges on the existence of a food web capable of supporting its considerable energetic needs and the evolution of specific physiological and morphological adaptations to survive extreme climatic conditions and challenging terrain.

In conclusion, while the dense, biodiverse forests of the tropics and temperate zones remain the most probable strongholds for a large, undiscovered primate, the possibility of such creatures persisting in mountainous and alpine regions cannot be entirely dismissed. These environments present significant ecological hurdles, particularly concerning food availability and climatic extremes. However, their inherent remoteness offers crucial sanctuary, and the adaptations seen in extant mountain fauna demonstrate the capacity of these landscapes to support substantial animal life. A hypothetical mountain primate would require specialized adaptations for cold tolerance, locomotion on steep terrain, and efficient foraging on tough, fibrous vegetation. Such a creature would likely occupy the lower, more vegetated slopes and valleys, perhaps descending to even lower altitudes during the harshest winter months, and would represent a remarkable example of evolutionary resilience in one of Earth's most challenging environments. The persistent myths and legends associated with these high-altitude realms, while not scientific proof, underscore the enduring allure and speculative potential of these wild, untamed landscapes as potential custodians of nature's grandest secrets.

The intricate tapestry of life within an ecosystem, known as biodiversity, plays a pivotal and often underestimated role in its ability to support substantial animal populations, particularly those of large body size. For a megafauna species, such as a hypothetical large primate, the sheer energetic and nutritional demands necessitate an environment rich in varied resources across multiple trophic levels. A simplistic ecosystem, with limited species diversity, would struggle to provide the consistent and diverse caloric intake required by a large animal. Conversely, a biodiverse habitat acts as a robust natural buffer, ensuring that if one food source experiences a seasonal decline or a localized impact, others remain available, thus averting widespread starvation or nutritional deficiency.

At the base of any food web lies primary productivity – the plants. A biodiverse plant community is not merely a collection of different species; it represents a mosaic of nutritional profiles, growth cycles, and accessibility. For a large primate, this translates into a varied menu throughout the year. It might involve the consumption of nutrient-dense fruits and berries during their fruiting seasons, the fibrous leaves and bark of certain trees and shrubs, succulent roots and tubers dug from the soil, or even the nectar and pollen from flowering plants. Different plant species offer distinct combinations of carbohydrates, proteins, fats, vitamins, and minerals. A broad spectrum of plant life ensures that the primate can meet all its nutritional requirements. For instance, some plants might be rich in essential amino acids, while others provide crucial micronutrients or easily digestible energy sources. The varied life cycles of plants also mean that food is available at different times of the year. Deciduous trees provide new growth in spring and fruits in summer and autumn, while evergreen species may offer edible leaves or bark during winter. Furthermore, different plant species occupy different microhabitats within an ecosystem. Some may thrive in sun-drenched clearings, others in shaded understory, and yet others in riparian zones. This spatial distribution means that a large primate, capable of traversing its territory, can exploit these varied resource patches as they become available.

Beyond the plant kingdom, biodiversity extends to the animal kingdom, forming the crucial intermediate trophic levels that large primates may also utilize. While many large primates are primarily herbivorous or omnivorous, few can subsist on plant matter alone, especially when considering the high energy demands of maintaining body mass and reproductive cycles. The availability of invertebrates – insects, larvae, earthworms, snails – can provide a vital source of protein, fats, and essential minerals like iron and zinc. A biodiverse insect population, comprising a multitude of species with different life stages and habits, ensures a continuous supply of these small but nutritionally significant food items. For example, a large primate might dig for grubs and earthworms in the soil, forage for ants and termites in decaying logs, or consume beetles and caterpillars found on foliage. The sheer abundance and diversity of insect life in a healthy ecosystem can be staggering, representing a readily accessible and energy-rich food source.

Moving up the food chain, larger invertebrates and smaller vertebrates can also play a role in the diet of an omnivorous or opportunistic large primate. This could include amphibians, reptiles, bird eggs, or even small mammals. The presence of a robust and diverse population of these animals is, in turn, dependent on the health of the lower trophic levels. A biodiverse ecosystem supports a greater variety of predators, each specializing in different prey items. This complex web ensures that populations of smaller animals do not become excessively large, which could destabilize the ecosystem, and also ensures that a substantial biomass of these potential food sources is available. A primate that can exploit these various animal resources gains a significant advantage, particularly during times when plant resources might be scarce. For example, during dry seasons or periods of drought, the availability of succulent insects or small vertebrates might become more critical for survival than fibrous plant matter.

The concept of niche partitioning is fundamental to understanding how biodiversity supports large populations. In a diverse ecosystem, different species evolve to exploit slightly different resources or the same resources at different times or in different ways. This reduces direct competition and allows a greater number of species, including larger ones, to coexist. For a large primate, this means it does not have to compete directly with every other herbivore or omnivore for every single food item. Instead, its foraging strategies and dietary preferences will occupy a unique niche. It might specialize in consuming larger, tougher leaves that smaller animals cannot digest, or it might have the dexterity and strength to access fruits from higher branches or to dig for deep roots. This specialization, enabled by the sheer variety of available resources, allows the primate to thrive without depleting any single resource to the point of collapse.

Furthermore, a biodiverse ecosystem is inherently more resilient to environmental perturbations. Whether it's a drought, a flood, a disease outbreak, or a localized pest infestation, a diverse ecosystem has a greater capacity to absorb the shock and recover. If a particular plant species is decimated by a disease, a biodiverse ecosystem will have other plant species that can fill the ecological gap. Similarly, if a particular insect population collapses, the primate can switch to other food sources. This resilience is crucial for the long-term survival of large animal populations, which often have slower reproductive rates and longer generation times, making them more vulnerable to sudden environmental changes. A species that relies on a single, narrow food source in a less biodiverse environment is at a much higher risk of extinction when that source is compromised.

Consider the contrast between a species-poor environment and a species-rich one. Imagine a vast, monoculture grassland. While it might support a large population of grazers that specialize in that particular grass, it would be vulnerable. A blight that affects that grass would be catastrophic. Now imagine a varied savannah landscape with numerous grass species, a variety of tree and shrub cover, different types of legumes, and a diverse insect and small animal population. This environment can support a wider array of herbivores, from specialized grazers to browsers and frugivores, as well as a richer predator community and a wider range of omnivorous species. A large primate in such a setting would have a more stable and reliable food supply, distributed across various plant types and trophic levels.

The role of biodiversity in supporting large primate populations is also intrinsically linked to habitat structure. A biodiverse ecosystem often translates to a complex and heterogeneous habitat. This might include a mix of forest types, open clearings, water bodies, and varied topography. Such structural diversity provides different foraging opportunities, shelter from predators and adverse weather, and suitable areas for nesting or denning. For example, a large primate might forage in the open grasslands for roots and succulent shoots, seek shelter and browse on leaves in dense thickets, and consume fruits from trees in the forest canopy. The complexity of the habitat, driven by the diversity of plant life and the resulting microhabitats, is essential for providing these varied needs. A simple, uniform habitat, even if it contains a single abundant food source, might not offer the necessary refuge or diversity of foraging substrates to support a large, sustainable population.

The presence of keystone species within a biodiverse ecosystem further amplifies its capacity to support large animals. Keystone species are those that have a disproportionately large effect on their environment relative to their abundance. They can shape the structure of the ecosystem and influence the populations of many other species. For instance, large herbivores can act as grazers that maintain grasslands, preventing them from being encroached by woody vegetation. Their grazing patterns can create diverse microhabitats beneficial to smaller animals and insects. Similarly, large predators, by keeping herbivore populations in check, prevent overgrazing and maintain the balance of plant communities. A large primate, potentially interacting with these keystone species either as a competitor, a predator, or even as prey for apex predators, exists within this intricate web. Its success is thus dependent on the overall health and diversity of the ecosystem, including the presence and function of its key components.

In essence, biodiversity acts as a biological insurance policy for large animal populations. It provides a rich and varied buffet of resources, a mosaic of microhabitats, and a resilient ecological structure that can withstand disturbances. For a large primate, with its significant resource demands, an ecosystem with low biodiversity would be a precarious place to exist. It would face a higher risk of starvation due to resource scarcity, increased competition, and greater vulnerability to environmental changes. The ability of an ecosystem to sustain a large population of any species, particularly one as energetically demanding as a large primate, is a direct reflection of its internal richness and complexity – its biodiversity. This richness ensures a continuous supply of energy and nutrients across multiple trophic levels, facilitating niche partitioning, providing habitat heterogeneity, and fostering ecological resilience, all of which are fundamental for the establishment and maintenance of substantial animal populations.

The increasing footprint of human civilization represents perhaps the most pervasive and detrimental force shaping the ecological landscape for any wildlife species, particularly those with extensive habitat requirements and specialized needs, such as large primates. The relentless progression of deforestation, driven by agricultural expansion, logging, and the demand for timber and land for development, systematically dismantles the contiguous tracts of forest that these animals depend upon for survival. As forests are cleared, the complex, multi-layered canopy that provides food, shelter, and protection is replaced by a mosaic of degraded secondary growth, open fields, or entirely human-dominated landscapes. This process directly reduces the total area of available habitat, but its consequences are far more profound.

One of the most significant impacts of human encroachment is habitat fragmentation. This is not merely a matter of shrinking territory; it is the dissection of once-continuous landscapes into smaller, isolated patches, often separated by a matrix of human-altered environments. For a large, often nomadic primate species, this fragmentation acts as a severe impediment to essential life processes. Movement corridors, vital for foraging across a wide home range, for finding mates, and for dispersing to new territories, become severed. Imagine a vast, unbroken forest, allowing a primate to roam freely, accessing diverse food sources as seasons change and its needs evolve. Now, visualize that same forest intersected by roads, agricultural fields, or villages. The once-seamless journey is now punctuated by dangerous crossings of open, predator-exposed areas, or by impassable artificial barriers.

These isolated habitat fragments create what are effectively islands of wilderness within a human-dominated sea. The consequences for the primate population are manifold and severe. Firstly, the size of these fragments directly limits the carrying capacity of the habitat. Smaller patches can simply support fewer individuals due to reduced resource availability and increased competition within the confined space. A large primate species, with its substantial dietary needs and potentially large home ranges, requires extensive areas. When these areas are broken into small pieces, the overall biomass of available food, water, and shelter is dramatically reduced, making it difficult to sustain even a small population.

Secondly, fragmentation isolates populations, preventing gene flow between them. This genetic isolation is a critical threat to long-term viability. Over generations, inbreeding can occur, leading to reduced genetic diversity. This makes the population more susceptible to diseases, less adaptable to environmental changes, and can result in a decline in overall fitness, reproductive success, and the survival rates of offspring. A small, isolated population in a fragment is like a single gene pool, vulnerable to collapse if a disease or environmental stressor affects that specific genetic makeup. Without the introduction of new genes from other populations, the species' ability to adapt and persist diminishes significantly.

The edges of these fragmented habitats also present unique challenges. Forest edges, where woodland meets farmland or open land, are often characterized by different vegetation composition and increased exposure to environmental factors like wind and sunlight. They can also be zones of increased predation risk. For a large primate, these edges might offer some supplementary resources, such as fruits from pioneer species or insects attracted to the varied vegetation. However, they also bring the species into closer proximity with human activities and domestic animals, increasing the potential for negative interactions. Furthermore, edge effects can alter the microclimate within the forest fragments, potentially affecting the availability and palatability of native food plants.

The construction of roads, railways, and other linear infrastructure is a particularly insidious form of fragmentation. These corridors not only divide habitats but also act as significant mortality factors. Primates, especially those that are arboreal but descend to the ground, are highly vulnerable to vehicle collisions. The noise and light pollution associated with these infrastructures can also disrupt their natural behavior, including foraging patterns, resting periods, and communication. For a species that relies on stealth and camouflage, the constant disruption of roads can be extremely stressful and energetically costly.

Urbanization and industrialization exacerbate these issues. The expansion of cities and industrial complexes leads to direct habitat destruction and the creation of large, impenetrable barriers. Pollution, both air and water, emanating from these areas can contaminate food and water sources, posing long-term health risks. Waste disposal sites, while potentially attracting some opportunistic foraging, can also introduce novel pathogens and expose animals to dangerous materials like plastics and chemicals.

The consequences of habitat fragmentation extend beyond the direct impact on the primates themselves. It disrupts the intricate ecological web within which they exist. For instance, if a primate species is a key seed disperser, its inability to move between fragmented forest patches can hinder forest regeneration and affect the distribution of plant species. Similarly, if it serves as prey for a predator, its reduced mobility and population size in fragmented landscapes can have cascading effects on predator populations.

Moreover, fragmentation can lead to an increase in human-wildlife conflict. As the available habitat shrinks and becomes discontinuous, primates may be forced to venture into agricultural areas or human settlements in search of food. This can result in crop raiding, damage to property, and, in some instances, direct confrontations between humans and primates. Such conflicts often lead to retaliatory killings, further decimating already vulnerable populations. The perception of these animals as pests can foster negative attitudes among local communities, undermining conservation efforts and creating a cycle of negative interactions.

The specific requirements of a large primate species, such as their dietary preferences for certain fruits or their need for old-growth forest structure for nesting or foraging, make them particularly susceptible to the effects of fragmentation. If a fragment is too small to contain the necessary diversity of food plants or the specific types of trees they rely on, the population within that fragment will struggle to survive, even if the fragment itself appears vegetated. The loss of connectivity means that individuals cannot simply move to a neighboring area that might better meet their needs.

Consider the ecological requirements of a hypothetical large primate that relies on the mature canopy of primary rainforest for sustenance and shelter. As logging operations slice through this forest, they create gaps and clearings, and the remaining forest is often left in a degraded state, with a reduced diversity of tree species and an altered microclimate. This not only reduces the availability of their preferred food sources but also exposes them to greater predation risk from terrestrial predators that can now more easily access the forest floor. If the logging continues, and the forest is replaced by plantations or secondary scrub, the conditions necessary for this primate may be lost entirely. Even if a corridor is left, its suitability as a functional link depends on its width, the quality of vegetation within it, and the absence of insurmountable barriers like major roads.

The challenge is compounded by the fact that many large primates, particularly those that are elusive and rare, may have low population densities to begin with. Fragmentation can push these already sparse populations below a critical threshold, making them more vulnerable to extinction. The interconnectedness of populations is vital for maintaining a healthy metapopulation – a group of spatially separated populations that interact through dispersal. Fragmentation severs these connections, turning a healthy metapopulation into a collection of isolated, struggling demes.

Furthermore, the matrix surrounding the habitat fragments plays a crucial role. If the matrix is composed of relatively benign land uses, such as uncultivated grasslands or sparsely populated woodlands, movement between fragments might still be possible, albeit with some risk. However, if the matrix consists of intensive agriculture, urban areas, or industrial zones, movement becomes virtually impossible, and the fragments are effectively rendered as isolated islands, leading to rapid declines in genetic diversity and population viability. The quality of the matrix can determine whether a fragmented landscape remains ecologically functional or becomes a complete barrier.

The long-term implications of habitat fragmentation are dire. It leads to a decrease in species richness and abundance, a simplification of ecological communities, and a loss of ecosystem services. For large primates, it represents a pathway towards endangerment and, ultimately, extinction. Conservation strategies must therefore prioritize not only the protection of existing habitat but also the restoration and creation of functional ecological corridors that can reconnect fragmented landscapes, allowing for movement, gene flow, and the maintenance of viable populations. Without addressing the fundamental issue of habitat connectivity, efforts to conserve these magnificent creatures will remain incomplete and ultimately unsustainable in the face of ongoing human expansion.

The very process of development, from the construction of a single new road to the establishment of a vast agricultural frontier, can have profound and lasting impacts on habitat connectivity. These human-induced barriers do not respect natural ecological boundaries. They cut across river valleys, ascend mountain slopes, and traverse ancient forest ecosystems, effectively dissecting landscapes that were once seamlessly connected. For a large primate, whose home range can span many square kilometers, the introduction of such linear features can transform a landscape from a continuous, resource-rich expanse into a checkerboard of isolated, potentially inadequate patches.

The psychological and behavioral impact of these barriers on primate populations cannot be overstated. Many species have evolved innate patterns of movement and exploration that are intrinsically linked to their perception of a safe and navigable environment. The presence of unfamiliar and potentially dangerous human structures, coupled with the altered sensory cues (noise, light, artificial scents), can induce stress, fear, and avoidance behaviors. This can lead to a behavioral bottleneck, where individuals become reluctant to leave the perceived safety of their fragment, even if resources are becoming depleted. They might exhibit increased vigilance, reduced foraging times, and altered social interactions, all of which can negatively impact their health and reproductive success.

The nature of the matrix also dictates the degree of isolation. A forest fragment surrounded by actively managed monoculture plantations, such as oil palm or rubber, offers very little in terms of supplementary resources and presents a highly inhospitable environment for a forest-dwelling primate. The dense canopy of the plantation might mimic some aspects of forest cover, but the lack of diverse food plants, the presence of pesticides, and the intensive human activity within them make them poor substitutes for natural forest. In contrast, a fragment bordered by regenerating secondary forest or traditional agroforestry systems might offer a more permeable boundary, allowing for some level of interaction and resource utilization.

The concept of "habitat quality" also becomes critically important in the context of fragmentation. A large fragment of degraded forest might be less ecologically valuable than a smaller, intact patch of primary forest. When fragmentation leads to a reduction in the overall quality of the remaining habitat – for example, through selective logging that removes key tree species or an increase in invasive plant species – the ability of the fragment to support a viable population is further compromised. The loss of old-growth trees, which may provide essential foraging substrates, nesting sites, or den locations, can be a particularly devastating consequence for species that rely on these specific structural elements.

Furthermore, the scale of fragmentation is a critical determinant of its impact. A single, minor road might cause localized disruptions and mortality, but a network of roads, settlements, and agricultural fields can effectively render an entire region uninhabitable for large, wide-ranging primates. The cumulative effect of multiple human intrusions creates a mosaic of barriers and resource-poor zones that can isolate populations to the point of no return. This is especially true for species that have slow dispersal rates or are inherently cautious in their movements.

The impact of human-induced fragmentation is often amplified by climate change. As global temperatures rise and weather patterns become more erratic, the ability of species to adapt is crucial. If populations are already confined to small, isolated fragments, their capacity to migrate to more suitable climates or to exploit newly available resources in adjacent areas is severely curtailed. They are trapped in suboptimal conditions, with limited options for adaptation. This makes them exceptionally vulnerable to the synergistic pressures of habitat loss and a changing environment.

In essence, human impact, particularly in the form of habitat fragmentation, transforms a species' ecological niche from a potentially continuous and diverse domain into a series of disconnected, often suboptimal, and increasingly risky enclaves. The ability of a large primate to fulfill its fundamental ecological role – to find food, reproduce, and avoid predation – is systematically undermined by these human-imposed alterations to the landscape. The creation of barriers, the reduction of resource availability, and the increase in human-animal conflict all converge to pose an existential threat, demanding comprehensive and landscape-level conservation strategies that prioritize connectivity and the restoration of ecological integrity.

4

Diet and Nutritional Demands

The possibility of a large primate, or megafauna primate, being primarily herbivorous hinges on a delicate interplay of ecological factors, foremost among them being the availability and nutritional quality of plant matter capable of sustaining such substantial energetic demands. Drawing parallels with extant large herbivores, most notably the gorilla, offers a valuable framework for understanding the potential dietary strategies of an undiscovered primate of similar stature. Gorillas, the largest of all primates, are overwhelmingly herbivorous, with leaves forming the bulk of their diet, supplemented by fruits, stems, and occasionally insects. This dietary specialization allows them to thrive in their forest environments, but it is also a testament to the rich botanical resources that these habitats can provide.

For a hypothetical megafauna primate, the cornerstone of a herbivorous diet would likely be foliage. Leaves, particularly young, tender shoots and buds, are a primary food source for many large herbivores. Their abundance in many forest ecosystems, especially tropical and subtropical regions, makes them a seemingly inexhaustible resource. However, leaves are not uniformly nutritious. They can be high in structural carbohydrates like cellulose and hemicellulose, which are difficult to digest, and often contain secondary compounds such as tannins and alkaloids that can be toxic or unpalatable. A truly herbivorous megafauna primate would need to possess a specialized digestive system, perhaps a sacculated stomach or enlarged hindgut, capable of breaking down these complex plant materials efficiently. Furthermore, they would likely exhibit a preference for species or specific plant parts that offer a higher nutrient-to-fiber ratio. This might involve selecting leaves from certain tree species, or focusing on the softer, more digestible new growth that emerges seasonally. The sheer volume of leaf matter required to meet the caloric needs of a large primate would necessitate significant foraging time and the ability to process large quantities of vegetation daily.

Fruits would undoubtedly play a crucial role, especially for a primate aiming for a primarily herbivorous diet. While fruits are typically more energy-dense than leaves, offering sugars and readily digestible carbohydrates, their availability is often highly seasonal. During fruiting seasons, a megafauna primate could capitalize on this rich energy source, building up fat reserves. However, reliance on fruits alone would be precarious. Periods of scarcity, known as inter-fruiting intervals, would necessitate a fallback diet, likely consisting of less palatable but more consistently available plant parts like leaves, stems, or roots. The ability to locate and exploit ephemeral fruit patches would require a sophisticated understanding of the forest ecosystem, including knowledge of ripening times and the spatial distribution of fruit-bearing trees. This could also drive dispersal patterns, as individuals or groups move across their home range in pursuit of fruiting trees.

Roots and tubers represent another potential, albeit more challenging, component of a herbivorous diet. These underground storage organs are often rich in carbohydrates and can be available year-round, providing a crucial food source during lean periods. However, accessing them requires significant effort and specialized adaptations. Digging through soil demands strength and dexterity, and some roots may contain high levels of indigestible fibers or even toxins that deter herbivores. A primate adapted to consuming roots would likely possess strong limbs and claws for excavation, and a digestive system capable of detoxifying or tolerating any noxious compounds. Furthermore, the energetic cost of digging might outweigh the caloric return unless the roots are particularly nutrient-dense or abundant. In environments where the soil is relatively loose and rich in organic matter, root foraging could become a viable strategy, especially for species that are already adapted to ground-dwelling.

Stems, including bark and pith, could also contribute to the diet, particularly the tender inner bark of certain trees or the succulent pith of herbaceous plants. These parts are generally fibrous and less palatable than leaves or fruits, but they can offer a consistent source of moisture and some carbohydrates. Bark consumption, for example, is a common strategy among large herbivores in environments where other food sources are scarce. However, this can have detrimental effects on the trees themselves, and a primate relying heavily on bark would need to be selective to avoid damaging its own habitat or consuming species with high concentrations of toxins. Young, tender stems and shoots, similar to new leaves, would be more digestible and palatable, offering a valuable nutritional supplement.

The energetic demands of a large primate are substantial. A gorilla, for instance, can weigh upwards of 150-200 kg and requires a considerable daily intake of food to maintain its body mass and energy levels, particularly for activities such as locomotion, social interaction, and reproduction. If a hypothetical megafauna primate were even larger, these demands would be proportionally higher. This necessitates a diet that is not only abundant but also possesses a sufficient caloric density. While leaves are abundant, their low energy content means that very large quantities must be consumed and processed. Fruits offer higher energy but are temporally and spatially variable. Therefore, a truly successful herbivorous megafauna primate would likely exhibit a mixed strategy, optimizing its intake by exploiting the most energy-rich and readily available resources at any given time. This might involve a daily or seasonal shift in dietary focus, moving from leaf-dominated foraging to fruit-heavy consumption, and potentially incorporating roots or stems as crucial fallback options.

The seasonality of food resources is a critical factor that would shape the foraging behavior and life history of a herbivorous megafauna primate. In environments with distinct wet and dry seasons, or temperate regions with pronounced seasonal changes, the availability of leaves, fruits, and young shoots would fluctuate dramatically. During the wet season, when plant growth is vigorous, food might be abundant and nutritious. However, during the dry season, leaves may become tougher and less digestible, fruits scarce, and water sources may diminish. A large herbivore would need to adapt to these fluctuations, perhaps by:

1. **Mobility:** Migrating to areas where food and water remain available, a strategy common among large terrestrial herbivores. This implies the need for large home ranges and migratory corridors, which are increasingly threatened by habitat fragmentation.
2. **Dietary Flexibility:** Shifting to more fibrous, less palatable food sources, or relying on stored energy reserves accumulated during periods of abundance.
3. **Physiological Adaptations:** Developing adaptations to conserve water, tolerate a wider range of food types, or to efficiently extract nutrients from low-quality forage.

For a primate species, intelligence and social learning would play a significant role in navigating these seasonal challenges. Knowledge about the timing of fruit ripening, the locations of reliable water sources, and the identification of safe and nutritious fallback foods would be passed down through generations, conferring a substantial survival advantage. Group foraging strategies might also evolve, with individuals collectively searching for food patches, defending resources, or sharing information about food availability.

The ecological context of potential habitats is also paramount. A megafauna primate would likely inhabit environments rich in diverse plant life, such as primary or mature secondary forests, savannas with significant tree cover, or montane woodlands. These habitats would need to support a high biomass of edible vegetation capable of sustaining a large animal's daily requirements. Areas dominated by a single plant species, or those with low overall plant diversity, would be less likely to support a large, primarily herbivorous primate due to insufficient nutritional variety and potential dietary limitations. The presence of various tree species that fruit at different times, or a variety of understory plants offering edible leaves and stems, would be essential for providing a consistent and balanced diet throughout the year.

Furthermore, the presence of essential micronutrients, beyond mere caloric content, would be a consideration. While fruits and leaves provide a spectrum of vitamins and minerals, deficiencies could arise. A varied diet, encompassing a range of plant types, would be crucial for ensuring the uptake of all necessary nutrients. It is also possible that such a primate might engage in geophagy, the consumption of soil or clay, to supplement mineral intake or to neutralize toxins present in certain plant foods.

In conclusion, the scenario of a primarily herbivorous megafauna primate is ecologically plausible, provided that its environment offers a continuous and abundant supply of nutrient-rich plant matter. The successful adaptation would necessitate specialized digestive physiology to process fibrous vegetation, behavioral flexibility to cope with seasonal variations in food availability, and potentially, the development of foraging strategies that exploit a diverse range of plant parts, from tender leaves and fruits to roots and stems. The parallels with gorillas highlight that such a niche exists within the primate order, demonstrating that even the largest primates can thrive on a plant-based diet when the ecological conditions are sufficiently supportive. The challenge for such a creature, as with many large herbivores, would be the scale of its caloric needs and the reliability of its food sources in a dynamic and often challenging environment.

The vibrant tapestry of a primate's diet, particularly for a species of significant size, is often interwoven with the sweet allure of fruits. These fleshy, carbohydrate-rich offerings from the plant kingdom represent a critical and highly sought-after resource, providing a concentrated source of energy that is often more readily digestible than the fibrous components of leaves. The evolutionary success of many primate species, from the smallest marmosets to the great apes, is inextricably linked to their ability to locate, consume, and derive sustenance from a diverse array of fruits. For a hypothetical megafauna primate, the role of frugivory would be even more pronounced, serving as a vital engine for energy acquisition, especially during specific periods within the annual cycle.

The nutritional profile of fruits is a complex subject, varying dramatically depending on the species, ripeness, and environmental conditions. Generally, ripe fruits are characterized by high sugar content, providing a quick and easily accessible source of calories. This sugar, primarily in the form of glucose, fructose, and sucrose, fuels the high metabolic rates of primates. Beyond simple sugars, fruits also contribute essential vitamins, minerals, and water. Ascorbic acid (Vitamin C), for example, is abundant in many fruits and plays a crucial role in immune function, collagen synthesis, and as an antioxidant. Beta-carotene, a precursor to Vitamin A, is also commonly found, vital for vision and cell growth. Trace minerals, though present in smaller quantities, are equally important for a myriad of physiological processes. The water content of fruits is also a significant consideration, especially in environments where access to free water might be limited. Many fruits can contribute substantially to an animal's daily hydration needs, reducing the reliance on distant or potentially unsafe water sources.

However, the bounty of fruits is rarely constant. The seasonality of fruiting is a fundamental ecological reality that profoundly shapes the foraging strategies and dietary patterns of frugivorous primates. Many plant species have evolved intricate mechanisms to synchronize their flowering and fruiting with favorable environmental conditions, such as the onset of the rainy season, to maximize pollination and seed dispersal success. This temporal variability means that fruit availability can fluctuate dramatically throughout the year. During peak fruiting seasons, a forest canopy can become a veritable larder, offering a rich and diverse selection of edible fruits. At such times, a large frugivore would likely shift its dietary focus, maximizing its intake of these energy-dense foods, potentially building up fat reserves for leaner periods. This can involve spending extended periods foraging within fruit-rich areas, consuming large quantities to capitalize on the ephemeral abundance.

The periods between peak fruiting seasons, often referred to as inter-fruiting intervals, present a significant challenge. During these times, the availability of preferred fruits diminishes, forcing primates to turn to alternative food sources. This is where the concept of a fallback diet becomes critical. For a large primate, this might involve a return to a more folivorous diet, consuming leaves, shoots, and buds. Alternatively, they might exploit less palatable but more reliable food items such as stems, bark, roots, or even insects. The ability to successfully navigate these seasonal shifts in food availability is a key determinant of a species' ecological success and its geographic distribution. A large primate with high energy demands would need a broad understanding of its environment, including the fruiting phenology of numerous plant species and the locations of reliable fallback food sources.

The nutritional quality of fruits also varies with ripeness. Unripe fruits are often more astringent and contain higher levels of tannins and other secondary compounds that can deter herbivores. As fruits ripen, these compounds typically decrease, while sugars increase, making them more palatable and digestible. However, overripe fruits can also present challenges, potentially becoming fermented and containing alcohol, or attracting undesirable pests and pathogens. Therefore, primates often exhibit a refined ability to select fruits at their optimal stage of ripeness, a process that requires learned experience and keen sensory perception. This selective feeding can further influence the effectiveness of seed dispersal, as fruits consumed at their peak ripeness are more likely to have viable seeds pass through the digestive tract intact.

The ecological impact of a large frugivore, particularly one that consumes a significant quantity of fruit, extends far beyond its own nutritional needs. These animals play an indispensable role as seed dispersers, a process that is fundamental to the regeneration and maintenance of forest ecosystems. As primates consume fruits, they ingest the seeds contained within. These seeds then travel through the digestive tract, often unharmed, and are eventually deposited in new locations via the animal's feces. This dispersal mechanism offers several critical advantages to plant species. Firstly, it helps to colonize new areas, increasing the geographic range of the plant and reducing competition with the parent plant. Secondly, the passage through the digestive tract can scarify the seed coat, improving germination rates for some species. Thirdly, the deposition of seeds within fecal matter can provide a readily available nutrient source for the germinating seedling, giving it a better chance of survival.

For a large primate, the sheer volume of fruit consumed translates into a significant capacity for seed dispersal. Their extensive home ranges and daily movements allow them to transport seeds over considerable distances, potentially establishing new plant populations in areas that would otherwise be inaccessible. This is particularly important for plant species with heavy or adhesive fruits that are not effectively dispersed by wind or gravity alone. The effectiveness of a primate as a seed disperser is influenced by several factors, including the size of its home range, its dietary preferences (i.e., which fruits it consumes), and the degree to which it processes the seeds during digestion. Large primates, with their considerable body mass and energetic requirements, are often capable of consuming a wide variety of fruits and have the mobility to move seeds across diverse landscapes.

The availability of fruiting trees is therefore a linchpin for the survival of large frugivorous primates. Their distribution, density, and the timing of their fruiting cycles directly dictate the carrying capacity of a habitat for such species. Forests that are rich in a diversity of fruit-bearing trees, with a staggered fruiting phenology that ensures a continuous supply of food throughout the year, are ideal environments. Such habitats would support a larger population of primates and allow them to maintain their energetic balance with less difficulty. Conversely, areas with a low density of fruit trees, or those dominated by species with short, overlapping fruiting periods, would present significant challenges, leading to periods of food scarcity and potentially limiting population size.

Understanding the specific fruit preferences of a hypothetical megafauna primate is crucial for predicting its ecological niche and potential distribution. Different primate species exhibit varying degrees of frugivory, with some being almost exclusively fruit-eaters (e.g., many species of New World monkeys) while others incorporate fruits as a significant but not sole component of their diet (e.g., chimpanzees and orangutans). A large primate might exhibit a preference for large, calorie-dense fruits, as these would provide the most efficient return on foraging effort. This could include fruits from canopy trees, which are often larger and more nutritious than those found in the understory. The ability to access these higher-canopy fruits would require specialized arboreal skills, such as powerful climbing abilities and a strong grip, or a terrestrial foraging strategy that exploits fallen fruits.

The ecological consequences of a large frugivore's presence are far-reaching. By dispersing seeds, they contribute to genetic diversity within plant populations, help maintain forest structure and composition, and influence the success of other species that depend on those plants for food or shelter. A decline in the population of a key seed disperser can have cascading effects throughout the ecosystem, potentially leading to reduced forest regeneration and altered community dynamics. Therefore, the ecological role of frugivory in the life of a large primate is not merely about sustenance; it is about actively participating in the intricate web of life that defines its forest habitat. The abundance and diversity of fruits are not just food for the primate; they are also a testament to the health and functioning of the ecosystem itself, an ecosystem that, in turn, relies on the primate for its continued regeneration. The primate, through its consumption and dispersal of fruits, becomes an active agent in shaping the future landscape, a vital cog in the grand machinery of forest ecology.

The nutritional demands of a large primate, especially one that relies heavily on fruits, are substantial and dynamic. While fruits offer readily available energy, their water content means that they can be less calorically dense than fat-rich foods. Therefore, to meet the caloric needs of a large body size, a significant volume of fruit must be consumed. This necessitates long foraging bouts and efficient processing capabilities. The sugars in fruits, while providing quick energy, can also lead to rapid fluctuations in blood glucose levels. A primate's physiology would need to be adapted to manage these energy spikes and dips, perhaps through efficient metabolic pathways for sugar utilization and storage.

Furthermore, the composition of fruits can vary seasonally, not just in quantity but also in nutritional quality. As fruits ripen, their sugar content increases, but other nutrients like proteins and fats might be more concentrated in specific species or at particular times. A varied diet of fruits from different plant species, each with its unique nutritional profile, would be essential for ensuring the intake of all necessary macronutrients and micronutrients. For example, some fruits are richer in fats, providing a more sustained energy source and essential fatty acids, while others might offer a higher protein content, crucial for tissue repair and growth. A large primate's ability to discern and select these diverse fruits based on their nutritional value would be a significant advantage, requiring a sophisticated understanding of plant biology and seasonal changes.

The energetic cost of foraging for fruits can also be considerable, particularly if the preferred fruits are located high in the canopy or in dispersed patches. Navigating complex forest terrain, climbing trees, and defending fruit resources from competitors all require significant energy expenditure. Therefore, a primate that relies heavily on frugivory must possess an efficient energy budget, balancing the calories gained from foraging with the calories expended in the process. This might involve strategies such as optimal foraging, where the animal seeks to maximize its net energy intake by selecting food items and foraging locations that offer the highest caloric return for the least energetic cost.

The presence of toxins in fruits, even when ripe, is another factor to consider. Some fruits contain natural chemical defenses to deter herbivores or parasites. A large primate would need to possess physiological mechanisms to detoxify or tolerate these compounds. This could involve specialized liver enzymes or a gut microbiome capable of breaking down toxic substances. The ability to process a wide range of fruits, including those with mild levels of toxins, would broaden the available food base and increase the primate's resilience to fluctuating food availability.

The social dynamics of foraging can also play a role in fruit acquisition. In some primate species, individuals forage solitarily, while in others, group foraging is common. For a large primate, group foraging could offer advantages such as increased efficiency in locating fruit patches through information sharing or enhanced defense of valuable food resources. However, larger groups also lead to increased competition for food, potentially leading to reduced individual intake, especially for less dominant individuals. The specific social structure and foraging strategy would therefore be closely linked to the spatial distribution and abundance of fruit resources.

In conclusion, frugivory is a pivotal dietary component for many primate species, and for a hypothetical megafauna primate, it would likely represent a critical strategy for meeting its substantial energetic requirements. The seasonal availability, nutritional variability, and ecological role as a seed disperser underscore the importance of fruits in the primate's life. A successful large frugivore would need to navigate the challenges of temporal and spatial variability in fruit abundance, possess adaptations for efficient energy acquisition and processing, and exhibit a sophisticated understanding of its environment to exploit this resource effectively. Its existence and ecological impact would be intrinsically linked to the health and diversity of the fruit-bearing flora within its habitat, highlighting the profound interdependence between primate frugivores and the forest ecosystems they inhabit. The primate, in turn, actively contributes to the perpetuation of these very ecosystems through its indispensable role as a seed disperser, a testament to the intricate and mutually beneficial relationships that shape the natural world.

While fruits form the cornerstone of a large primate's sustenance, providing readily accessible carbohydrates and vital hydration, a truly resilient and ecologically versatile species would likely exhibit opportunistic omnivorous tendencies. This means venturing beyond the predictable bounty of the botanical realm to exploit alternative food sources when they become available or when preferred fruits are scarce. The inclusion of animal matter, even in small quantities, can offer significant nutritional advantages, particularly in terms of concentrated protein and essential fats, which are often less abundant in a purely frugivorous diet. Such supplemental feeding would not necessarily represent a fundamental shift in dietary strategy but rather a pragmatic adaptation to maximize nutritional intake and energy efficiency across varying environmental conditions.

The most accessible and likely animal-based food sources for such a primate would be invertebrates. The forest floor, decaying logs, and the bark of trees are teeming with a diverse array of insects, larvae, and grubs. For a large primate, extracting these small creatures might require specialized foraging techniques. Imagine a powerful hand, capable of tearing apart rotting wood or stripping bark with surprising dexterity, revealing hidden caches of protein-rich sustenance. The mandibles and teeth, already adapted for processing tough plant material, could also be utilized to crush the exoskeletons of beetles or the chitinous casings of insect larvae. This kind of foraging would demand a degree of manual dexterity and patience, as well as a keen sensory awareness to locate these often inconspicuous food items. The energetic return from consuming a handful of plump grubs or a cluster of ants could be significant, offering a rapid influx of protein and fats that are crucial for muscle maintenance, growth, and overall metabolic function.

Furthermore, the availability of these invertebrate resources often follows predictable ecological patterns. Many insect species undergo population booms during specific seasons, such as the warmer, wetter months. This temporal abundance could coincide with periods when fruit availability might be lower, thus presenting a valuable fallback option. A large primate that can exploit these invertebrate pulses would have a distinct advantage, buffering its nutritional intake during critical times. The social dynamics of foraging for insects could also be interesting. While solitary foraging is common, a large primate might engage in cooperative behaviors, such as driving insect swarms towards a gathering point, or perhaps using its strength to disturb large areas of leaf litter to flush out hidden prey. The olfactory senses would also play a crucial role, allowing the primate to detect the scent of insect colonies or the subtle aroma of decaying organic matter that signals the presence of grubs.

Beyond insects and their larval forms, the possibility of consuming small vertebrates cannot be entirely dismissed. While a large primate might not be an apex predator in the conventional sense, opportunistic predation on slow-moving or vulnerable prey could supplement its diet. This could include amphibians, reptiles, or even small mammals encountered during foraging excursions. The acquisition of such prey would likely be a matter of chance rather than targeted hunting. A primate might stumble upon a basking lizard, a slow-moving frog, or a young, uncoordinated rodent. The ability to react quickly and decisively, utilizing its powerful limbs and jaws, would be key. The nutritional value of such prey would be considerably higher than that of invertebrates, offering a rich source of protein, fats, and essential micronutrients like iron and B vitamins.

The energetic cost of pursuing and capturing larger prey might outweigh the potential benefits for a primate whose primary adaptations are for arboreal locomotion and fruit consumption. However, for a truly large species, even a moderate success rate in acquiring small vertebrates could contribute significantly to its overall nutritional balance. Consider the possibility of raiding bird nests. Many primates are known to raid nests for eggs, and this hypothetical megafauna primate would likely be no different. Bird eggs are a concentrated source of protein, fats, and essential vitamins. The dexterity required to reach nests, often located precariously in tree branches or cliff faces, would be essential. The ability to carefully extract eggs without breaking them would be a learned behavior, honed through experience. The size of the primate would also play a role; a larger animal might be able to access nests that are out of reach for smaller species, or even deter adult birds from defending their nests.

The physiological adaptations necessary to digest and process animal matter would also be a consideration. While primate digestive systems are primarily adapted for plant material, many species possess the capability to digest proteins and fats. The gut microbiome would likely play a crucial role in breaking down complex animal tissues and neutralizing any potential pathogens. The presence of specific enzymes in the digestive tract would also be important for efficient nutrient extraction. A primate that frequently consumes animal matter might develop a more robust digestive system, capable of handling a wider range of food types. This would involve not only the ability to break down proteins and fats but also to metabolize any residual toxins or indigestible components.

The social implications of incorporating animal matter into the diet could also be significant. If group foraging for larger prey or nest raiding becomes a common behavior, it could foster increased cooperation and social bonding within the primate community. Competition for these high-value food resources could also lead to more complex social hierarchies and dominance displays. The ability to defend a kill or a raided nest would require significant strength and potentially coordinated effort from multiple individuals. This could lead to the development of specialized roles within the group, with some individuals acting as scouts, others as defenders, and others as processors of the food.

The occasional consumption of carrion, while perhaps unappealing from a human perspective, is another potential avenue for nutrient acquisition in the wild. Large primates, with their powerful jaws and robust digestive systems, might be capable of scavenging on the carcasses of deceased animals. This would require a tolerance for decaying flesh and a lack of aversion to the associated smells and textures. The ability to detect the scent of carrion from a distance would be a valuable adaptation. While scavenging is often associated with obligate carnivores, many omnivorous species will opportunistically consume carrion when it is available. For a large primate, this could represent a windfall of nutrients, particularly protein and fats, that would be difficult to obtain otherwise. The main challenge with consuming carrion would be the risk of contracting diseases from the decomposing animal. A strong immune system and efficient detoxification mechanisms would be essential for survival.

The behavioral repertoire of a large omnivorous primate would be incredibly diverse. It would need to possess the agility and arboreal skills to forage for fruits in the canopy, the ground-level foraging abilities to find insects and grubs, and potentially the stealth and opportunistic predatory instincts to capture small vertebrates. The ability to switch between these foraging strategies based on resource availability would be a key indicator of its ecological success. This flexibility would allow it to thrive in a wider range of habitats and to persist through periods of seasonal scarcity. The learning capacity of such a primate would also be paramount, as it would need to learn which food sources are safe and nutritious, where and when they are available, and the most effective techniques for acquiring them. This knowledge would be passed down through generations, shaping the foraging traditions of the species.

The energetic benefits of incorporating animal matter, even in small quantities, can be substantial. Proteins are essential for building and repairing tissues, and fats provide a concentrated source of energy and are crucial for hormone production and nutrient absorption. While fruits provide carbohydrates, they are often low in protein and fat. Therefore, a diet that includes animal matter can offer a more balanced nutritional profile, supporting greater muscle mass, bone density, and overall physiological function. This is particularly important for large animals with high metabolic demands. The increased protein intake could also aid in wound healing and immune system function, making the primate more resilient to disease and injury.

The question of feasibility hinges on the balance between the energetic costs and benefits of acquiring these supplemental food sources. Foraging for insects might be relatively low-cost, requiring only a short period of focused effort. However, the caloric return per unit of effort might also be low. Conversely, hunting small vertebrates or raiding nests could offer a higher caloric return but would involve a greater energetic investment in terms of pursuit, capture, and potential conflict. A successful omnivorous strategy would involve optimizing this balance, capitalizing on readily available and high-yield food sources while minimizing energy expenditure and risk. This might involve utilizing existing tools or skills, such as using a strong limb to break open a rotten log to access grubs or employing a rapid pounce to catch a slow-moving reptile.

The role of the gut microbiome in a hypothetical omnivorous primate cannot be overstated. The diverse array of microorganisms residing in the gut plays a critical role in digesting complex food materials, synthesizing essential vitamins, and detoxifying harmful compounds. A primate that consumes both plant and animal matter would likely have a more diverse and specialized gut microbiome, capable of handling a wider range of substrates. For instance, bacteria capable of breaking down chitin from insect exoskeletons or metabolizing the fats and proteins in animal tissues would be crucial. Research on extant omnivorous primates, such as chimpanzees and baboons, highlights the importance of their gut microbiomes in processing a varied diet. Extrapolating this to a larger primate species suggests that a robust and adaptable microbiome would be a key element of its omnivorous success.

The development of specific foraging tools, while not typical for primates beyond the use of simple probes for insects, could be a fascinating avenue to consider for a highly intelligent, large-bodied omnivore.

While complex tool manufacture is rare, the opportunistic modification of existing materials could occur. For example, a primate might learn to use a sturdy stick to pry open tough seed pods or to dig for tubers. It might also use its hands and feet in conjunction with the environment, such as wedging a branch into a crevice to leverage open a source of grubs, or using a large leaf to scoop up water or collect smaller food items. The evolutionary pressures driving such behaviors would be directly linked to the availability and importance of these supplemental food sources.

In conclusion, the possibility of omnivorous tendencies in a large primate is not merely speculative but a logical extension of adaptive foraging strategies seen in many extant primate species. The exploitation of invertebrates, small vertebrates, and bird eggs would provide crucial supplemental nutrients, enhancing the species' resilience to seasonal fluctuations in fruit availability and supporting its substantial energetic demands. The feasibility of such a diet relies on a combination of behavioral adaptations, including specialized foraging techniques, opportunistic predation, and potentially rudimentary tool use, alongside physiological adaptations such as a robust digestive system and a diverse gut microbiome. This dietary flexibility would not only contribute to the individual's survival and reproductive success but would also shape its ecological role, influencing the populations of its prey species and contributing to the overall dynamics of its ecosystem. The omnivorous primate, in this context, becomes a more integrated and impactful component of its environment, a testament to the power of adaptive plasticity in the face of ecological challenges.

The sheer scale of a hypothetical giant primate immediately brings into focus one of the most fundamental biological constraints: energy requirements. Simply put, a creature of immense size demands an immense amount of fuel to power its biological machinery. Estimating these caloric needs is not merely an academic exercise; it is crucial for understanding the ecological footprint such an animal would possess and the carrying capacity of its environment. To approach this, we can draw upon established principles of metabolic scaling and comparative physiology, examining how energy expenditure relates to body mass in the animal kingdom, particularly among mammals and, more specifically, other large primates.

Metabolic scaling laws, most famously embodied by Kleiber's law, describe the relationship between an animal's metabolic rate and its body mass. Kleiber's law posits that an animal's basal metabolic rate (BMR) scales to the three-quarters power of its body mass (M^(3/4)). This means that as an animal gets larger, its metabolic rate increases, but not in direct proportion to its mass. A larger animal has a lower surface area to volume ratio, which helps it conserve heat, meaning it doesn't need to generate as much heat metabolically per unit of mass as a smaller animal. However, a creature weighing hundreds of kilograms, as our hypothetical giant primate might, would still have a remarkably high total metabolic demand.

Let's consider a hypothetical giant primate weighing, for instance, 500 kilograms (approximately 1100 pounds). Using Kleiber's law as a starting point, we can estimate its basal metabolic rate. While precise species-specific coefficients exist, a general approximation for mammals suggests a BMR of roughly 70 kilocalories per kilogram to the power of 0.75 (70 *M^0.75). For our 500 kg primate, this calculation would be: 70* (500)^0.75. Calculating (500)^0.75 gives us approximately 151.5. Therefore, the BMR would be around 70 151.5 = 10,605 kilocalories per day. This is the energy required to sustain basic life functions at rest – breathing, circulation, cell production, and so on.

However, basal metabolic rate is only a fraction of an animal's total daily energy expenditure. We must also account for the thermic effect of food (TEF), the energy expended during digestion and absorption, and, most significantly, the energy spent on activity. For a large, active primate, activity will likely constitute a substantial portion of its daily caloric needs. This includes everything from foraging, traveling, social interactions, grooming, and any potential defensive or offensive behaviors.

Primate activity levels vary greatly, but it's reasonable to assume a large primate would need to exert considerable energy simply to move its massive body through its environment, especially if it is arboreal or semi-arboreal. If we apply an activity factor, typically ranging from 1.5 to 3.0 or even higher for very active animals, to the BMR, we can arrive at an estimate for total daily energy expenditure. For a moderately active large primate, an activity factor of 2.0 might be a conservative estimate. This would bring the daily caloric requirement to approximately 10,605 kcal 2.0 = 21,210 kilocalories. If the primate is highly active, perhaps spending much of its day traversing large territories or engaging in strenuous social activities, the factor could easily climb to 2.5 or 3.0, pushing the daily requirement to over 25,000 or even 30,000 kilocalories.

To put these numbers into perspective, consider known large mammals. An adult male elephant, weighing between 4,000 and 6,000 kg, can consume between 100,000 and 200,000 kilocalories per day. A rhinoceros, weighing around 1,500-2,500 kg, might consume 30,000-50,000 kilocalories daily. Even a large gorilla, typically weighing between 135 and 220 kg, has a daily caloric requirement estimated to be in the range of 3,000 to 5,000 kilocalories, although this can vary significantly with activity and diet. Our hypothetical 500 kg primate's needs of 20,000-30,000+ kilocalories per day place it in a significant position in terms of energy consumption, surpassing that of a gorilla by a considerable margin and approaching the needs of some smaller rhinoceros species, despite being considerably lighter.

The dietary implications of such a massive caloric demand are profound. If our giant primate is a frugivore, as many large primates are, the sheer volume of fruit required would be staggering. A kilogram of fruit typically yields between 0.5 and 1.5 kilocalories per gram (or 500 to 1500 kcal per kg). To meet a daily requirement of 25,000 kilocalories, the primate would need to consume between 16.7 kg (25,000 kcal / 1500 kcal/kg) and 50 kg (25,000 kcal / 500 kcal/kg) of fruit daily. This vast quantity necessitates not only an abundant food source but also a significant digestive capacity and considerable foraging time. The energetic cost of simply processing and metabolizing such a large volume of food must also be factored into the overall equation, though it is partially accounted for in the activity factor and TEF.

If the primate is indeed omnivorous, as suggested by previous discussions, the caloric density of its diet becomes more manageable. Animal matter, while likely consumed in smaller quantities than fruits, offers a much higher energy return per unit mass. For example, 100 grams of lean meat can provide around 150-200 kilocalories, while 100 grams of high-fat meat could yield upwards of 300-400 kilocalories. Insects and larvae are also nutrient-dense, with some estimates suggesting they can provide significantly more calories and protein per unit of weight than many fruits. Therefore, a diversified diet would likely be more efficient in meeting the substantial energy requirements of a large primate, reducing the sheer volume of food that needs to be ingested daily.

Furthermore, the energy expenditure associated with foraging is a critical consideration. Traveling long distances to find food, climbing trees, tearing apart logs, or even passively waiting for prey all consume energy. For a large primate, even moving from one feeding site to another can represent a significant caloric expenditure. The morphology of the animal plays a role here; a bipedal primate might have different energetic costs associated with locomotion compared to a quadrupedal or brachiating species. The height of the canopy and the density of the forest would also influence the energy required for arboreal foraging.

The thermoregulatory aspect of a large body mass also needs to be considered in the context of energy expenditure. While larger animals are more efficient at retaining heat due to their lower surface area-to-volume ratio, they can also face challenges with dissipating excess heat, particularly in warm climates or during strenuous activity. This thermoregulatory burden can contribute to overall energy expenditure, as the animal may need to engage in behaviors to cool down, such as seeking shade, resting, or even panting, which itself requires energy.

The calculation of these caloric requirements also has direct implications for population dynamics and social structure. A species that requires tens of thousands of kilocalories per day per individual would necessitate vast territories or highly productive feeding grounds to support even a small population. This could lead to intense competition for resources, driving the evolution of sophisticated foraging strategies, territorial defense, and complex social hierarchies. The sheer energetic cost of raising offspring would also be substantial, requiring mothers to consume even more to support lactation and provide for their young.

The estimations provided, while based on established scientific principles, are inherently approximations. The precise caloric requirements would depend on a multitude of factors specific to the individual and the environment: age, sex, reproductive status, health, ambient temperature, habitat type, food availability and quality, and the specific behavioral repertoire of the animal. However, the magnitude of the estimated daily caloric intake – in the tens of thousands of kilocalories – underscores the immense energetic burden of maintaining such a large primate. This highlights the critical need for a stable and abundant food supply, making the ecological niche and resource availability paramount in determining the viability of such a species. The vastness of its caloric needs firmly places this hypothetical creature as a significant consumer within its ecosystem, demanding a proportionate supply of resources to sustain its existence.

The fundamental requirement for life, beyond the substantial caloric intake discussed previously, is access to potable water. For any primate, and particularly for a hypothetical giant species with its proportionally greater physiological demands, a consistent and reliable supply of fresh water is not merely a convenience but an absolute necessity. Dehydration can rapidly lead to a cascade of physiological failures, impacting everything from nutrient transport and waste elimination to thermoregulation and cognitive function. Therefore, the presence and accessibility of suitable water sources would be a paramount factor in defining the geographical distribution and ecological niche of such a creature.

Potential habitats for a large primate would need to be characterized by a network of perennial water sources. Rivers, streams, and natural springs are the most obvious candidates. The volume and flow rate of these sources would be critical. A small, intermittent trickle might suffice for smaller animals, but a large primate would require a consistent, significant volume to drink from. The depth of the water is also a consideration; while a large animal might not require deep water for drinking, access to the water's edge would need to be unimpeded by steep banks or dense undergrowth that could impede movement or pose a risk. Natural springs, often bubbling up from underground aquifers, provide a continuous supply of fresh, often cool, water and would be highly prized locations. The geological conditions that give rise to springs are also often associated with lush vegetation, potentially indicating a rich food environment as well.

Lakes and large ponds, where they exist, would also serve as vital hydration points. The presence of these larger bodies of water can create microclimates that support diverse flora and fauna, potentially making them central hubs of ecological activity. The edges of these water bodies are often rich in vegetation, providing a food source, and can also offer opportunities for foraging on aquatic plants or the invertebrates and small vertebrates that inhabit them. For a large primate, the ability to stand in shallow water to drink and perhaps even cool down during hot periods would be a significant advantage.

Beyond these obvious sources, more subtle forms of hydration must also be considered. The moisture content of the food itself is a crucial, albeit often overlooked, aspect of an animal's overall water balance. Many fruits, leaves, and even some roots can contain a substantial percentage of water. As previously discussed, a frugivorous or herbivorous diet, likely to be a significant component of a large primate's sustenance, would contribute a considerable amount of water to its daily intake. A diet composed primarily of dry matter, such as seeds or tough, fibrous vegetation, would place an even greater reliance on direct drinking water. The ecological assessment of a potential habitat must therefore not only map water bodies but also consider the water content of the available food resources.

In environments where surface water is scarce or seasonally absent, such as arid or semi-arid regions, or even in certain high-altitude environments with frozen water sources for much of the year, specialized hydration strategies would become paramount. Animals in such environments often exhibit remarkable adaptations to conserve water and maximize intake from limited sources. A large primate, with its proportionally larger water needs, would face significant challenges in these scenarios.

One key behavioral strategy in water-scarce environments is to concentrate drinking at times when water is most available or most easily accessed. This might mean visiting water sources only during specific times of day, such as dawn or dusk, when temperatures are cooler, reducing evaporative water loss through respiration and skin. It could also involve exploiting seasonal rains or the melting of snow and ice at higher elevations. During prolonged dry periods, the primate might need to travel considerable distances to reach the last remaining waterholes, a behavior that itself carries significant energetic costs and risks.

The physiological adaptations for water conservation are also critical. Many desert-dwelling mammals have highly efficient kidneys that produce concentrated urine, minimizing water loss through excretion. While primate kidneys are generally efficient, a species adapted to arid conditions might exhibit even more specialized adaptations in renal function. Furthermore, the ability to tolerate a certain degree of dehydration, allowing body water levels to drop slightly below optimal without severe impairment, could be a vital survival trait in environments with unpredictable water availability.

Another crucial behavioral adaptation is the exploitation of indirect water sources. This can include drinking dew that has condensed on vegetation in the early morning. While the amount of water obtained this way might be small per plant, by visiting a large number of dew-laden plants, a significant quantity could be accumulated. Similarly, certain plants, particularly succulent species, store water within their tissues. While the nutritional value of such plants might be lower, they could serve as a vital source of moisture during dry spells. Accessing water from the prey consumed would also be a significant factor for any omnivorous or carnivorous component of the diet. The blood and tissues of hunted animals are rich in water.

The social structure of a large primate population could also influence hydration strategies. If water sources are limited and highly contested, social dynamics might revolve around access to these vital resources. This could lead to territoriality, with dominant individuals or groups controlling access to waterholes. Conversely, in environments where water is more abundant but dispersed, individuals or smaller foraging groups might move more independently. The presence of young, dependent offspring would add another layer of complexity, as mothers would need to ensure a consistent water supply not only for themselves but also for their highly vulnerable young, who are more susceptible to dehydration.

Consider the specific challenges faced by a large primate in a montane environment. High altitudes are often characterized by cooler temperatures, but also by thinner air and potentially less readily available surface water. Snow and ice might be present seasonally or year-round at the highest elevations. While melting snow can provide water, the process of melting and consuming it requires energy. Furthermore, navigating steep, rugged terrain to reach these sources can be physically demanding. The primate might need to adapt its locomotion and foraging patterns to cope with such an environment. The water content of alpine vegetation also tends to be lower than in lowland forests, meaning more effort might be required to extract sufficient moisture from food.

In contrast, consider a primate inhabiting a tropical rainforest. While these environments are typically characterized by high rainfall and abundant surface water, the sheer volume of water present can also present challenges. Fast-flowing rivers can be difficult to cross, and localized flash floods can be a hazard. The dense canopy can limit the amount of direct sunlight reaching the forest floor, potentially reducing dew formation. However, the consistent availability of numerous small streams, rainwater collected in epiphytic plants, and the high water content of fruits and leaves generally make hydration less of a limiting factor in these ecosystems compared to drier regions.

The role of thermoregulation in water balance is also inextricably linked. In hot climates, or during periods of high activity, animals lose water through evaporation to maintain a stable body temperature. A large primate, with its lower surface area to volume ratio, might be more efficient at conserving heat but could face challenges dissipating excess heat. This would necessitate behavioral responses such as seeking shade, reducing activity during the hottest parts of the day, or immersing itself in water if available. The ability to access water for drinking and potentially for cooling would therefore be a crucial aspect of its thermoregulatory strategy.

For a hypothetical giant primate, the sheer volume of water required to meet daily needs would be substantial. While specific figures are difficult to ascertain without knowing the exact physiology and activity levels, it is safe to assume that it would far exceed that of a human or even a gorilla. This high demand implies that the primate would likely be found in close proximity to reliable water sources. Its home range would almost certainly encompass rivers, large streams, or areas with consistent rainfall and groundwater availability. Its daily movements might be dictated, in part, by the need to visit these water sources.

The relationship between diet and hydration is also dynamic. If a primate's diet shifts seasonally, with periods of high fruit consumption (high water content) followed by periods where it relies more on dry matter or hunted prey (lower water content), its reliance on direct water sources would also fluctuate. Understanding these dietary shifts is therefore essential for comprehending the primate's hydration strategy throughout the year. For instance, a primate that primarily consumes fruits during the wet season might shift to consuming more browse or roots during the dry season, necessitating increased visits to water sources.

Furthermore, the social and ecological implications of water availability are profound. Areas with reliable water sources are often biodiversity hotspots, attracting a wide array of fauna and flora. These areas would likely become focal points for any large primate population, potentially leading to higher population densities in the vicinity of water compared to more arid or less hydrated regions. This concentration of animals could, in turn, increase competition for both food and water, shaping the primate's social dynamics and territorial behavior. The presence of predators, which also rely on water sources, would further influence the primate's behavior around these vital sites, necessitating vigilance and perhaps even group defense strategies.

In summary, the availability and accessibility of water sources, combined with the primate's physiological and behavioral adaptations for hydration, would be a critical determinant of its survival and ecological success. Whether through direct drinking from rivers and springs, or indirectly from its food and meteorological phenomena like dew, ensuring adequate water intake would be a constant, fundamental priority. The strategies employed to achieve this would be as varied and complex as the environments it inhabits, underscoring the intricate interplay between a large animal and its aquatic resources.

5

Social Structure and Behavior

The intricate tapestry of primate social life offers a rich wellspring of data for understanding the potential social dynamics of any large ape, especially one whose existence remains within the realm of hypothesis. Our closest living relatives – chimpanzees, gorillas, orangutans, and bonobos – exhibit a remarkable array of social structures, each shaped by evolutionary pressures, ecological constraints, and specific behavioral adaptations. By dissecting these established models, we can begin to construct plausible scenarios for the social organization of a hypothetical giant ape, considering factors such as group size, composition, mating systems, and the complex web of inter-group relations.

Among the great apes, group size and composition vary considerably. Gorillas, for instance, typically live in stable, multi-male, multi-female groups, often referred to as "harem" groups, though this term can be misleading as it implies a strict, exclusive pairing that isn't always the case. These groups are usually led by a dominant silverback male, who is responsible for the safety and cohesion of the troop. He typically associates with a number of adult females and their offspring. The silverback's leadership is crucial; he mediates conflicts within the group, leads foraging expeditions, and defends his females and their young from rival males and predators. While the silverback is the primary breeder, other subordinate males may sometimes be present in the group, though their reproductive opportunities are significantly limited. The size of these gorilla groups can vary, typically ranging from five to twenty individuals, though larger aggregations have been observed, particularly at rich food sites. This social structure is thought to be an adaptation to the gorillas' herbivorous diet, which requires them to spend a significant amount of time foraging, and the need for protection against predators like leopards, a threat that a powerful silverback can effectively deter. The stability of these groups provides consistent social bonds and cooperative benefits, such as shared vigilance and the protection of young.

Chimpanzees, in contrast, exhibit a fission-fusion social system. This means that while they belong to a larger community, their daily social lives are characterized by individuals forming temporary subgroups of varying size and composition. These subgroups often form and dissolve throughout the day based on foraging opportunities, social interactions, and individual preferences. A typical chimpanzee community might consist of anywhere from twenty to over a hundred individuals. Within this larger community, males tend to form strong, lifelong bonds and typically remain in their natal group. They often cooperate in hunting, territorial defense, and aggressive displays against neighboring communities. Female chimpanzees, on the other hand, tend to be more solitary, especially when they have infants, and often transfer to different communities upon reaching sexual maturity. This fission-fusion structure allows chimpanzees to exploit dispersed food resources efficiently. When food is abundant, they can form larger foraging parties, but when resources are scarce, they can break into smaller, more manageable groups to reduce competition. The strong male bonds are crucial for defending their large, shared territories, which are vital for accessing adequate food resources throughout the year.

Orangutans represent a more solitary extreme among the great apes, though this is more a reflection of their arboreal lifestyle and the dispersed nature of their primary food sources – fruit. While orangutans are not strictly solitary, adult males and females generally interact only briefly for mating. Females live with their dependent offspring for an extended period, sometimes up to eight years, and these maternal bonds are incredibly strong. They maintain overlapping home ranges, and while direct social interactions are infrequent, there is a degree of awareness and tolerance between individuals whose ranges coincide. Males, particularly mature individuals, tend to be more territorial and may actively avoid each other, though disputes can arise. The solitary nature of orangutans is likely an adaptation to their frugally dispersed fruit diet, which makes large, cohesive groups energetically inefficient. Their arboreal locomotion also limits the size of groups that can move together through the canopy without significant disruption.

Bonobos, the closest living relatives to chimpanzees, also possess a fission-fusion society, but their social dynamics are notably different, characterized by a more peaceful and egalitarian social structure. Female bonobos form strong bonds, often forming alliances that can rival or even dominate the males. While male chimpanzees are often aggressive and hierarchical, male bonobos tend to be less dominant, and social conflict is often resolved through sexual activity, which plays a far more prominent role in bonobo social life than in chimpanzees, serving to reduce tension, build bonds, and facilitate social interactions between individuals of all ages and sexes. This unique social dynamic, with its emphasis on female bonding and socio-sexual behavior, creates a more cohesive and less conflict-ridden community structure compared to chimpanzees.

Considering these diverse models, we can begin to hypothesize about the social structure of a hypothetical giant ape. Several factors would influence its organization: body size, diet, locomotion, predator-prey dynamics, and resource distribution.

If our hypothetical giant ape were predominantly herbivorous and arboreal, its social structure might resemble that of orangutans, with individuals or mothers and their dependent young maintaining large, overlapping home ranges. The sheer mass of such an animal would make arboreal travel difficult and energetically costly for large groups. Foraging on dispersed fruit or leaves would favor a more solitary existence to minimize competition. However, if this giant ape possessed a more robust, terrestrial or semi-terrestrial lifestyle and a diet that included more readily available resources such as grasses, tubers, or communal fruiting trees, then a more cohesive group structure, perhaps akin to gorillas or even a modified chimpanzee fission-fusion system, might evolve.

The size of a potential giant ape would also impact its social dynamics in terms of defense. A large, powerful animal might have fewer natural predators, potentially reducing the need for large, cohesive groups for protection. However, even apex predators can pose a threat to younger or weaker individuals, or in situations of resource scarcity. If a significant predator threat exists, then a gorilla-like group structure with a dominant, protective male could be advantageous. This would provide a powerful deterrent and coordinated defense against threats. The energy expenditure required to maintain such a group, including the dominance struggles and territorial defense by the dominant male, would need to be offset by reliable and abundant food resources.

The mating system is another critical aspect of social organization. Among the great apes, polygyny (one male mating with multiple females) is common, particularly where males can monopolize access to females, as seen with the silverback gorilla. Promiscuity, as observed in chimpanzees and bonobos, can also occur, especially in fission-fusion societies where individuals have multiple mating partners throughout their reproductive lives. For a giant ape, the mating system would likely be influenced by its social structure. If it lived in stable, multi-male groups, a form of polygyny or promiscuity, potentially with significant male-male competition, would be expected. If it were more solitary, mating encounters would likely be brief and opportunistic, with males seeking out receptive females. The intensity of male-male competition would be a key determinant of social hierarchy and the degree of aggression within the species.

Inter-group relations are also crucial. Many primate species exhibit territoriality, with groups defending their home ranges against incursions by conspecifics. This defense can range from passive avoidance to active, often aggressive, patrolling and conflict. For chimpanzees, inter-group aggression is a significant aspect of their social lives, with coordinated raids on neighboring territories and lethal aggression sometimes occurring. Gorillas, while defending their groups, are less prone to inter-group warfare. Orangutans, being largely solitary, exhibit overlapping ranges, with interactions typically limited to males avoiding each other or females tolerating overlapping territories. A hypothetical giant ape's inter-group relations would depend heavily on its resource needs and social structure. If resources were concentrated and highly valuable, territorial defense would likely be intense, potentially leading to behaviors similar to chimpanzees, especially if the animals were capable of coordinated group action. Conversely, if resources were more evenly distributed, or if the animals' large size and power made individual encounters sufficient for defense, then territoriality might be less pronounced.

The influence of ecological factors cannot be overstated. The availability and distribution of food and water, as discussed previously, would fundamentally shape social dynamics. A species that relies on dispersed, ephemeral food sources, such as ripe fruit that appears seasonally and in scattered locations, would likely favor a fission-fusion system or a more solitary existence. This allows individuals or small groups to efficiently exploit these patchy resources without intense direct competition. In contrast, species that subsist on more uniformly distributed resources, like grasses or leaves, might be able to support larger, more stable social groups. The energy demands of a large animal would necessitate consistent access to substantial quantities of food, which in turn would influence how individuals organize themselves to acquire it.

Furthermore, the physical environment itself can dictate social structures. A species living in dense forests might be more inclined towards smaller, less visible groups, while one inhabiting more open savannas might benefit from larger groups for enhanced vigilance against predators. The ability to navigate the terrain would also play a role. Arboreal apes are limited by the structure of the canopy, while terrestrial apes have more freedom of movement but are more exposed. A giant ape, whether arboreal or terrestrial, would have unique considerations regarding its locomotion and how its size impacts group movement.

The communication methods of a hypothetical giant ape would also be intrinsically linked to its social structure. Complex social lives require sophisticated communication. Apes utilize a combination of vocalizations, gestures, facial expressions, and scent marking. For a large, potentially dispersed animal, long-distance communication would be particularly important. Deep, resonant vocalizations could travel far through forests or across open landscapes, facilitating contact between individuals or groups. Visual signals might be less effective in dense vegetation, making auditory and perhaps olfactory cues more critical.

The development of a complex repertoire of calls could signal presence, alarm, mating readiness, or social status.

Considering all these factors, several plausible social structures for a hypothetical giant ape emerge:

1. **Large, Stable, Multi-Male, Multi-Female Groups:** Similar to gorillas, but potentially larger, with a dominant male or a coalition of males leading the group. This structure would be favored if the species were primarily herbivorous, terrestrial, and faced significant predation pressure, requiring strong group defense. The reproductive success of subordinate males would likely be limited, and social hierarchy would be pronounced. Resource needs would need to be substantial to support such a large group.

2. **Fission-Fusion Society:** Analogous to chimpanzees or bonobos, with individuals belonging to a larger community but forming fluid, temporary subgroups. This structure would be advantageous if the species had a diverse diet, including fruit that ripens sporadically, or if it occupied an environment with patchy resources. It would allow for efficient exploitation of resources while minimizing direct competition. The relative dominance of males versus females could vary, leading to structures more akin to either chimpanzees (male-dominated) or bonobos (female-dominated).

3. **Solitary or Matriarchal Groups:** Similar to orangutans, where adults generally live alone or females live with their dependent young. This would be favored if the species were arboreal, fruit-dependent, or

occupied a niche where resources are highly dispersed and competition is best avoided through isolation. Interactions would be limited to mating and perhaps brief territorial encounters between males.

4.

Hybrid Systems: It is also possible that a giant ape could exhibit a hybrid social structure, with different patterns emerging based on sex, age, or ecological conditions. For example, females might form more cohesive units, while males are more solitary or nomadic, or group structures might change seasonally in response to resource availability.

Understanding these variations among living great apes provides a vital framework for predicting the social life of a hypothetical giant primate. The complexity of primate sociality underscores the intricate interplay between evolutionary history, ecological pressures, and the adaptive advantages conferred by different social strategies. For a creature of immense size, these dynamics would be amplified, making its social organization a critical area of inquiry for understanding its ecological role and evolutionary trajectory. The research on chimpanzees, gorillas, orangutans, and bonobos serves not just as a comparative study but as a foundational blueprint, allowing us to project plausible social architectures onto the unknown. The lessons learned from their intricate social dances – the alliances, the rivalries, the bonds, and the separations – offer a powerful lens through which to view the potential social fabric of a truly monumental ape. The size and intelligence attributed to such a creature would undoubtedly lead to sophisticated social behaviors, where cooperation, competition, and communication would form the bedrock of its existence, echoing the complex realities observed in its smaller, yet equally fascinating, primate cousins.

The question of whether a large, hypothetical primate would thrive in solitude or within the dynamic of a social group is a fundamental one, touching upon the very core of evolutionary strategy. Both solitary living and complex group dynamics offer distinct advantages and disadvantages, each shaped by the crucible of natural selection and heavily influenced by ecological pressures. For a creature of megafauna proportions, these considerations are magnified, with profound implications for its survival, reproduction, and ecological impact.

Solitary existence, as exemplified by the orangutan, presents a compelling strategy for certain ecological niches. The primary benefit of solitary living is the reduction of direct competition for resources. When an animal forages alone, it does not have to contend with the immediate presence of conspecifics vying for the same food items. For a large primate whose dietary needs would be substantial, this can be a significant advantage, particularly if its food sources are dispersed, ephemeral, or of low density. Imagine a solitary giant ape roaming a vast territory, its large frame requiring a considerable caloric intake. By avoiding other members of its species, it can maximize its personal foraging efficiency, dedicating its time and energy to locating and consuming food without the constant pressure of sharing or competing with others. This is especially relevant if the primate's diet consists heavily of fruits, which often ripen asynchronously and are scattered across the landscape. A solitary forager can exploit these fleeting opportunities without the logistical challenges and increased competition that a larger group would introduce. This efficiency extends beyond food; a solitary individual can also conserve energy by not having to navigate the social complexities and potential conflicts inherent in group living. The absence of constant social interaction means less energy expended on maintaining dominance hierarchies, mediating disputes, or engaging in complex social rituals.

However, the solitary life is not without its considerable drawbacks, particularly concerning predator avoidance. While a large primate may possess fewer natural predators than smaller ape species, it is not entirely immune. Young, old, or injured individuals can be vulnerable. In a solitary existence, the burden of vigilance rests entirely on the individual. There is no "many eyes" effect, where a group can collectively scan the environment for threats, providing earlier warnings and a greater likelihood of detecting an approaching danger. A lone primate, however large, might be caught unaware by a coordinated predator attack, especially if it is ambushed or if its attention is solely focused on foraging. This lack of collective defense makes solitary living a riskier proposition in environments with significant predation pressure. Reproduction also presents a challenge for solitary species. Finding a mate can be a difficult and time-consuming endeavor when individuals are dispersed across large territories. This often leads to brief, opportunistic encounters, and the competition among males to find and secure receptive females can be intense, potentially leading to heightened aggression during mating periods. Furthermore, the extended period during which many primate mothers care for their young means that a female must either remain solitary with her offspring, potentially limiting her own foraging efficiency, or find a way to integrate her offspring into a social group, which may not be an option in a strictly solitary species.

The advantages of group living, conversely, are most pronounced in the realms of predator defense and resource acquisition through cooperation. For a large primate, the presence of a group can act as a powerful deterrent to potential predators. A single, large animal is a significant meal, but a group of such animals presents a much more formidable challenge. The collective strength, size, and vigilance of a group can make them an unappealing target for most predators. Furthermore, groups can mount coordinated defenses, with individuals working together to drive away threats. This shared vigilance is a cornerstone of group living, significantly increasing the chances of early detection of danger. Imagine a troop of giant apes moving through the savanna; any lurking predator would be made acutely aware of their presence, and the sheer number of potential defenders would likely dissuade an attack.

This benefit is particularly crucial for protecting vulnerable juveniles and offspring, who benefit immensely from the collective protection of the group.

Group living also facilitates more efficient foraging and resource acquisition in certain scenarios. While it increases direct competition, it can also enable cooperative hunting, as seen in chimpanzees, where the success rate of bringing down large prey is higher in coordinated groups. Even in herbivorous species, groups may be able to defend rich food patches from competitors, whether they are other primate groups or different species entirely. The ability to collectively defend a prime foraging territory can ensure a more stable and reliable food supply for all group members. Moreover, social learning plays a significant role. Young primates learn foraging techniques, identify edible versus toxic plants, and understand predator avoidance strategies by observing and interacting with older, more experienced individuals within the group. This accelerated learning curve can be critical for survival, especially in a species with a long developmental period.

Reproduction is often more straightforward in group-living species, although it can also be more complex due to social dynamics. In species with polygynous mating systems, such as gorillas, the group structure inherently supports a male monopolizing access to multiple females. This ensures consistent mating opportunities for the dominant male and provides a stable social unit for raising offspring. Even in more promiscuous systems, like those of chimpanzees, the larger community provides ample opportunities for finding mates. Furthermore, within groups, there can be significant social support for mothers. Alloparenting, where individuals other than the mother assist in caring for infants, can reduce the burden on the mother, allowing her to forage more effectively and increasing the survival rate of her offspring.

The primary disadvantage of group living is the inherent increase in competition for resources. A larger group means more individuals vying for the same food, water, and prime resting or nesting sites. This competition can lead to social stress, increased aggression, and a potential decrease in individual foraging success if resources are scarce or unevenly distributed. Dominance hierarchies, while potentially reducing overt conflict, require energy expenditure for their maintenance and can disadvantage subordinate individuals, who may receive less food or fewer mating opportunities. The risk of disease transmission is also higher in dense groups.

For a hypothetical megafauna primate, the balance between these pros and cons would be heavily influenced by its specific ecological context. If its primary food source were abundant, widely distributed, and relatively easy to access, such as vast grasslands or extensive forests with consistently available foliage, then the benefits of group living—enhanced predator defense and potential for cooperative resource defense—might outweigh the increased competition. A large, herbivorous terrestrial ape, for instance, might benefit from the "safety in numbers" principle, especially if it inhabits an environment with significant large predator activity. The energy expended in maintaining group cohesion and navigating social hierarchies would be justified by the increased security and access to reliable food sources.

Conversely, if the primate's diet were highly specialized, relying on dispersed, seasonal fruits or other ephemeral resources, a more solitary or loosely structured social system might be favored. In such a scenario, the high costs of competition within a large group would likely exceed the benefits of collective defense. The orangutan's strategy of wide-ranging solitary foraging, punctuated by brief mating encounters, is a testament to the success of this model when resources are patchy and dispersed. For a giant ape with a similar dietary specialization and arboreal or semi-arboreal lifestyle, this would likely be the most energetically efficient path.

The specific type of predator threat would also be a crucial factor. If predators were large, powerful, and capable of tackling solitary individuals, group living would be a strong evolutionary advantage. However, if the primary threats were less formidable, or if the primate's sheer size and strength made it largely invulnerable as an adult, the pressure to form large, cohesive groups for defense would be reduced. This might allow for smaller, more fluid social units, or even a transition towards greater solitary tendencies among adults.

Ultimately, the most successful social strategy for a hypothetical megafauna primate would be a finely tuned adaptation to its unique ecological circumstances. It is not a simple dichotomy of solitary versus group living, but rather a spectrum of possibilities. Some species might exhibit flexible social structures, shifting between solitary foraging and temporary aggregations based on resource availability or seasonal changes. For instance, females with dependent young might remain more solitary for safety and efficient foraging, while juvenile groups might form for social learning and play, and adult males might form bachelor groups or patrol territories independently. The complex interplay of diet, habitat, predation pressure, and reproductive strategies would determine whether our hypothetical giant ape ultimately embraced the communal strengths of a troop or the independent resilience of a solitary wanderer. The evolutionary trajectory of such a creature would likely have favored the social system that most effectively minimized energetic costs while maximizing opportunities for survival and successful reproduction within its specific environmental context.

The establishment and maintenance of territory represent a fundamental aspect of the behavioral ecology of many large mammals, and a hypothetical megafauna primate would likely be no exception. Territoriality, the defense of a specific area against intrusion by conspecifics, is driven by a complex interplay of resource availability, reproductive needs, and predator avoidance strategies. For an animal of considerable size and potentially significant resource requirements, the spatial dynamics of its existence would be a critical determinant of its survival and reproductive success. The concept of a "home range"—the area an animal regularly traverses in its daily activities of foraging, mating, and sheltering—is central to understanding territorial behavior. However, not all home ranges are actively defended. Some are merely areas of use, while true territories are actively patrolled and defended against rivals.

The size of a home range for a megafauna primate would be dictated by a multitude of factors, primarily its metabolic rate, dietary habits, and the spatial distribution and predictability of its food resources. Given the substantial energetic requirements of large-bodied animals, their home ranges would, by necessity, be considerably larger than those of their smaller ape relatives. Consider the dietary needs of a creature weighing several hundred kilograms or even a ton. To sustain such a mass, it would require access to a continuous and abundant supply of calories. If its diet were primarily frugivorous, relying on fruits that often ripen asynchronously and are dispersed across the landscape, its home range would need to encompass a diverse array of fruiting trees and patches, requiring extensive daily or seasonal travel to exploit these ephemeral food sources. This would necessitate a large geographical area, potentially spanning many square kilometers, to ensure a reliable food supply throughout the year. The more dispersed and unpredictable the food, the larger the home range must be. For example, if a particular type of nutrient-rich fruit only ripens seasonally in scattered locations, the primate would need access to multiple such areas across its territory to bridge the gaps between fruiting seasons.

In contrast, if the megafauna primate were primarily herbivorous, feeding on abundant and widespread vegetation such as grasses or leaves, its home range might be comparatively smaller, provided that the vegetation was consistently available and of sufficient nutritional quality. Large herbivores, like elephants or rhinoceroses, often maintain extensive ranges, but their territoriality can be more fluid, particularly when following seasonal vegetation growth or water sources. However, even with abundant vegetation, a large primate would still require substantial amounts of food, meaning its range would need to be large enough to support the sheer volume of biomass consumed daily without depleting local resources too rapidly. The concept of carrying capacity would be paramount; the home range must be large enough to sustain the individual or social group without leading to habitat degradation and starvation.

The nature of the habitat itself would also play a crucial role in defining home range size. An animal inhabiting a dense forest might have a more restricted home range in terms of linear distance covered but a higher density of resources within that area. Conversely, an open savanna dweller might traverse vast distances daily to find scattered food patches and water. The presence of geographical barriers such as rivers, mountains, or impassable terrain would also influence the shape and size of a home range, creating natural boundaries and forcing the animal to adapt its movements accordingly. Arboreal or semi-arboreal species, while potentially having more localized resource patches within trees, might still require extensive arboreal pathways and interconnected canopies, leading to large, complex home ranges that also encompass terrestrial elements for inter-forest travel or access to specific ground-level resources.

Beyond resource needs, reproductive strategies significantly influence territorial behavior. In many primate species, males establish and defend territories that encompass the home ranges of multiple females. This polygynous mating system ensures that the dominant male has preferential access to receptive females within his territory, thereby maximizing his reproductive success. For a megafauna primate, such a system would imply that a male's territory would need to be sufficiently large to encompass the home ranges of several females, each of whom would also require substantial foraging areas. This would result in very large male territories, potentially several times the size of the individual female home ranges. The defense of such a vast area would demand considerable energy expenditure and vigilance. Alternatively, in more monogamous or pair-bonded species, the territories might be more co-defended by both the male and female, or the pair might share a common home range. In species with a more egalitarian social structure, home ranges might be more fluid, with individuals or small groups of related individuals sharing overlapping ranges.

The implications of territoriality for population density and distribution are profound. Active territorial defense typically limits the number of individuals that can occupy a given area, leading to lower population densities compared to non-territorial species. If a megafauna primate were highly territorial, especially among males competing for resources and mates, its population would likely be sparse, with individuals or small groups spread out over vast tracts of land. This low density can have cascading effects on social dynamics, mating opportunities, and the ability to maintain social bonds. Finding mates could become a significant challenge, necessitating long-distance travel or specific signaling behaviors to attract partners from afar.

The boundaries of these territories are often marked through a variety of methods. Scent marking, through urine and feces, is a common and effective way for large mammals to communicate their presence and ownership of an area. Vocalizations, such as roars or bellows, can also serve to announce territorial claims and warn off rivals, especially in species with large home ranges where visual confirmation of intruders might be difficult. Visual displays, such as the erection of fur or the display of physical prowess, can also play a role in territorial disputes, though direct physical confrontations are often avoided due to the high risk of injury for large animals. The establishment and maintenance of these boundaries require constant patrolling and reinforcement. A male might routinely traverse the perimeter of his territory, reinforcing his scent marks and inspecting for signs of intrusion by rivals. Similarly, females might defend their core foraging areas against other females, especially if resources are localized and critical for survival, such as specific water sources or prime feeding grounds.

The density of population is intrinsically linked to resource availability and the degree of territoriality. In areas where resources are exceptionally abundant and predictable, territories might be smaller, allowing for higher population densities. Conversely, in resource-poor or highly variable environments, territories would need to be larger to ensure survival, resulting in much lower population densities. This would mean that for a megafauna primate, vast, contiguous tracts of suitable habitat would be necessary to support even a moderate population. Fragmentation of habitat due to human encroachment or natural events would thus pose a severe threat, as it could break up territories, isolate individuals, and reduce access to essential resources, leading to population decline.

The social structure of the species would also influence how territories are managed. In a species where males hold large territories encompassing multiple females, there might be a hierarchy among subordinate males who are tolerated within the territory but do not have direct mating access to the females. These subordinate males might live on the fringes of the dominant male's territory or occupy smaller, less resource-rich areas within it, waiting for an opportunity to challenge the dominant individual. In species where territories are more fluid and overlapping, the concept of strict territorial defense might be replaced by a system of dominance hierarchies within larger social groups, where access to resources and mates is determined by social status rather than exclusive territorial ownership.

The energy expenditure associated with territorial defense cannot be overstated. Patrolling, marking, and confronting intruders all require significant caloric investment. This means that the benefits of territoriality—primarily securing adequate resources and reproductive opportunities—must outweigh these costs. If the food resources within a large home range are scarce or unpredictable, the energy spent defending that range might be more than the energy gained from the food found within it. This is a crucial trade-off that evolutionary pressures would have finely tuned. For a megafauna primate, the size of its territory would represent a delicate balance between ensuring survival and reproduction and minimizing the costs of defense.

Furthermore, the reproductive status of individuals can influence their territorial behavior. Pregnant or lactating females, with their increased energetic demands and the need to protect vulnerable offspring, might establish and defend smaller, core areas rich in essential resources. Males, particularly during the mating season, might intensify their territorial patrols and displays, becoming more aggressive in defending their claimed areas. Juveniles, while often roaming within their mother's home range or within a group's territory, might also begin to establish their own nascent territories as they approach independence, potentially engaging in exploratory forays and testing the boundaries of established territories.

The concept of "resource defense polygyny" is highly relevant here. If certain critical resources, such as prime feeding sites or reliable water sources, are spatially concentrated, a male might defend access to these resources, thereby controlling access to the females who rely on them. This would lead to territories being defined not just by the overall extent of travel but by the critical nodes of resource availability within that area. The distribution of these critical resources would therefore directly shape the size, shape, and defensibility of territories. If these resources are scattered, the territories would be larger and potentially more challenging to defend comprehensively. If they are concentrated, territories might be smaller but fiercely contested.

The presence of other species can also influence territoriality. A megafauna primate might compete with other large herbivores for vegetation, or with large carnivores for prey. This interspecific competition could either expand or contract the effective size of its territory. For example, if a rival herbivore species is a strong competitor for a particular food source, the primate might need to expand its range to find sufficient quantities of that food, or it might need to defend its preferred feeding grounds more vigorously against interlopers. Similarly, the presence of dangerous predators might influence whether a primate chooses to defend a fixed territory or to adopt a more nomadic lifestyle, following safer routes or relying on group defense.

Ultimately, the territoriality and range use of a hypothetical megafauna primate would be a complex mosaic reflecting its unique ecological niche. It would be a system shaped by the fundamental drivers of survival and reproduction, finely tuned by the pressures of resource availability, predation, and social dynamics. The scale of such a creature would magnify these factors, demanding vast expanses of habitat and intricate behavioral strategies to navigate the spatial challenges of its existence, defining a significant footprint on the landscape it inhabits.

The ability to communicate is a cornerstone of social behavior, influencing everything from individual interactions to the dynamics of entire populations. For a hypothetical megafauna primate, sophisticated communication systems would be essential for navigating its complex social world and interacting with its environment. These communication strategies would likely encompass a rich repertoire of vocalizations, scent markings, and visual signals, each serving distinct purposes and conveying nuanced information.

Vocalizations would undoubtedly play a critical role, especially for a creature potentially inhabiting vast territories where visual cues might be limited by dense foliage or distance. The sheer size and robust anatomy of such a primate would likely allow for the production of powerful and far-carrying sounds. Imagine deep, resonant calls, perhaps akin to the roars or bellows of other large mammals, used for territorial demarcation. These vocalizations could serve as a primary means of announcing an individual's presence and claim to an area, acting as a constant reminder to conspecifics to maintain a respectful distance. Such calls would be particularly effective at dawn and dusk, times when many animals are most active and territorial interactions are frequent. The frequency, duration, and pitch of these territorial roars could convey vital information about the caller's age, size, and dominance status, allowing potential rivals to assess the threat without engaging in direct confrontation.

Beyond simple territorial announcements, a complex vocal repertoire could facilitate a wide array of social interactions. Alarm calls are a prime example. A distinct sound might signal the presence of a predator, prompting immediate evasion or defensive behaviors from other group members. Different calls could be associated with different types of threats—a rapid, high-pitched series of shrieks for a stealthy ground predator versus a deep, guttural rumble for a large aerial threat. The nuances of these calls could even indicate the severity of the danger, influencing whether individuals flee, freeze, or prepare to defend themselves.

Social bonding within groups would also likely be facilitated by vocalizations. Soft, low-intensity calls, perhaps chirps, grunts, or coos, could be used during grooming sessions, infant care, or simply while resting together. These affiliative vocalizations would reinforce social bonds, reduce tension, and signal friendly intentions. They would be crucial for maintaining group cohesion, especially in a species where individuals might spend significant amounts of time foraging or traveling apart. The ability to locate and stay in contact with group members over distances, particularly in dense forests or during periods of low visibility, would be greatly enhanced by these soft vocalizations.

Mating behavior would undoubtedly involve specialized vocalizations. Males might emit specific calls to attract females, signaling their readiness to mate and advertising their fitness. These calls could be part of elaborate courtship rituals, designed to impress potential mates and ward off rivals. Females, in turn, might vocalize to signal their receptivity or to attract the attention of a preferred male. The timing of these vocalizations would be critical, likely coinciding with periods of peak fertility, and the unique acoustic properties of each individual's voice could contribute to mate recognition.

The ecological context would heavily influence the specific nature of these vocalizations. For instance, a species inhabiting a noisy environment, such as a rainforest with constant insect and bird activity, would need calls that are distinct and can cut through the background noise. This might involve higher frequencies, pulsed calls, or even mimicry of other sounds to avoid being masked. Conversely, in open savanna environments, where sound travels freely, deeper, more resonant calls might be more effective for long-distance communication. The anatomy of the vocal tract, including the larynx and nasal cavities, would evolve to support these specific acoustic requirements, potentially leading to unique vocal organs adapted for producing these distinctive sounds. The size of the primate would certainly contribute to the volume and depth of its vocalizations, creating a sonic landscape that would be both impressive and informative.

Scent marking represents another powerful modality for communication, particularly for large mammals with extensive territories. The production and deposition of scent are metabolically costly, indicating the importance of the information being conveyed. Scent glands, strategically located on the body, would likely be utilized to leave olfactory messages on vegetation, rocks, or the ground. Urine and feces would also serve as significant scent markers, imbued with pheromones and other chemical signals.

Males, in particular, would likely employ scent marking extensively to advertise their presence and territorial ownership. Regular patrols of their boundaries would involve depositing scent at prominent locations—trees, prominent rocks, or crossing points—creating a mosaic of olfactory signals that define their domain. This constant "re-signing" of the territory would inform rivals of occupancy and deter them from encroaching. The intensity and frequency of marking could also convey information about the male's condition and mating motivation. A male in prime condition might mark more frequently and vigorously than a subordinate or an individual in poor health.

Females would also utilize scent marking, though perhaps with different motivations. They might mark to indicate their reproductive status, signaling to males that they are receptive or, conversely, that they are not currently a viable mating option. Scent marking could also play a role in maintaining social bonds within a family group or for mothers to keep track of their offspring. The scent of a mother could be a comforting presence for a young primate, and infants might also deposit their scent to facilitate reunion after periods of separation.

The chemical composition of these scents would be incredibly complex, containing a wealth of information. Pheromones, airborne chemical signals that elicit specific physiological or behavioral responses in conspecifics, would be key. These could include signals related to reproduction, such as estrus signals from females or aggression-inhibiting signals from males. Other volatile organic compounds could indicate diet, health status, or even stress levels. The persistence of these scent marks would vary depending on environmental factors like rain and wind, and the primate might adapt its marking behavior accordingly, perhaps marking more frequently in areas exposed to the elements.

Visual signals, while potentially less dominant in dense environments, would still be crucial for close-range communication and for specific contexts. Body posture, facial expressions, and gestures would all contribute to the visual communication system. A dominant individual might stand tall, chest puffed out, and stare directly at a subordinate, conveying a clear message of hierarchy and discouraging challenge. Conversely, a subordinate might adopt a submissive posture, averting its gaze, lowering its head, or exposing its vulnerable underbelly as a sign of appeasement.

Facial expressions would be particularly important for conveying emotional states. Wide eyes might indicate fear or alertness, while bared teeth could signify aggression. Subtle changes in ear position or the erection of fur could also communicate mood and intention. For instance, an animal with its fur bristling and ears laid back might be signaling intense agitation or preparing for a fight.

Gestures, such as a head bob, a tail flick, or a specific limb movement, could also be used to convey specific messages. These might be part of courtship displays, aggressive challenges, or invitations to play. The physical size and impressive musculature of a megafauna primate would lend significant weight to its visual displays, making them powerful tools for social signaling. Imagine a male primate performing a charging display, not to attack, but to intimidate rivals and impress females, showcasing its strength and agility.

Furthermore, some primates might utilize specific colored or patterned body parts for visual signaling. Although less common in primates compared to birds, certain species, like the mandrill with its vibrant facial coloration, demonstrate the potential for pigment-based visual communication. If a hypothetical megafauna primate possessed distinct markings, these could be used for individual recognition, signaling species identity, or conveying social status. For example, a particular pattern of fur on the face or chest might become associated with dominant individuals.

The interplay between these different communication modalities would be critical. A vocalization might be accompanied by a specific visual display or a strong scent mark, amplifying the message and ensuring its clear reception. For instance, a territorial roar might be reinforced by a visible scent marking on a nearby tree. Similarly, an aggressive visual display might be accompanied by a growl or hiss. This multi-modal communication would provide redundancy, ensuring that important information is not lost, even if one channel is compromised. Understanding these integrated communication systems is paramount to deciphering the social dynamics and ecological strategies of such a creature, offering glimpses into its world even in the absence of direct observation, by inferring the purpose and meaning behind these complex signals. The richness and sophistication of these communication channels would directly influence the species' ability to coordinate social activities, manage conflicts, reproduce successfully, and ultimately, to thrive in its environment.

The reproductive strategies and patterns of parental care exhibited by a hypothetical large primate would be deeply intertwined with its ecological niche, social structure, and evolutionary history, as is characteristic of primate species. Given their size and likely complex social dynamics, we can anticipate a suite of adaptations geared towards ensuring the successful rearing of offspring in a challenging environment. Primates, as a rule, invest heavily in their young, and this species would be no exception. This investment typically manifests as prolonged gestation periods, extended infant dependency, and significant contributions to nurturing from various group members.

The gestation period for a large primate would likely be substantial, reflecting the developmental demands of producing a relatively large and neurologically complex infant. Comparable to existing large primate species such as gorillas and orangutans, gestation could range from seven to ten months, or potentially even longer, given the sheer scale of the developing fetus. This extended period within the womb allows for significant fetal development, equipping the infant with a degree of maturity at birth that aids in its initial survival. However, even with a lengthy gestation, the infant would still be born in a highly altricial state, dependent on its mother for virtually all aspects of survival. The energetic demands on the mother during this phase are immense, requiring a diet rich in nutrients to support both her own metabolic needs and the rapid growth of the fetus.

Following birth, the infant's dependency on its mother would extend for an exceptionally long duration, a hallmark of primate life history. This extended period of dependence, often spanning several years, is critical for several reasons. Firstly, it allows for the gradual maturation of the infant's motor skills, sensory systems, and cognitive abilities. Learning to navigate the complex arboreal or terrestrial environment, to identify edible foods, to recognize social cues, and to avoid predators are all skills that require considerable time and practice. The mother serves as the primary tutor and protector during this formative stage, guiding the infant through these essential learning processes. The physical size of the infant at birth and its rate of growth would also dictate the duration of this dependency. A larger infant, requiring more calories, would necessitate a longer period of nursing and protection.

Weaning would therefore be a gradual process, not an abrupt cessation of milk feeding. As the infant's dentition develops and its digestive system matures, it would begin to explore solid foods, initially under the watchful eye of its mother. This transition would likely be staggered, with the infant still relying on its mother's milk for a significant portion of its nutritional needs for an extended period. The mother's milk composition would likely evolve to meet the specific dietary requirements of a rapidly growing primate, perhaps being particularly rich in fats and proteins. The age at which full weaning occurs would be a crucial factor influencing interbirth intervals, as a mother typically cannot conceive again until her current offspring is at least partially weaned.

Parental care, particularly maternal care, would be the cornerstone of infant survival. The mother's role extends far beyond simply providing nourishment. She would be responsible for grooming the infant, keeping it clean and free from parasites, which is not only a hygienic practice but also a vital component of social bonding. Grooming reinforces the mother-infant bond, reducing stress for both individuals and fostering a sense of security for the young primate. The mother would also provide warmth and protection, carrying the infant almost constantly during its early months. This close physical contact, often in the form of ventral carrying (infant clinging to the mother's underside) or dorsal carrying (infant riding on her back), allows the infant to remain close and secure while the mother forages, travels, or defends herself.

The physical demands on the mother during this period are considerable. Carrying a growing infant, which can become quite heavy, would necessitate significant energy expenditure. This would likely influence her foraging behavior, requiring her to consume more food and potentially travel longer distances. The risk of predation also increases for a mother with an infant, as the infant's cries or movements could attract predators, and the mother's movements might be encumbered by the need to protect her young. Therefore, the mother's vigilance and defensive capabilities are paramount.

The social structure of the species would also profoundly influence parental care. In many primate species, care is not solely the responsibility of the mother. While mothers bear the primary burden, other group members, particularly females, can play significant roles in alloparenting, or helping to care for infants that are not their own. In a large primate species with a cohesive social group, older females, perhaps related females like sisters or aunts, might assist in caring for infants. This could involve holding, grooming, or playing with the infant, allowing the mother a brief respite for foraging or rest. Such alloparenting can be beneficial for both the helper and the mother. The helper gains valuable experience in infant care, which can be advantageous when they eventually have their own offspring, and the mother benefits from reduced energetic costs and increased safety. However, alloparenting can also carry risks for the infant, as inexperienced or overly enthusiastic helpers might inadvertently harm a young primate. The degree of alloparenting would likely depend on the specific social dynamics, the relatedness of individuals within the group, and the perceived benefits and costs to each participant.

The role of males in parental care, if any, would be highly variable and depend on the mating system. In species with strong pair bonds or highly cohesive multi-male, multi-female groups where paternal certainty is high, males might engage in some level of infant protection or even direct care. This could involve defending the infant from predators or rivals, or even carrying the infant during travel. However, in species where male parental investment is low, males might be more focused on territorial defense and competition for mating opportunities. Given the potential for large social groups, it is plausible that males would play a protective role for the group as a whole, thereby indirectly benefiting the infants by reducing overall predation risk. They might also exhibit tolerance towards infants, allowing them to approach and even interact, which can be a sign of affiliation within the group.

Infanticide, the killing of infants by adults, is a tragic but not uncommon phenomenon in the animal kingdom, including among some primate species. It is often driven by males seeking to bring females back into estrus more quickly or by unrelated individuals competing for resources. A large primate species might evolve specific strategies to mitigate this risk. These could include strong group defense, where adult males or other females actively protect infants from potential infanticides. The close social bonds within the group, reinforced through grooming and other affiliative behaviors, could also create a social environment where infanticidal acts are less likely or are met with swift group retaliation. The mother's own vigilance and her ability to defend her infant are also crucial deterrents.

The nutritional requirements of a large, growing primate infant would be substantial, and this would influence both maternal foraging strategies and the overall diet of the group. The mother would need to locate and consume calorie-dense foods to produce sufficient milk, and as the infant begins to consume solid food, it would also need access to nutritious items. This could lead to specialized foraging behaviors, such as focusing on high-energy fruits, seeds, or protein-rich insects or small vertebrates, depending on the species' overall diet. The mother's ability to effectively forage while simultaneously protecting and caring for her infant would be a key determinant of infant survival.

The physical demands of carrying and protecting a large infant would likely influence the mother's own body condition. Mothers often experience a decline in body mass during lactation and infant care, and their ability to regain this lost condition would be crucial for subsequent reproductive success. This highlights the importance of ample food resources and safe foraging areas for the long-term viability of the population. The energy expenditure associated with reproduction and parental care is a significant factor in the life history of primates, and for a large species, these costs would be amplified.

The weaning process would likely be accompanied by a gradual shift in the infant's social interactions. As it becomes more mobile and independent, it would likely engage in more play behavior with its peers, as well as with other adults in the group. Play is essential for developing motor skills, social competence, and cognitive abilities. Through play, young primates learn about social hierarchies, practice communication signals, and explore their environment in a safe context. The tolerance and participation of adult group members in these play activities would be indicative of the group's social cohesion and the general level of care afforded to the young.

The age at which young primates become fully independent would mark a significant transition. This independence does not necessarily mean leaving the social group, but rather the ability to fend for themselves in terms of foraging, finding shelter, and avoiding immediate threats. Even after becoming independent, however, social bonds established during infancy and juvenile periods would continue to play a crucial role in an individual's social life and reproductive success. Young adults would likely remain within their natal group, or a closely related one, and their integration into the adult social structure would be a gradual process, influenced by dominance hierarchies, social alliances, and mating opportunities.

The long period of infant and juvenile dependency has significant implications for population dynamics. Because females invest so much time and energy into each offspring, and because interbirth intervals are likely to be long, population growth rates for such a large primate species would be relatively slow. This makes them particularly vulnerable to environmental changes, habitat loss, and hunting pressure. A slow reproductive rate means that populations take a long time to recover from declines. Therefore, understanding the nuances of their reproductive strategies and the extended duration of parental care is not just an academic exercise; it is fundamental to comprehending their ecological role and their conservation needs. The success of each individual, and by extension the species, hinges on the intricate and often demanding processes of reproduction and dedicated parental investment. The prolonged developmental trajectory is a testament to the complexity of primate life and the significant evolutionary advantages conferred by extensive learning and social development facilitated by dedicated parental figures.

6

Intelligence, Tool Use, and Adaptability

The cognitive landscape of primates, particularly the great apes, presents a fascinating spectrum of intellectual prowess, demonstrating capacities that extend far beyond simple instinctual responses. This section delves into these sophisticated abilities, examining problem-solving, memory, and social intelligence as observed in our closest living relatives. By understanding these existing capabilities, we can then extrapolate and hypothesize about how such faculties might manifest and potentially amplify in a hypothetical larger primate species, pushing the boundaries of what we currently understand about animal cognition.

One of the most striking cognitive abilities observed in great apes is their capacity for insightful problem-solving. This is not merely a matter of trial and error, but often involves understanding the underlying mechanics of a situation and devising novel solutions. Classic experiments with chimpanzees, for instance, have shown their ability to use tools in complex ways, not just as they find them, but by modifying them to suit a particular task. This can include stripping leaves from branches to create fishing probes for termites or using stones as hammers to crack nuts. These are not learned behaviors passed down rigidly, but demonstrate a flexible understanding of object properties and their potential applications. When faced with a novel problem, such as reaching a reward that is just out of grasp, chimpanzees might spontaneously stack boxes or use a stick to extend their reach, exhibiting a foresight and planning capacity that is truly remarkable. Orangutans, known for their solitary nature, also display impressive problem-solving skills, often related to their arboreal lifestyle. They can devise strategies to navigate complex environments, select appropriate materials for nest construction, and even exhibit an understanding of basic physics, such as how to maintain balance on unstable branches. Gorillas, while often perceived as less tool-oriented than chimpanzees, also demonstrate problem-solving in their daily lives, particularly in their intricate social interactions and foraging strategies. The ability to understand cause and effect, to anticipate consequences, and to adapt strategies based on environmental feedback are all hallmarks of their cognitive sophistication.

Beyond immediate problem-solving, the memory capacities of great apes are equally impressive and crucial for their survival. They possess excellent long-term spatial memory, enabling them to recall the locations of fruiting trees, water sources, and safe sleeping sites over vast territories, sometimes weeks or months after their last visit. This spatial knowledge is vital for efficient foraging and for navigating the complex social and ecological landscapes they inhabit. Furthermore, their social memory is profound. Apes can recognize and remember a large number of individuals, not just within their own social group, but also individuals from neighboring groups and even those they have encountered only fleetingly. This social recognition is fundamental for maintaining complex social hierarchies, forming alliances, and understanding the potential threats or benefits posed by others. They can recall past interactions, distinguishing between cooperative partners and rivals, and remembering the social status of individuals. This allows them to adjust their behavior accordingly, engaging in reconciliation after conflicts or seeking out allies when needed. Studies have also indicated that some apes possess a form of episodic-like memory, meaning they can recall "what, where, and when" specific events occurred, although this area is still under active investigation. This capacity for detailed recall of past events is critical for learning from experience, avoiding past mistakes, and anticipating future needs.

The realm of social intelligence in great apes is perhaps the most complex and well-studied aspect of their cognition. Their lives are intricately woven into a tapestry of social relationships, and their cognitive abilities are heavily geared towards navigating this intricate web. They exhibit a profound understanding of social dynamics, including dominance hierarchies, kinship ties, and alliances. They can accurately assess the intentions and emotional states of others, a phenomenon known as theory of mind, though the extent of this capacity is a subject of ongoing debate. Evidence for this includes their ability to deceive others for personal gain, such as hiding food from a dominant individual or feigning ignorance. They also demonstrate empathy, comfort distressed individuals, and engage in complex forms of cooperation. Grooming, for example, is not merely a hygienic practice but a vital social currency, used to build and maintain relationships, resolve tensions, and solidify alliances. Apes engage in tactical deception, manipulate social relationships, and display remarkable foresight in their social interactions, planning their movements and behaviors to maximize their social standing and reproductive success. Their communication systems, while not as complex as human language, are sophisticated, involving a range of vocalizations, gestures, and facial expressions that convey a wide array of information about their emotional state, intentions, and the environment. They can learn to understand symbolic communication, as demonstrated by apes trained in sign language or using lexigrams. This capacity for understanding symbols and abstract concepts suggests a level of cognitive flexibility that is truly impressive.

Now, let us consider how these remarkable cognitive abilities might scale and adapt in a hypothetical larger primate species. Size itself can be a significant factor influencing cognitive demands and expression. A larger body mass often correlates with a larger brain size, although the relationship is not always linear and the encephalization quotient (EQ) – the ratio of actual brain size to expected brain size for a given body mass – is a more informative metric. If our hypothetical large primate possesses an EQ comparable to or exceeding that of existing great apes, we might anticipate an amplification of their cognitive faculties.

The problem-solving abilities of a larger primate could be even more sophisticated, driven by more complex ecological challenges and social needs. Imagine an environment where food sources are more dispersed or require more intricate extraction methods. A larger primate, with a larger brain and potentially greater physical strength, might develop even more elaborate tool use. This could involve creating more complex tools, perhaps multi-component tools that require sequential assembly, or tools designed for more specialized purposes such as digging for roots in harder soil or processing tougher plant materials that are inaccessible to smaller species. Their understanding of physical properties might extend to a more nuanced grasp of leverage, trajectory, and material resistance, enabling them to overcome obstacles that would stymie their smaller relatives. Consider the possibility of innovative methods for accessing water during dry seasons or for constructing more robust shelters capable of withstanding extreme weather. Their capacity for planning could also be enhanced, allowing them to develop long-term foraging strategies that span across seasons, anticipating shifts in food availability and preparing accordingly. This might involve caching food in ways that are protected from scavengers or developing complex migration routes based on seasonal resource cycles.

Memory, too, would likely be enhanced. A larger primate inhabiting a more expansive territory would require an even more robust spatial memory to navigate and exploit its resources effectively. This could involve mental maps of territories that are orders of magnitude larger than those of current great apes, with detailed recall of a vast array of food locations, water sources, predator territories, and safe havens. Their social memory could also be expanded to encompass larger social groups or more fluid social networks. If their species lives in larger, more complex social structures, they would need to remember and track relationships with a greater number of individuals, understanding subtle social nuances and shifting alliances within a much larger community. The ability to recall past social transgressions or triumphs, and to use this information to inform present-day interactions, would be crucial for maintaining stability and cooperation within a larger group. Furthermore, if they engage in inter-group interactions, their social memory would need to extend to recognizing and remembering individuals from multiple distinct groups, understanding their respective social dynamics and potential affiliations.

The social intelligence of a large primate could manifest in ways that are currently unparalleled. In larger social groups, the complexity of social dynamics increases exponentially. Navigating these intricate social structures would require advanced capacities for social reasoning, strategic maneuvering, and conflict resolution. They might develop more sophisticated forms of communication, perhaps incorporating a greater range of vocalizations or more nuanced gestural repertoires, to convey complex social information. Their understanding of intentions and emotions could be refined, allowing for more subtle forms of manipulation, empathy, and cooperation. Consider the potential for advanced forms of coalition building, where individuals form complex, multilayered alliances to achieve common goals, whether it be for defense, resource acquisition, or social dominance. The ability to predict the actions of multiple individuals simultaneously, and to strategize within this complex social milieu, would represent a significant leap in social cognition. They might also develop more intricate forms of social learning, where information is not just transmitted through observation but through more deliberate teaching or mentoring. This could lead to the development of cultural traditions, where specific behaviors, tool-use techniques, or social norms are passed down through generations, leading to distinct cultural variations between different groups.

The ability to plan for the future would be a key differentiator. While great apes show evidence of planning, a larger primate might exhibit more long-term foresight. This could include anticipating future needs for food and shelter not just for days or weeks, but for months or even years. This might manifest in the development of complex storage strategies, the creation of more permanent or semi-permanent shelters, or even rudimentary forms of landscape modification to improve resource availability. The concept of "delayed gratification" could be more pronounced, enabling them to forgo immediate rewards for greater future benefits. For instance, they might invest significant effort in cultivating or managing a food source, knowing that the rewards will be reaped over a much longer period.

Moreover, the potential for abstract thought and symbolic representation in a larger primate warrants consideration. While current evidence for abstract reasoning in apes is intriguing, a species with greater cognitive capacity might push these boundaries further. This could involve understanding abstract concepts like quantity, causality, or even rudimentary forms of self-awareness that extend beyond the present moment. If they develop more complex communication systems, it's conceivable that these systems could evolve towards symbolic representation, where specific sounds or gestures represent not just immediate needs or emotions, but abstract ideas or past events. This would be a significant step towards a more generalized intelligence, moving beyond task-specific problem-solving towards a capacity for understanding and manipulating abstract information.

The development of advanced cognitive abilities in a larger primate would also have profound implications for their ecological impact. A species with sophisticated problem-solving, planning, and social intelligence would be exceptionally adaptable, capable of thriving in a wider range of environments and exploiting a broader spectrum of resources. Their ability to learn from experience and to innovate would allow them to respond effectively to environmental challenges, whether they be climate change, natural disasters, or shifts in prey availability. This adaptability could also lead to them becoming apex predators or dominant herbivores in their ecosystems, shaping the environment through their activities.

In conclusion, the cognitive abilities of extant great apes provide a strong foundation for hypothesizing about the potential intelligence of a larger primate. Their capacities for problem-solving, memory, and social intelligence, when scaled up and potentially amplified by a larger brain, suggest a being capable of sophisticated planning, intricate social navigation, and a profound understanding of its environment. The emergence of such advanced cognitive functions would not only be a testament to evolutionary ingenuity but would also have far-reaching implications for their ecological role and their place within the broader tapestry of life. Exploring these possibilities allows us to appreciate the spectrum of intelligence in the primate order and to consider the theoretical limits of cognitive evolution.

The intricate relationship between intelligence and tool use is a cornerstone of primate evolution, and nowhere is this more evident than in the great apes. Chimpanzees, our closest living relatives, exhibit a remarkable array of tool-using behaviors that go beyond simple manipulation of found objects. Their skills are not merely instinctual; they often involve a sophisticated understanding of the properties of materials and the physics of their environment. For instance, chimpanzees have been observed modifying branches by stripping away leaves and twigs to create probes for extracting termites or ants from their nests. This act of modification demonstrates foresight and an understanding of how to optimize a tool for a specific task. Further complexity arises when they use stones or pieces of wood as hammers and anvils to crack open hard-shelled nuts, a behavior that requires not only the selection of appropriate tools but also an understanding of force, trajectory, and the structural integrity of the target object. These behaviors are not universally distributed across all chimpanzee populations, suggesting a degree of learned behavior and cultural transmission, indicating that these skills can be taught and refined within social groups. The development of such tool-use traditions further underscores their cognitive sophistication, hinting at a capacity for cumulative innovation.

Orangutans, though often perceived as more solitary than chimpanzees, also display impressive ingenuity in their use of tools. Their arboreal lifestyle presents unique challenges, and they have evolved to meet them with clever adaptations. For example, orangutans have been observed using leaves as improvised tools – not just as simple coverings, but as multi-purpose instruments. They might use a large leaf as an umbrella to shield themselves from rain, or fold it to create a scoop for collecting water. In some instances, they have been documented using sticks to explore potential food sources in crevices or to extract honey from beehives. This suggests a flexible cognitive approach, where an object's primary function can be readily repurposed based on situational needs. Their ability to gauge the necessary force and angle to extract food from difficult locations, or to construct simple tools for reaching otherwise inaccessible resources, points to a nuanced understanding of their physical world. The selection of specific types of wood or fiber for certain tasks, and the modification of these materials to improve their efficacy, further highlight their problem-solving capabilities and their understanding of material properties.

These observed behaviors in chimpanzees and orangutans provide a crucial foundation for hypothesizing about the potential for tool use and even tool manufacture in a hypothetical megafauna primate. If such a creature, with its proportionally larger brain and enhanced cognitive capacities, were to exist, the complexity and sophistication of its tool use could be significantly amplified.

Consider the foraging needs of a large primate. While current great apes might use simple probes for insects or hammers for nuts, a megafauna primate would likely face different, and perhaps more significant, challenges in acquiring sustenance. If its diet included tough roots, tubers, or fibrous plant matter that are difficult to process with bare hands or teeth alone, the development of specialized tools would be highly advantageous. We might envision tools designed for digging more effectively. These could range from sharpened sticks, similar to those used by humans for millennia, to more robust implements fashioned from bone or stone, capable of breaking through harder soil or even prying loose larger root systems. The manufacture of such digging tools would require a sophisticated understanding of material properties – selecting appropriate woods that are strong yet pliable, or identifying suitable stones that can be shaped and sharpened. The process would involve multiple steps: sourcing the raw material, shaping it through abrasion or percussive force, and potentially hardening it through fire if such a technology were within their grasp.

Furthermore, if this large primate species were to engage in hunting, or even scavenging, for larger prey, the demands on tool use would escalate dramatically. The creation of cutting tools, such as sharp flakes or blades, would be essential for butchering carcasses and processing meat. This would necessitate a deeper understanding of lithic technology – the art of stone tool manufacture. This involves not only identifying suitable types of stone like flint or obsidian, which fracture predictably, but also mastering the techniques of knapping, striking the stone with another object (a hammerstone) to detach sharp flakes. The creation of a functional cutting edge would require precision and knowledge of angles, force, and the inherent weaknesses within the stone. Beyond simple cutting, there's the potential for projectile tools. While the exact mechanisms would depend on their physical capabilities, imagine the development of spears, or even more rudimentary throwing tools, to incapacitate prey from a distance. The manufacture of a spear would involve not just shaping a pointed tip, but also hafting it to a shaft – attaching the stone or sharpened wooden point securely to a long handle, likely using natural adhesives like tree sap and binding materials such as plant fibers. This multi-component tool requires a higher level of planning and execution, involving different materials and assembly techniques.

Defense would also be a significant driver of tool use. A large primate, while possessing formidable physical strength, might still face threats from other large predators or environmental hazards. Tools could serve as deterrents or weapons. This might include the creation of defensive clubs, perhaps made from dense wood or even incorporating stone elements for added impact. The manufacture of such a club would involve selecting a suitable piece of wood, shaping it for optimal grip and swing, and potentially attaching sharpened stones or bone fragments to its striking surface. The ability to fashion such implements would require not only strength but also a degree of dexterity and foresight in planning the tool's defensive capabilities.

Shelter construction presents another avenue for advanced tool use. While many primates build rudimentary nests, a megafauna primate might require more substantial shelters to protect itself from the elements or to create a secure environment for raising young. This could involve using tools to gather and manipulate larger materials. Imagine using sharpened poles to dig foundations for a more permanent structure, or employing crudely fashioned levers to move heavy branches or stones into place. The manufacture of tools for cutting and shaping wood, such as rudimentary axes or adzes, would be a significant leap forward, enabling them to process larger timbers and construct more elaborate shelters. This would involve a deep understanding of how to apply force to wood to break or shape it, and how to create tools that can withstand repeated use against tough materials.

The neural underpinnings for such advanced tool use and manufacture are intrinsically linked to brain structure and function. The development of these skills would necessitate significant expansion and refinement in several key cognitive areas. Firstly, enhanced fine motor control and dexterity would be paramount. The precise manipulation required for shaping stone, hafting components, or carefully assembling parts of a tool demands sophisticated neural pathways connecting the brain to the hands. This suggests a larger representation of the hands and fingers in the primary motor cortex and sensory cortex, allowing for a higher degree of sensory feedback and volitional control.

Secondly, a heightened capacity for planning and sequential processing would be critical. Tool manufacture is rarely a single, immediate action. It involves a series of steps, from identifying the need for a tool, to selecting raw materials, to shaping, assembling, and finally using the tool. This requires an ability to hold multiple steps in working memory, to execute them in the correct order, and to adapt the plan as needed based on feedback. The prefrontal cortex, known for its roles in executive functions such as planning, decision-making, and working memory, would likely be significantly enlarged and more interconnected in such a species.

Thirdly, spatial reasoning and a sophisticated understanding of physical properties would be essential. To shape a stone, one must understand its fracture patterns. To build a spear, one needs to gauge the correct length, weight distribution, and balance. To construct a shelter, one needs to understand principles of leverage, load-bearing, and stability. This implies a well-developed parietal lobe, involved in spatial processing and integrating sensory information, and a more refined understanding of physics, even if not consciously articulated, that guides their actions.

Furthermore, the ability to learn from observation and to transmit knowledge across generations, a hallmark of cumulative culture, would be crucial. This requires not only observational learning but also a capacity for imitation and social learning that allows for the faithful replication of complex skills. This social learning capacity is often linked to specialized brain regions and networks that facilitate understanding the intentions and actions of others, such as mirror neuron systems, and the ability to generalize learned behaviors to new contexts.

The very act of tool manufacture, especially in its more complex forms like lithic technology, also implies a degree of foresight and an understanding of abstract concepts. The knapper does not just see a rock; they envision the sharp flakes that can be produced from it. They understand that a specific strike at a specific angle will yield a desired result. This move from immediate perception to abstract representation and planned action is a significant cognitive leap. This could also involve a form of proto-symbolic thought, where the tool itself, or the process of its creation, gains a meaning beyond its immediate utility.

The potential for such advanced tool use and manufacture in a megafauna primate is not merely an extrapolation of current primate behavior, but a logical consequence of evolutionary pressures that favor intelligence and adaptability. If such a creature were to exist, its capacity to manipulate and shape its environment through tool use would undoubtedly have profound implications for its ecological niche, its social structure, and its evolutionary trajectory. It would represent a significant departure from the tool use observed in extant primates, moving from opportunistic modification of found objects to deliberate, complex manufacture, potentially laying the groundwork for a form of technological development unparalleled in the animal kingdom outside of humans. The development of tools for foraging, defense, and shelter would allow this hypothetical primate to exploit a wider range of environments, to overcome physical limitations, and to enhance its survival and reproductive success, ultimately shaping its own evolutionary path in ways that are both fascinating and profoundly significant.

The capacity for environmental adaptability is a defining characteristic of successful species, and primates, with their inherent intelligence and behavioral plasticity, are no exception. While the previous sections have explored the cognitive underpinnings of tool use, it is this very adaptability that allows such complex behaviors to be deployed effectively across a spectrum of ecological challenges. For a hypothetical megafauna primate, thriving in a dynamic world would necessitate a profound ability to adjust to shifting environmental conditions, whether these be the predictable rhythms of seasons or the more unpredictable forces of climate change and habitat alteration. This resilience, honed by millions of years of evolution, would be key to its survival and a significant factor in explaining its potential presence across diverse geographical and climatic zones.

Consider the dramatic seasonal variations that characterize many terrestrial ecosystems. Temperate regions, for instance, experience stark contrasts between warm, verdant summers and cold, resource-scarce winters. A large primate species would need a robust strategy to navigate these cycles. This might involve a shift in diet, a move towards more energy-dense but less palatable food sources during lean times, or the development of physiological adaptations to conserve energy, such as periods of reduced activity or even a form of torpor, though this is less common in primates than in other large mammals. For instance, during the summer months, they might capitalize on the abundance of fruits, berries, and tender leaves, supplementing their diet with insects and perhaps small vertebrates. As autumn approaches, with the decline of readily available fruits, their foraging would need to adapt. This could mean a greater reliance on stored foods, like nuts and seeds, or an increased focus on tougher, more fibrous plant matter that persists through the colder months. Their intelligence would play a crucial role here, enabling them to remember the locations of food caches, to predict the ripening and fruiting times of various plants, and to efficiently exploit the available resources. The ability to process tougher vegetation, perhaps through specialized dental morphology or the use of tools to break down fibrous material, would be a significant advantage.

Furthermore, adaptability to seasonal changes can also manifest in migratory patterns, albeit on a smaller scale than seen in many avian or ungulate species. A megafauna primate might not undertake transcontinental journeys, but localized movements to exploit seasonal food sources or to find more favorable microclimates could be crucial. They might move from higher elevations, where vegetation becomes sparse in winter, down to valleys where food is more abundant and temperatures are milder. This requires a sophisticated understanding of their territory, an ability to navigate complex landscapes, and an innate sense of when and where to move. Such movements would also necessitate social coordination, with groups likely moving together to ensure safety and to share knowledge about routes and resources.

Beyond the predictable cycles of seasons, primates, like all life, are subject to broader environmental shifts. Climate change, whether driven by natural cycles or anthropogenic influences, presents a significant challenge. A large primate species, with its potentially extensive home range and specific habitat requirements, would be particularly vulnerable to large-scale alterations in temperature, rainfall patterns, and vegetation composition. Their adaptability would be tested in their ability to cope with drought, increased frequency of extreme weather events, or the encroachment of unfamiliar flora and fauna.

One of the most critical aspects of adapting to changing climates is dietary flexibility. If a primary food source, such as a specific type of fruit or nut, becomes scarce due to altered rainfall or temperature, a resilient species must be able to switch to alternatives. This requires a broad knowledge of edible plants and an ability to quickly assess the nutritional value and edibility of new food items. Primates are known for their omnivorous tendencies, and a megafauna species would likely exhibit an even greater breadth of dietary opportunism. This could involve exploiting a wider array of insects, tubers, roots, or even carrion if the opportunity arises. Their intelligence would allow them to learn about these new food sources, to understand how to process them effectively (for instance, removing toxins from certain plants), and to incorporate them into their established foraging strategies.

Habitat alteration, whether due to natural events like wildfires or volcanic activity, or human-induced deforestation and land development, also demands significant adaptability. A large primate with a substantial territory would need to be able to adjust its behavior and movements in response to these changes. This might involve expanding its range into less disturbed areas, adapting to fragmented habitats, or even developing novel ways to utilize resources in altered environments. For example, if forest cover diminishes, they might need to spend more time in more open woodlands or grasslands, relying on different foraging strategies and increasing their vigilance against predators. Their intelligence would be vital in assessing risks and opportunities in these new settings, and their social structure could play a role in cooperative foraging or defense in unfamiliar terrain.

The resilience of other large mammals provides a valuable comparative framework for understanding the potential adaptability of a megafauna primate. Consider, for instance, bears. Many bear species exhibit remarkable adaptability, inhabiting a wide range of environments from arctic tundra to temperate forests. They achieve this through a combination of physiological adaptations (like hibernation), dietary plasticity, and opportunistic foraging. Grizzlies, for example, are known to consume berries, nuts, roots, insects, fish, and even carrion, allowing them to thrive across diverse ecosystems. Similarly, elephants, while requiring significant resources, demonstrate adaptability by migrating long distances to find water and food, and by modifying their environment through foraging to create pathways or access resources.

A megafauna primate might draw upon similar strategies. Their intelligence would allow for more sophisticated planning and problem-solving than seen in many other large mammals. Instead of merely reacting to environmental changes, they could anticipate them and proactively adjust their behavior. For example, they might learn to predict the impact of droughts on water sources and strategically move towards areas where water is more likely to be available. Their tool-use capabilities, as discussed previously, could also be instrumental in adaptation. If a particular food source becomes difficult to access due to changes in vegetation structure, they might develop or adapt tools to reach it. Imagine using a sharpened stick to dig for tubers exposed by drought, or using a stone to break open a hardy seed pod that has become a staple during lean times.

The ability to learn from experience and to transmit that knowledge socially is paramount for adaptability. A young primate would learn foraging techniques, migratory routes, and survival strategies by observing and imitating its elders. This cultural transmission allows for the rapid dissemination of adaptive behaviors throughout a population, enabling them to respond to environmental challenges more effectively than if each individual had to learn everything anew. This is particularly important in the face of rapid environmental change, where established behaviors might become obsolete. The capacity for flexible learning and innovation within social groups would therefore be a critical factor in their long-term survival.

Furthermore, the concept of "niche construction" – the ability of an organism to modify its own environment – could also play a role in adaptability. While not on the scale of beavers building dams, a large primate might subtly alter its surroundings in ways that enhance its survival.

This could involve creating well-worn paths through dense vegetation, clearing areas for better visibility, or even building more robust shelters that offer protection from extreme weather. These modifications, while seemingly small, could cumulatively create a more favorable microenvironment, allowing them to persist in areas that might otherwise be marginal.

The potential for a megafauna primate to inhabit diverse terrains, as suggested by occasional unexplained sightings, is directly linked to this environmental adaptability. It implies that such a creature would not be confined to a single, specialized niche but would possess the behavioral and perhaps even physiological flexibility to thrive in a variety of conditions. This could range from dense, humid rainforests to more open woodlands, savannah edges, or even mountainous regions, provided that essential resources like food and water are available. Their intelligence would allow them to exploit the unique opportunities presented by each environment, while their adaptability would enable them to cope with the challenges.

For instance, in a tropical rainforest, they might utilize their arboreal skills to access fruits and leaves high in the canopy, while also foraging on the forest floor for roots and invertebrates. In a more open woodland, they might adopt a more terrestrial lifestyle, relying on foraging strategies suitable for sparser vegetation and increased predator awareness. Their social structure would likely adapt to the demands of the environment; a more nomadic lifestyle in response to seasonal resource availability might necessitate larger, more coordinated groups, while a stable, resource-rich environment might support smaller, more territorial units.

The capacity to adjust to fluctuating resource availability is another critical facet of adaptability. Many ecosystems experience boom-and-bust cycles, where certain food sources are abundant for a period and then disappear. A large primate would need strategies to cope with these fluctuations. This might involve developing methods for storing food, such as caching nuts or fruits for later consumption, or having the ability to digest a wide variety of foodstuffs to buffer against the scarcity of any single item. Their intelligence would enable them to remember the patterns of resource availability and to plan their foraging accordingly.

In essence, the adaptability of a hypothetical megafauna primate is not a single trait but a complex interplay of cognitive abilities, behavioral flexibility, and potentially even physiological adaptations. It is this suite of characteristics that would allow such a creature to navigate the ever-changing tapestry of the natural world, to persist in diverse environments, and to potentially evade consistent detection. Their ability to learn, to innovate, to transmit knowledge, and to adjust their behavior in response to environmental pressures would be the ultimate testament to their evolutionary success and the key to their enduring presence, however elusive, in the wild. This inherent resilience makes the prospect of their existence in varied terrains not just plausible, but a logical outcome of evolutionary pressures that favor intelligence and adaptability in large-bodied organisms.

The intricate tapestry of primate intelligence is woven not only from an individual's capacity to learn and problem-solve but also, crucially, from the ability to pass this learned knowledge down through generations. This phenomenon, often referred to as cultural transmission, is a cornerstone of behavioral evolution and is particularly vital when considering the survival and ecological success of a hypothetical megafauna primate. While individual learning allows for immediate adaptation to novel circumstances, cultural transmission enables the accumulation and refinement of behaviors over time, creating what can be termed "primate culture." This cultural inheritance acts as a powerful buffer against environmental stochasticity and resource variability, allowing a species to maintain complex adaptive strategies even if individuals within the population are widely dispersed.

The significance of cultural transmission in a sparsely distributed species like a hypothetical megafauna primate cannot be overstated. Unlike highly gregarious primates, where knowledge can be exchanged with relative ease within a cohesive group, a large, solitary, or semi-solitary primate would rely heavily on intergenerational learning to maintain sophisticated behaviors. Imagine a young primate reaching maturity. Its innate abilities are enhanced and guided by observing its mother, or other experienced individuals, during critical learning periods. These observations are not passive; they involve focused attention, imitation, and often trial-and-error guided by the subtle cues of elders. The foraging techniques that allow these animals to exploit specific food sources, the intricate methods for processing difficult-to-access nutrients, or the nuanced understanding of the local landscape's resources are all prime candidates for cultural transmission.

Consider the complexity of foraging strategies. Different primate species, and indeed different populations of the same species, can exhibit remarkable variation in what they eat and how they obtain it. This variation is rarely solely due to genetic predisposition; a substantial component is learned. For our hypothetical megafauna primate, this could translate into the inheritance of specialized diets. For instance, one population might have learned to extract nutrient-rich grubs from specific types of decaying wood using carefully crafted probes, while another, living in a different habitat, might have developed a sophisticated understanding of how to locate and excavate tubers from deep within the soil, possibly aided by rudimentary digging tools. The knowledge of which tubers are edible, their seasonal availability, and the most effective methods for extraction would be a valuable piece of information passed from generation to generation. This is not merely instinct; it is learned expertise, honed by generations of trial and error and refined through observation and practice.

Tool use, as previously discussed, is a prime example of a behavior that is heavily reliant on cultural transmission. The manufacture and application of tools are not typically innate behaviors. A young primate might be born with the motor skills to manipulate objects, but the specific techniques for selecting the right branch, stripping it of leaves, shaping its end for optimal probing, or using a stone to crack a hard-shelled fruit are acquired through observation and practice. If a particular tool-use technique proves successful, it can become a cornerstone of a population's behavioral repertoire, passed down through successive generations. For a megafauna primate, the development and perpetuation of tool-use traditions could be crucial for accessing food sources that are otherwise inaccessible, or for processing food in ways that enhance its digestibility and nutritional value. The discovery of a new tool-making technique or a novel application for an existing tool could significantly improve survival rates and would undoubtedly be a strong candidate for cultural transmission. The persistence of such complex behaviors, even across vast distances and between individuals who may rarely interact, suggests a powerful mechanism of knowledge transfer at play.

The ability to learn and transmit information is not limited to direct imitation. Social learning also encompasses observational learning, where individuals learn by watching others, and instructive learning, where experienced individuals actively teach novices. For a large, potentially solitary primate, the opportunities for direct instruction might be limited, making observational learning even more critical. A young primate might spend considerable time observing its mother or other adults from a safe distance, internalizing their foraging patterns, their routes through the territory, and their responses to environmental cues and potential threats. This passive observation, coupled with an innate curiosity and propensity for exploration, allows for the gradual acquisition of complex skills.

The concept of "primate culture" extends beyond just foraging and tool use. It can encompass a range of learned behaviors, including social interactions, alarm calls, and even rudimentary forms of communication. While a megafauna primate might not possess the complex vocalizations of some smaller primate species, learned nuances in body language, scent marking, or subtle postural cues could convey vital information about territory, mating readiness, or the presence of danger. The transmission of such subtle signals would allow for a degree of social coordination and understanding, even among individuals who are not in constant contact. For example, if a particular alarm call or a specific evasive maneuver is learned and consistently demonstrated in response to a specific predator, its perpetuation through observation ensures that new generations are equipped to deal with that threat effectively.

The sparse distribution of a hypothetical megafauna primate presents a unique challenge and opportunity for cultural transmission. While it might seem that isolation would hinder the spread of knowledge, it is precisely in such scenarios that the efficiency of cultural transmission becomes paramount. Instead of relying on frequent social interactions, the knowledge is embedded within the parent-offspring lineage. A mother teaches her offspring, who then teaches their own offspring, creating a lineage of learned behaviors that can persist for generations. This vertical transmission (from parent to offspring) is the most direct form of cultural transmission and is a robust mechanism for ensuring the continuity of learned skills.

Furthermore, even in a sparsely distributed population, occasional encounters between individuals can lead to the diffusion of cultural traits. If one individual possesses a particularly effective foraging technique or a novel tool-use innovation that is observed by another, that knowledge can then be carried back to the observer's lineage, leading to horizontal or oblique transmission (from non-parent to non-offspring). While less common than vertical transmission in solitary species, these inter-individual transmissions can act as catalysts for spreading new innovations or for reintroducing lost knowledge into different parts of the population. The cumulative effect of these transmission pathways ensures that adaptive behaviors are not lost and can, in fact, spread and become incorporated into the broader population's behavioral repertoire.

The persistence of certain behaviors across vast geographical areas, even where individuals are widely separated, can be strong evidence for cultural transmission. If a specific, complex foraging technique or tool-use method is observed consistently across different regions inhabited by this hypothetical primate, it strongly suggests that the behavior is learned and passed down, rather than being a purely genetic predisposition. For instance, if multiple populations are observed to use a specific type of stone to crack open a particular type of nut, and the technique for selecting and using that stone is detailed and not universally obvious, this points towards a learned tradition. The absence of genetic variation that would explain such specific behavioral differences further strengthens the argument for cultural inheritance.

The role of innovation and the capacity for "teaching" or facilitating learning in such a species is also an important consideration. While active teaching, as seen in some bird species or even human children, might be less pronounced in a solitary primate, behaviors that facilitate learning in others would be highly adaptive. This could include a mother deliberately exposing her young to food sources or tool-making opportunities, or a more experienced individual allowing a younger one to observe its actions up close. Such facilitative behaviors, if they improve the learning efficiency of the offspring, would be strongly selected for.

The longevity of a megafauna primate would also play a significant role in the transmission of culture. Longer lifespans provide more opportunities for learning and for the transmission of knowledge from experienced individuals to younger ones. A primate that lives for several decades would have ample time to master a wide range of skills and to impart this knowledge to multiple offspring or observers over its lifetime. This extended period of learning and teaching allows for the perpetuation of complex behavioral traditions, ensuring that even highly specialized skills are not lost due to the premature death of a knowledgeable individual.

The study of cultural transmission in non-human primates, particularly in species like chimpanzees and orangutans, provides valuable analogies. Chimpanzees, for instance, exhibit distinct regional "cultures," with variations in tool use, grooming styles, and social rituals that are specific to different communities and are transmitted across generations. The same can be said for orangutans, where mothers patiently teach their offspring the intricate skills needed to forage and build sleeping nests. These examples highlight that the capacity for cultural transmission is a well-established phenomenon within the primate order and is likely to be a significant factor in the evolutionary success of any large-bodied, intelligent primate.

For a hypothetical megafauna primate, the implications of cultural transmission are profound. It means that the species' behavioral repertoire is not static, dictated solely by genes, but is dynamic and cumulative. This allows for adaptation to changing environments and novel challenges in ways that genetic evolution alone could not achieve. A behavior that was adaptive a thousand years ago might be obsolete today, but the capacity for learning and cultural transmission allows the species to discard outdated practices and develop new ones as needed. This makes them exceptionally resilient and adaptable.

The existence of learned traditions also raises interesting questions about the cognitive architecture of such a species. It implies a level of social cognition that allows for the accurate observation, interpretation, and replication of complex behaviors. It suggests a capacity for understanding the intentions or goals of others, even if those intentions are not explicitly communicated. This sophisticated cognitive machinery is essential for the fine-tuning of learned skills, from the precise angle at which to strike a nut with a stone to the subtle cues that indicate the ripeness of a fruit.

Furthermore, the transmission of knowledge can also lead to the emergence of sophisticated problem-solving abilities at the population level. While an individual primate might solve a particular problem, the sharing of that solution through cultural transmission means that the entire population benefits. Over time, this can lead to the development of increasingly complex and effective strategies for dealing with environmental pressures. It creates a cumulative cultural evolution, where each generation builds upon the knowledge and innovations of the previous ones, leading to a sophisticated and adaptable behavioral repertoire that is far greater than what any single individual could achieve alone. This allows for a dynamic and ongoing process of adaptation, ensuring the long-term survival and success of the species in an ever-changing world. The intelligence and adaptability of such a creature would thus be a product of both its innate cognitive capacities and the rich tapestry of learned behaviors passed down through the generations, forming a distinct and enduring "culture" of survival.

The very nature of our subject, a hypothetical megafauna primate, inherently suggests a profound understanding and mastery of concealment. The evolutionary pressures that favor survival in a world populated by formidable predators and, more recently, an increasingly intrusive human presence, would undoubtedly select for individuals and populations adept at vanishing. This isn't merely a passive trait; it's an active, sophisticated art form, a testament to their intelligence and adaptability. The elusive nature of such a creature, often described as being more ghost than animal, is not a matter of chance but a finely honed survival strategy.

One of the primary pillars of this cryptic behavior is the adoption of a nocturnal or crepuscular lifestyle. While many primate species are diurnal, relying on daylight for foraging and social activity, a large, potentially conspicuous primate would find significant advantage in operating under the cloak of darkness. The reduced visibility afforded by dawn, dusk, and the deep night offers a natural shield against detection. Predators, including humans with their keen visual senses, are significantly hampered in their ability to spot and track prey when light is scarce. This temporal shift in activity patterns allows our hypothetical primate to move, forage, and engage in social interactions with a considerably lower risk of being observed. Imagine the sheer advantage of a creature that can navigate dense forests, traverse open glades, or even approach human settlements under the cover of moonless nights, leaving behind only the faintest of traces – a broken twig, a displaced leaf, a scent that dissipates quickly on the night air. The temporal segregation of its activities from those of its primary threats provides a fundamental layer of evasion.

Furthermore, the adoption of a nocturnal rhythm often correlates with a quieter, more deliberate mode of locomotion. While diurnal primates might exhibit more boisterous movements, vocalizations, and social interactions, a nocturnal species would likely move with an almost preternatural silence. Every step would be carefully placed, every branch tested before full weight is applied, every rustle of leaves minimized. This isn't just about being quiet; it's about understanding the acoustics of their environment and moving in harmony with it, rather than against it. Their large size, which might otherwise be a liability, could be transformed into an advantage through sheer precision and controlled movement. Think of a great cat moving through tall grass; it doesn't crash and thrash, but flows, each muscle controlled, each movement deliberate. Our hypothetical primate would likely employ a similar, albeit primate-specific, mastery of silent passage.

Beyond temporal and behavioral modifications, physical adaptations for camouflage would also play a crucial role. While primates are not typically known for their vibrant, easily spotted coloration, a megafauna primate would likely possess pelage that blends seamlessly with its environment. This could manifest in several ways. In dense forest environments, their fur might be a dappled pattern of greens, browns, and greys, mimicking the interplay of sunlight and shadow filtering through the canopy. For those inhabiting more open or rocky terrains, a coloration that breaks up their outline against cliffs or scrubland would be advantageous. The texture and length of their fur could also be an adaptive trait, allowing them to appear less solid and more like natural features of the landscape. Imagine a large primate with fur that, from a distance, resembles moss-covered branches or lichen-draped rocks. This isn't just about passive coloration; it's about an active integration into the visual tapestry of their habitat.

The ability to remain motionless for extended periods, a behavior known as crypsis, is another critical component of their evasive strategy. When detection is imminent, or when simply resting during daylight hours, these animals would possess an extraordinary capacity to freeze, becoming virtually indistinguishable from their surroundings. This is a behavior observed in many prey species, but for a large primate, the level of control and stillness required would be exceptional. It would involve not just a physical cessation of movement but also a suppression of physiological cues. Their breathing would become shallow and slow, their heart rate subtly regulated, and their gaze fixed, perhaps blending with the natural patterns of the vegetation. They would essentially become part of the landscape, a living statue that only springs to life when the perceived threat has truly passed. This capacity for immobility is particularly important when considering how they might avoid detection by human observers, who often rely on movement to spot wildlife.

Sophisticated avoidance strategies extend beyond simply hiding. These primates would possess an acute awareness of their environment and a deep understanding of predator behavior. They would learn the patrol routes of dangerous predators, the times of day when humans are most active in their territory, and the specific sounds or scents that signal danger. This knowledge would be accumulated over a lifetime and, as discussed previously, passed down through generations. Their ability to perceive threats from a considerable distance, utilizing keen senses of hearing, smell, and potentially even a rudimentary sense of seismic vibration through the ground, would allow them to preemptively alter their course or retreat long before they are directly observed. This proactive approach to threat avoidance is a hallmark of intelligent evasion.

The concept of "defensive departure" is also central to their cryptic behavior. Rather than confronting a perceived threat, their primary instinct would be to withdraw. However, this withdrawal would not be a panicked flight but a calculated retreat. They would likely employ a variety of maneuvers to break line of sight and confound pursuers. This might involve moving swiftly through dense undergrowth, utilizing natural features like rivers or steep ravines as barriers, or even employing deceptive tactics such as doubling back on their tracks or creating false trails. The sheer intelligence of these animals would enable them to assess the nature of the threat and select the most appropriate escape route and method. A human hunter with a firearm might elicit a different response than a pack of canids or a solitary large cat.

Furthermore, their social structure, or lack thereof, would also contribute to their elusiveness. While some primates are highly social and live in large, easily detectable groups, a megafauna primate might be solitary or live in very small, dispersed family units. Solitary individuals are inherently harder to track and monitor than large troops. The absence of constant social interaction and vocalization reduces the overall auditory and visual signature of the population. Even within family units, interactions would likely be kept to a minimum when in areas of higher risk. This isolation, while perhaps posing challenges in other areas, serves as a powerful tool for remaining undetected.

The strategic use of terrain is another element of their sophisticated evasion. These animals would likely favor habitats that offer a multitude of hiding places and escape routes. This could include dense primary forests with thick undergrowth, rugged mountainous regions with extensive cave systems and precipitous cliffs, or even complex wetland environments. They would know these landscapes intimately, understanding not only the best foraging spots but also the most secure locations for resting, sleeping, and raising young. These chosen refuges would be exceptionally difficult for outsiders, especially humans, to access or navigate, further enhancing their ability to remain hidden.

Their ability to minimize their environmental impact is also crucial. Beyond silent movement, this includes a conscious effort to leave as little trace as possible. This could manifest in their foraging habits, such as consuming food items completely or disposing of waste in ways that are not easily discoverable. Their waste disposal strategies might involve burying excrement or dispersing it over a wide area, making it difficult to pinpoint their presence through scat analysis. Even their nesting habits, if they build arboreal nests, would likely be designed to be inconspicuous, perhaps blending with natural formations or being constructed in locations that are difficult to access from the ground.

The psychological aspect of their cryptic behavior cannot be overlooked. These animals would possess an innate wariness and a deep-seated aversion to being observed. This isn't just fear; it's a highly developed survival instinct. They would be attuned to the subtlest signs of human presence – the distant snap of a twig, the unusual scent of smoke, the faint echo of voices. Upon detecting such indicators, their immediate reaction would be to cease all activity and melt away, becoming unseen. This constant vigilance, combined with their physical and behavioral adaptations, creates a formidable barrier to detection.

The persistence of reports of such a creature, often dismissed as folklore or misidentification, is itself evidence of their extraordinary ability to evade definitive proof. Consider the sheer number of expeditions, scientific surveys, and anecdotal accounts that have failed to produce conclusive evidence. This lack of concrete proof is not necessarily a reflection of the creature's non-existence, but rather a testament to its success in remaining hidden. Their cryptic behavior, honed over millennia of evolutionary pressure, ensures that they exist in the spaces between our observations, in the shadows of our knowledge.

The ability to adapt their evasive strategies based on the specific threats present in their environment is also a key element. If a particular region is experiencing increased human activity, these primates might shift their foraging times further into the night, travel through less accessible terrain, or even develop new avoidance techniques. This adaptability, driven by their intelligence and learned experiences, ensures that their cryptic behavior remains effective even in the face of changing environmental conditions or novel threats. They are not static in their elusiveness; they are dynamic and responsive.

The very act of scientific observation, which often relies on direct sightings and clear photographic or video evidence, is fundamentally challenged by the existence of such a creature. Their intelligence is demonstrated not only in their ability to solve problems or use tools but also, and perhaps most importantly for their survival, in their capacity to actively avoid being studied. They understand, on an instinctual and learned level, that observation by humans often precedes capture, exploitation, or even harm. Therefore, their most sophisticated tool might be their ability to remain an enigma. Their cryptic behavior is an active, intelligent, and deeply ingrained aspect of their evolutionary success, a masterpiece of natural evasion that allows them to persist, unseen and unknown, in the wilder corners of our world. This art of evasion is not merely a trait; it is their primary defense, their enduring legacy, and the reason why definitive proof remains so tantalizingly out of reach.

7

Population Dynamics and Viability

The long-term persistence of any species hinges critically on its ability to maintain a sufficiently large and genetically diverse population. For megafauna primates, given their presumed rarity and the potential challenges associated with their life history traits, understanding the concept of a Minimum Viable Population (MVP) is paramount. The MVP is not a fixed number but rather a dynamic threshold representing the smallest isolated population capable of persisting over a defined period (typically 100 to 1000 years) despite the impacts of demographic, environmental, and genetic stochasticity. Essentially, it is the absolute smallest number of individuals that can maintain a healthy, self-sustaining population in the long run.

Applying the MVP concept to hypothetical megafauna primates requires us to consider several intertwined factors. Foremost among these is population size. A larger population inherently buffers against random fluctuations. Small populations are far more vulnerable to extinction due to chance events. For instance, a few consecutive years of poor breeding success, a localized disease outbreak, or a severe weather event could decimate a small population, pushing it below a critical threshold from which recovery is unlikely. In contrast, a larger population can absorb such losses and continue to reproduce. While specific figures for megafauna primates are speculative, drawing parallels with other large mammals suggests that viable populations would likely require hundreds, if not thousands, of breeding individuals to buffer against such random demographic fluctuations. This isn't just about the total number of individuals, but the number of breeding adults capable of contributing to the next generation. A population skewed heavily towards juveniles or older, non-reproductive individuals would be far less viable than one with a robust breeding core.

Crucially, genetic diversity is the bedrock of long-term viability. As populations shrink, they are prone to inbreeding depression, a phenomenon where the offspring of closely related individuals exhibit reduced biological fitness. This can manifest as decreased fertility, increased susceptibility to disease, developmental abnormalities, and reduced lifespan. In a small, isolated population of megafauna primates, individuals would inevitably mate with relatives. Over generations, this leads to a loss of heterozygosity – the presence of different versions of genes – which can expose deleterious recessive alleles and reduce the population's overall vigor and adaptive capacity. Maintaining adequate genetic variation is therefore essential for the species' ability to adapt to changing environmental conditions, resist novel pathogens, and simply to reproduce successfully. The MVP must be large enough to retain sufficient genetic diversity, a figure often estimated to be in the low thousands of individuals for larger mammals to avoid significant inbreeding effects over evolutionary timescales.

The age and sex structure of a population also plays a critical role in its viability. A healthy population needs a balanced age distribution, with a sufficient number of individuals in their reproductive prime. If a population consists mainly of very young or very old individuals, reproductive output will be severely compromised, regardless of the total population size. Similarly, a skewed sex ratio can limit breeding opportunities. While some primate species exhibit fluctuating sex ratios, a consistent and significant bias towards one sex in a small population can dramatically reduce the effective breeding population size (the number of individuals contributing genes to the next generation), thereby accelerating the loss of genetic diversity and increasing the risk of inbreeding. For megafauna primates, a stable and relatively balanced sex ratio among reproductive-aged adults would be a crucial component of their MVP.

Environmental stochasticity – unpredictable changes in the environment such as fluctuations in food availability, extreme weather events, or natural disasters – poses a significant threat, especially to populations with limited ranges. A megafauna primate population confined to a single valley, for example, could be wiped out by a single volcanic eruption or a prolonged drought. The MVP calculation must account for the probability of such events and the population's capacity to recover. This implies that viable populations would likely need to exist across multiple, geographically distinct subpopulations. This metapopulation structure, where groups of populations are linked by occasional migration, provides a critical buffer against localized extinctions. If one subpopulation is wiped out by a stochastic event, others may survive, and recolonization from surviving populations can occur. The MVP, therefore, might be better understood as a collection of interconnected subpopulations, each of which is large enough to survive for a period, with enough connectivity to allow for rescue effects.

Demographic stochasticity, the random variation in birth and death rates, is particularly potent in small populations. In a population of 10 individuals, if two individuals fail to reproduce in a given year, that represents a 20% reduction in reproductive output. In a population of 1000, if 20 individuals fail to reproduce, that's only a 2% reduction. This highlights how small population sizes amplify the impact of random demographic events. For a species like a hypothetical megafauna primate, with potentially long generation times and perhaps specific breeding requirements, demographic fluctuations could be even more pronounced. Ensuring an MVP that is large enough to make the impact of random variations in birth and death rates negligible is thus fundamental.

Genetic stochasticity, encompassing factors like genetic drift and mutation, also influences long-term viability. Genetic drift, the random fluctuation of gene frequencies from one generation to the next, is far more pronounced in small populations. This can lead to the loss of beneficial alleles or the fixation of neutral or even slightly deleterious ones, reducing the population's adaptive potential. While mutation introduces new genetic variation, its rate is generally too slow to counteract the rapid loss of diversity through drift in small populations. Therefore, the MVP must be sufficiently large to minimize the effects of genetic drift and allow for the accumulation of advantageous mutations over time.

In considering the MVP for hypothetical megafauna primates, it is also important to acknowledge that this is an ongoing scientific endeavor. Estimates of MVP are refined as more data becomes available on a species' life history, habitat requirements, and the specific threats it faces. For a cryptic, potentially rare species, estimating these parameters is itself a significant challenge. Researchers would employ sophisticated modeling techniques, such as Population Viability Analysis (PVA), which integrate demographic data, environmental variability, and genetic considerations to predict the probability of a population's persistence under different scenarios. These models allow scientists to explore various management strategies and determine the minimum population size or habitat area required to achieve a desired level of conservation assurance. The MVP thus serves not only as a benchmark for assessing extinction risk but also as a crucial target for conservation efforts, guiding decisions about habitat protection, translocation, and captive breeding programs aimed at ensuring the long-term survival of these extraordinary, elusive creatures. Without a robust understanding of the minimum population size required for their enduring existence, any conservation efforts would be based on guesswork, significantly diminishing their chances of success.

Understanding the reproductive rates and generation times of any species is fundamental to grasping its population dynamics. For hypothetical megafauna primates, this becomes an even more intricate puzzle, demanding careful extrapolation from known primate biology and ecological principles. These life history traits dictate a population's inherent capacity for growth, its resilience to perturbations, and the speed at which genetic diversity can be replenished or lost. Without a clear grasp of these parameters, our models of population viability, and consequently our conservation strategies, would be built on shaky ground.

The typical primate reproductive cycle provides a valuable starting point. Most primate species exhibit a relatively slow reproductive tempo compared to many other mammals. This is often characterized by a prolonged gestation period, a significant period of infant dependency, and a delayed age at which individuals first reproduce. For instance, many Old World monkeys and apes have gestations lasting anywhere from five to nine months. Following birth, infants are typically altricial, meaning they are born in a relatively undeveloped state and are completely dependent on their mothers for sustenance, protection, and thermoregulation. This dependency period can extend for many months, and in larger species, often for several years. During this time, the mother's reproductive output is effectively paused, as she dedicates significant energy and resources to the survival of her single offspring. The interval between births, therefore, is often quite long, frequently exceeding one year and sometimes stretching to three or four years, or even longer, in species like gorillas and orangutans.

When we extrapolate these trends to the concept of megafauna primates – beings of significantly larger size and presumably greater energetic demands – we can anticipate an even more protracted reproductive schedule. Imagine a primate species weighing several hundred kilograms. Gestation in such an animal would likely be considerably longer than in smaller primates, potentially extending beyond a year, especially if the fetus requires substantial development before birth to cope with the rig أكبر physical demands of its existence. The energetic cost of carrying and nourishing such a large fetus would be immense, likely requiring the mother to shift her diet to higher-energy food sources and potentially impacting her own body condition and ability to forage.

Following birth, the challenges of raising a young megafauna primate would be equally, if not more, demanding. A single infant would represent a substantial nutritional and energetic investment for the mother. The extended period of nursing and the prolonged need for protection from predators or environmental hazards would necessitate a lengthy period of maternal care. We might conservatively estimate an infant dependency period of at least three to five years, and possibly even longer, during which the juvenile learns crucial survival skills, social behaviors, and foraging techniques from its mother and other group members. This prolonged dependency has profound implications for population growth. If a female can only successfully raise one offspring every five to ten years, the population's ability to rebound from declines or to expand its range will be severely limited. This contrasts sharply with species that can produce multiple offspring per year, which can achieve rapid population increases.

The age at which these megafauna primates reach sexual maturity and successfully reproduce for the first time would also likely be quite late. In many large-bodied primates, individuals may not become reproductively active until they are between eight and fifteen years old, and in some cases, even later. For a hypothetical megafauna primate, reaching sexual maturity might be delayed further, perhaps not occurring until the late teenage years or even early twenties. This delay is often linked to the time required for growth, maturation of social rank, and the accumulation of sufficient body mass and physiological readiness for the rigors of reproduction and parental care. A late age at first reproduction means that the generation time – the average time between the birth of an individual and the birth of its offspring – will be significantly extended. If individuals reproduce for the first time at, say, 15 years old, and have a lifespan of 50-60 years, their generation time might be in the range of 20-30 years, or even more. This is a fundamental constraint on population growth. A population with a short generation time can replace itself and grow quickly; a population with a long generation time does so much more slowly.

Litter size is another critical component of reproductive rate. While most primates are K-selected strategists, meaning they produce few offspring but invest heavily in their survival, the norm for primates is a single offspring per birth event. Twins are rare and often have lower survival rates, especially in species where infant dependency is prolonged. For a megafauna primate, the sheer size of the fetus and the energetic demands of gestation and lactation would almost certainly preclude the production of multiple offspring simultaneously. Therefore, we can reasonably assume a litter size of one for our hypothetical species. This further reinforces the slow reproductive tempo, as each reproductive cycle yields only a single potential new individual.

The implications of these extended reproductive timelines for population dynamics are profound. A population with a low intrinsic rate of natural increase (r) will struggle to recover from population declines caused by disease, habitat loss, or predation. If a catastrophic event reduces the population by half, it could take many decades, or even centuries, for it to return to its previous numbers, assuming favorable conditions persist. This slow recovery makes such populations exceptionally vulnerable to extinction. Even minor setbacks can have disproportionately large impacts on their long-term survival.

Furthermore, the extended generation time means that the effects of genetic drift are more pronounced. Genetic drift, the random fluctuation of allele frequencies, has a greater impact in smaller populations and over longer timescales. With each generation representing many years, the random loss or fixation of genes can occur more rapidly, leading to a decrease in genetic diversity. This slower turnover of generations also means that beneficial mutations, which could help the population adapt to changing environments, take longer to spread through the population. Conversely, deleterious mutations can also persist for longer periods.

The concept of reproductive senescence – the decline in reproductive capacity with age – also needs consideration, although its impact might be less significant for species with shorter lifespans or where individuals reproduce for a relatively short period within their overall lifespan. In many primates, fertility can decline in older individuals, and the risks associated with pregnancy and childbirth may increase. For a megafauna primate, the physiological stresses of reproduction could lead to a more pronounced decline in reproductive success in later life, further limiting the effective reproductive output of older individuals.

Estimating these parameters precisely for a hypothetical species is inherently challenging. We rely on comparative data from existing large primates and our understanding of life history trade-offs. For example, comparing the reproductive rates of gorillas, chimpanzees, and orangutans, all of which are large-bodied and exhibit slow reproductive strategies, provides a valuable baseline. Gorillas, for instance, have a gestation period of about 8.5 months, an interbirth interval of around 3-4 years, and females typically reach sexual maturity between 7 and 10 years of age. Chimpanzees have similar gestation and interbirth intervals, with sexual maturity reached around 8-13 years. Orangutans exhibit even longer intervals, with females giving birth every 6-8 years, and sexual maturity occurring between 7 and 15 years.

Given these figures, it is reasonable to infer that a megafauna primate, being larger and presumably facing greater energetic and physiological challenges, would operate at the extreme end of this spectrum. A gestation period approaching a full year, an interbirth interval of at least 5-7 years, and an age at first reproduction of 15-20 years would not be unreasonable estimates. This would result in a generation time of potentially 25-35 years. If such a species has a maximum lifespan of, say, 60 years, this would allow for perhaps only 2-3 successful reproductive events per female in her lifetime. This is a stark illustration of the incredibly slow reproductive potential of such creatures.

These very low reproductive rates have critical implications for conservation. Even if a population of megafauna primates is reduced to a few dozen individuals, the time required for it to recover to a self-sustaining level could be measured in centuries, not decades. This means that immediate, drastic conservation interventions would be necessary to prevent extinction. Simply protecting habitat and hoping the population will rebound on its own would likely be an insufficient strategy. Active management, potentially including captive breeding and reintroduction programs, might be essential, but even these endeavors would be hampered by the slow reproductive rates. Producing enough individuals in captivity to establish a viable wild population would be a long and resource-intensive undertaking.

Furthermore, the slow pace of reproduction means that the genetic consequences of inbreeding and genetic drift would accumulate more rapidly relative to the population's ability to replenish genetic diversity. If a population becomes isolated and small, the loss of heterozygosity due to genetic drift and inbreeding can become significant within just a few generations. Given the long generation times, "a few generations" could represent many decades or even a century. This slow replenishment rate makes genetic rescue or adaptation to new environmental pressures exceedingly difficult.

The ecological context also plays a role. For species that rely on specific, long-lived food sources or require large, stable territories, the slow reproductive rate is mirrored by a slow ecological turnover. Changes in habitat, availability of food resources, or the emergence of novel diseases could have long-lasting impacts on a population that reproduces so slowly. Recovery from such ecological disruptions would be a protracted affair.

In summary, the extrapolated reproductive rates and generation times for hypothetical megafauna primates are characterized by extremely long gestation periods, prolonged infant dependency, late age at sexual maturity, and a litter size of one. These factors combine to produce a very low intrinsic rate of natural increase and a long generation time, potentially in the range of 25-35 years. This slow reproductive tempo renders such species exceptionally vulnerable to extinction, making rapid population recovery after declines improbable and highlighting the urgent need for proactive and intensive conservation measures. The implications for genetic diversity, adaptation, and the practical challenges of conservation breeding are substantial and must be central to any assessment of their population viability.

The long-term survival of any species, especially one as potentially imposing as a megafauna primate, is intricately tied to the myriad of factors that contribute to mortality within its population. While the sheer size and presumed formidable physical capabilities of these creatures might suggest a relative immunity to predation, especially in adulthood, a comprehensive ecological assessment must delve deeper to uncover the subtler, yet equally potent, forces that shape population numbers. These mortality factors can range from the direct and immediate, such as disease outbreaks or environmental catastrophes, to the insidious and chronic, like chronic malnutrition or interspecific competition. Understanding the specific agents of death and their relative impact is not merely an academic exercise; it forms a critical bedrock for any attempt to model population viability and to devise effective conservation strategies for species that reproduce as slowly as we have extrapolated for these hypothetical megafauna primates.

One of the most significant, yet often underestimated, mortality factors for any large mammal, including our hypothetical megafauna primates, is the threat posed by disease. Pathogens, whether viral, bacterial, or parasitic, can sweep through a population with devastating effect, particularly when individuals are densely packed or when their immune systems are compromised by other stressors such as poor nutrition or environmental degradation. For a species with a very slow reproductive rate and a long generation time, a single severe epidemic could potentially decimate a significant portion of the population, setting back recovery for decades, if not centuries. The susceptibility of a population to disease is often correlated with its genetic diversity. If the population has experienced prolonged periods of small size or isolation, leading to reduced genetic variation, it may possess a narrower range of immune responses, making it more vulnerable to novel pathogens or existing ones that mutate. For instance, species like gorillas have shown extreme susceptibility to human-borne respiratory pathogens, leading to significant mortality events in both wild and captive populations. Extrapolating this, our megafauna primates, if they share certain physiological or immunological traits with existing primates, could be at considerable risk from diseases that they have no pre-existing immunity against. The potential for zoonotic transmission, where diseases jump from humans or other animals to the megafauna primates, or vice-versa, also represents a critical area of concern. As human encroachment into their presumed territories increases, the interface for such pathogen exchange becomes more frequent and more dangerous. The impact of parasitic loads, while often less dramatic than acute disease outbreaks, can also be substantial. Chronic parasitic infections can weaken individuals, reduce their foraging efficiency, impair their reproductive success, and ultimately increase their susceptibility to other mortality factors. The sheer energetic cost of fighting off a heavy parasite burden, especially for an animal with high metabolic demands, could be a significant drain on resources, leading to a slow decline in condition and eventual death.

Starvation, or more broadly, nutritional stress, is another paramount mortality factor. While large animals often have broad diets, the availability and quality of their food resources can fluctuate dramatically depending on season, climate, and ecological disturbances. For a megafauna primate, which likely requires vast quantities of nutrient-rich food to sustain its immense body mass and energetic needs, periods of scarcity could be exceptionally perilous. If their diet is specialized, relying on specific fruits, tubers, or leaves that are only available at certain times of the year or in particular locations, then any disruption to these food sources – such as prolonged drought, unseasonal frosts, or habitat destruction – could lead to widespread starvation. The impact of nutritional stress is often amplified in vulnerable age classes. Infants and juveniles, still growing and developing, are particularly susceptible to the detrimental effects of insufficient food. Their smaller body size means they have less energy reserve to draw upon during lean periods, and inadequate nutrition during critical developmental windows can lead to stunted growth, weakened immune systems, and long-term health problems, thereby increasing their mortality risk even when food becomes more abundant again. Older individuals, whose metabolic rates might be changing or who may have accumulated health issues, can also be more vulnerable. Furthermore, the link between nutrition and reproduction is undeniable. Females in poor body condition will have lower conception rates, higher rates of miscarriage, and produce smaller, weaker offspring. The prolonged periods of infant dependency we've discussed mean that maternal nutritional status is critical not just for her own survival, but for the survival of her offspring for several years. A mother who is undernourished will struggle to provide sufficient milk and protection, placing her infant at extreme risk.

Environmental hazards represent a broad category of mortality factors that can affect populations with little regard for age or physical condition. These can include extreme weather events such as severe storms, prolonged droughts, or unseasonal cold snaps. For a large, terrestrial animal, events like flash floods, landslides, or wildfires could pose direct threats. Imagine a large primate population occupying a valley that is suddenly inundated by floodwaters, or an area that becomes the path of a devastating wildfire. The ability of such animals to escape these events might be limited by their relatively slow speed compared to smaller, more agile creatures, or by their need to navigate challenging terrain.

Even avalanches in mountainous regions or the risk of falling from high perches or cliffs, if they inhabit such environments, could contribute to mortality. Natural geological events, like earthquakes, could also indirectly lead to mortality by causing habitat destruction or creating unstable conditions. For instance, a habitat reliant on specific geological formations or water sources could be irrevocably altered by seismic activity. The impact of these events is often exacerbated by habitat fragmentation. If a population is confined to smaller, isolated patches of suitable habitat, their ability to disperse and escape a localized hazard is severely curtailed, making them far more susceptible to the devastating consequences of a single extreme event.

While adult megafauna primates might be formidable opponents, the possibility of predation cannot be entirely dismissed, especially for younger or infirm individuals, or in specific ecological contexts. In most modern ecosystems, the apex predators capable of taking down such large animals are rare, if they exist at all. However, in the hypothetical scenario where these megafauna primates evolved, their suite of predators would have been shaped by co-evolutionary pressures. If they existed alongside large felids, such as saber-toothed cats, or powerful ursids, then these would undoubtedly have been significant predators, particularly targeting juveniles or subadults. Even in the absence of an apex predator capable of killing a healthy adult, communal hunting by packs of large carnivores could potentially pose a threat. Scavenging also plays a role; a predator that has successfully brought down other large prey might be challenged or ambushed by other opportunistic carnivores if the primary predator is not a particularly dominant one. Furthermore, the interaction with other large herbivores or omnivores could also lead to mortality. Competition for resources, particularly during times of scarcity, could escalate into aggressive encounters, resulting in injuries that prove fatal. While direct predation might be limited for adults, the cumulative impact of attacks on young, weakened, or sick individuals could be a significant factor in population regulation.

The increasing presence and influence of humans represent a potent and often multifaceted mortality factor for many large animal species, and it is highly probable that our hypothetical megafauna primates would be no exception. Direct killing through hunting, whether for subsistence, sport, or in retaliation for perceived threats to human life or property, is a well-documented cause of population decline in large mammals. The effectiveness of firearms means that even a few determined hunters can have a disproportionate impact on a slow-reproducing population. Beyond direct killing, human activities lead to habitat loss and fragmentation. The conversion of natural landscapes for agriculture, infrastructure development, or resource extraction inevitably reduces the available living space, food resources, and dispersal corridors for large animals. This fragmentation can isolate populations, leading to inbreeding depression and reduced genetic diversity, while also increasing the likelihood of human-wildlife conflict as animals venture into human-dominated areas in search of resources. Road construction is a particularly insidious threat, creating barriers to movement and a constant source of mortality through vehicle collisions. Even indirect impacts, such as pollution of water sources or soil, can affect the health and reproductive success of these animals. Furthermore, the introduction of novel diseases through human or domestic animal contact, as mentioned earlier, is a significant concern. The cumulative effect of these human-induced pressures can be devastating, often pushing populations towards extinction even when direct hunting is not the primary cause of decline.

Finally, interspecific competition, while often subtle, can contribute to mortality. If the megafauna primate occupies a niche that overlaps with other large herbivores or omnivores, intense competition for critical resources such as food, water, or denning sites during periods of scarcity can weaken individuals. This weakening can make them more susceptible to disease, reduce their reproductive output, and ultimately increase their mortality rate. While direct lethal competition might be rare, the chronic stress and reduced condition resulting from such competition can have a cumulative, significant impact on population health and survival. This is particularly true if the megafauna primates are less agile or efficient foragers than their competitors, or if they have a more specialized dietary requirement that becomes a bottleneck.

In summation, the mortality factors influencing a hypothetical megafauna primate population are diverse and interconnected. While their adult size may confer some protection against predation, threats from disease, starvation, environmental hazards, and particularly human activities, remain significant. For a species characterized by slow reproduction, any factor that increases mortality, especially among juveniles and reproductive females, has profound implications for population sustainability. A deep understanding of these specific threats, their relative magnitudes, and how they interact is essential for accurately assessing the viability of such populations and for designing effective, targeted conservation interventions that address the most critical sources of mortality.

The intricate tapestry of population dynamics is woven not only by the threads of birth and death but also by the vital processes of movement and exchange. For any species, and particularly for a large primate with potentially extensive territorial needs, the phenomenon of dispersal—the outward movement of individuals from their natal area—is a cornerstone of long-term viability. This section delves into the critical roles of dispersal and subsequent gene flow in shaping the population structure and genetic health of our hypothetical megafauna primates, examining the factors that drive these movements and their profound implications for preventing isolation and maintaining adaptive potential.

Dispersal is most pronounced in younger individuals, as they seek to establish their own territories, find mates, and avoid the intensifying competition and potential aggression from established adults within their birth group or home range. For males of many primate species, dispersal is often obligate, meaning they *must* leave their natal group to reproduce. This behavior is crucial for avoiding inbreeding and for reducing direct competition with their father and other older males who hold dominance and mating privileges. In our megafauna primates, assuming a social structure with some degree of group living or at least overlapping home ranges, young males would likely face significant pressure to venture out. The sheer energetic costs and inherent risks associated with traversing unknown and potentially dangerous territories would necessitate strong selective pressures favoring dispersal. Imagine a young, fully grown male primate, perhaps weighing several hundred kilograms, leaving the familiar confines of his birthplace. He would be venturing into a world of uncertain food sources, unfamiliar predators, and the constant threat of encountering hostile conspecifics. The distances covered could be substantial, driven by the need to find unoccupied optimal habitat and receptive females.

The timing and triggers for dispersal can be influenced by a confluence of factors. In many primate species, dispersal is often linked to reaching sexual maturity. However, social dynamics within the natal group can also play a significant role. As younger males mature, their interactions with older, dominant males may become increasingly tense, leading to aggressive encounters that effectively push them out. The availability of resources within the natal area is also a key determinant. If the home range is already saturated with individuals, or if food and water become scarce, the incentive to leave in search of better conditions intensifies. For a large-bodied primate, the requirement for substantial caloric intake means that even moderate resource scarcity can act as a powerful impetus for dispersal, especially for individuals who are not yet established as dominant or who lack strong social alliances.

Female dispersal, while often less consistently obligate than male dispersal in many primate species, can also be a significant factor influencing gene flow. In some species, females may disperse to avoid inbreeding with close kin, particularly if their natal group is small, or they might disperse to join groups with higher quality resources or a more favorable sex ratio. The decision for a female to disperse might be more influenced by the quality of the habitat and the presence of suitable mates than by direct social pressure, although competition for resources and social status within a group can also play a role. If our megafauna primates exhibit a more fluid social structure, with individuals or small family units forming and reforming, female movement could be equally important in connecting disparate populations.

The success of dispersal is heavily reliant on the connectivity of the landscape. A contiguous, unbroken expanse of suitable habitat allows individuals to move relatively unimpeded. However, the reality for many large animals, both in the past and increasingly in the present, is that landscapes are often dissected by natural or anthropogenic barriers. Rivers, mountains, vast open plains, or even dense forests can act as formidable obstacles. For our hypothetical megafauna primates, the effectiveness of these barriers would depend on their specific ecological tolerances and physical capabilities. A wide, fast-flowing river might be an impassable barrier for some, while a species adept at swimming or with the ability to find narrow crossing points might be less affected.

Similarly, mountain ranges could isolate populations for millennia, leading to distinct evolutionary trajectories.

Habitat fragmentation, driven by human activities such as deforestation, agriculture, and infrastructure development, poses perhaps the most significant challenge to dispersal and gene flow in the modern era. When large, continuous habitats are broken down into smaller, isolated patches, dispersal routes are severed. Individuals attempting to move between these patches may be forced to traverse dangerous, human-dominated landscapes—crossing roads, farmlands, or urban areas—where the risks of mortality from vehicle collisions, hunting, or disease transmission are exceptionally high. This not only reduces the rate at which individuals can move between populations but also significantly increases the selective pressure against dispersal, as the probability of successful journey and establishment in a new area plummets.

The influence of social structure on dispersal patterns is profound. If the megafauna primates live in strictly hierarchical, stable social groups with limited turnover of individuals, dispersal might be infrequent and highly regulated by dominant members. Conversely, in more fluid or fission-fusion societies, dispersal might be more common, with individuals moving between subgroups or even entirely different social units. The presence of strong kinship bonds within a group could either facilitate or inhibit dispersal. On one hand, strong familial ties might encourage individuals to remain close to relatives, limiting outward movement. On the other hand, the need to avoid inbreeding might necessitate dispersal, especially in smaller, more isolated groups.

The concept of "metapopulation dynamics" is highly relevant here. A metapopulation is a network of distinct populations that are connected by dispersal. Each local population may experience extinction and recolonization events. Gene flow, facilitated by successful dispersal, is the lifeblood that sustains these metapopulations. It rescues declining or isolated populations from the brink of extinction by introducing new genetic variation, thereby increasing their adaptive capacity and resilience to environmental change and disease. Without sufficient gene flow, local populations become increasingly genetically uniform, susceptible to inbreeding depression (reduced fitness due to mating between related individuals), and less able to adapt to novel challenges. For a species with a slow reproductive rate, like our hypothetical megafauna primates, the loss of genetic diversity can be a particularly insidious threat, as it can take many generations to recover.

The specific patterns of dispersal can also have implications for social structure in recipient populations. When a dispersed individual, particularly a male, enters a new area, they must negotiate their integration into an existing social structure. This could involve challenging resident males, forming alliances with females, or seeking out solitary individuals. The success of such integration can be influenced by the genetic relatedness of the newcomer to the resident population. Individuals arriving from genetically distant populations might be more readily accepted or might integrate more successfully if they bring novel genetic material that enhances the overall fitness of the local gene pool. Conversely, if the recipient population is already genetically robust and socially stable, a lone immigrant might face significant resistance.

Understanding the geographical scale of dispersal is critical for conservation. If dispersal distances are typically short, then conservation efforts must focus on maintaining connectivity between closely adjacent habitat patches. If, however, individuals are capable of traversing vast distances, then larger-scale landscape planning and the preservation of wide dispersal corridors become paramount. For a megafauna primate, which would likely require extensive foraging ranges, the capacity for long-distance dispersal would be a significant asset, allowing them to exploit a wider array of resources and to track shifting environmental conditions. However, this also means that they are vulnerable to fragmentation on a much larger scale.

The genetic consequences of restricted dispersal and gene flow are far-reaching. Inbreeding depression manifests in reduced fertility, increased infant mortality, weakened immune systems, and a general decline in overall vigor. Over time, a population that is genetically isolated can become so inbred that its ability to reproduce and thrive is severely compromised. This process is often exacerbated by environmental stressors. A population already weakened by inbreeding will be far less resilient to disease outbreaks, food shortages, or climate fluctuations. The slow generation time of our hypothetical primates means that the genetic consequences of isolation can accumulate over centuries, making recovery exceptionally difficult.

Conversely, high levels of gene flow, facilitated by consistent and successful dispersal, can act as a buffer against such genetic deterioration. It introduces novel alleles and combinations of genes, increasing heterozygosity and thus the overall genetic diversity within a population. This enhanced genetic variation provides the raw material for natural selection, increasing the population's capacity to adapt to changing environmental conditions and to resist novel pathogens. A genetically diverse population is a more resilient population.

Ecological factors beyond simple landscape connectivity also influence dispersal. The distribution and abundance of food resources, the presence of suitable denning or nesting sites, and the density of predators all play a role. If resources are clumped in specific areas, dispersal might be directed towards these resource hotspots. Conversely, if predators are concentrated in certain regions, dispersal might actively avoid these zones. The social dynamics of the species itself are also a major driver. If individuals within a group are highly cooperative, it might lead to longer periods of natal association, delaying dispersal. If competition for status and mates is fierce, it will likely accelerate the drive to disperse.

Furthermore, the sensory capabilities of the megafauna primates would influence their ability to navigate and find suitable dispersal routes. Do they rely primarily on sight, smell, or sound to orient themselves and detect potential dangers or resources? If they have a highly developed sense of smell, for example, they might be able to detect the scent of water or food sources from considerable distances, influencing their dispersal paths. Similarly, if they possess acute vision, they might be able to survey vast landscapes from elevated positions, identifying potential corridors or areas of suitable habitat.

The role of social learning in dispersal also warrants consideration. Young individuals often learn foraging techniques, predator avoidance strategies, and social behaviors from their mothers and other members of their natal group. This learned repertoire can influence their confidence and success when they eventually disperse. An individual who has acquired a broader set of learned skills may be more likely to successfully navigate unfamiliar territory and establish itself in a new environment. Conversely, individuals raised in less stimulating or socially restrictive environments might be less prepared for the challenges of dispersal.

The impact of climate change on dispersal patterns is a growing concern for all species. As habitats shift and environmental conditions change, the need for dispersal to track suitable climatic envelopes will increase. However, habitat fragmentation can create barriers that prevent species from migrating to more favorable regions, effectively trapping them in deteriorating environments. For our megafauna primates, this could mean that established dispersal routes become obsolete, and new ones are needed, but the ability to establish these new routes may be severely hampered by human land use.

In conclusion, dispersal and gene flow are not merely passive movements of individuals but active, dynamic processes that are fundamental to the ecological and genetic health of any population. For hypothetical megafauna primates, the ability of young individuals, particularly males, to successfully disperse and establish themselves in new territories is paramount for avoiding inbreeding, maintaining genetic diversity, and ensuring the long-term viability of the species across its range. Landscape features, habitat fragmentation, social structure, and environmental conditions all interact to shape dispersal patterns. The preservation of connected landscapes and the mitigation of human-induced fragmentation are therefore critical conservation imperatives, ensuring that the vital exchange of genes that sustains resilient and adaptive populations can continue. The success or failure of dispersal ultimately dictates whether a species can adapt to a changing world or whether it is doomed to genetic stagnation and eventual decline.

The ephemeral nature of elusive populations, particularly those hypothesized to exist in remote and inaccessible locales, necessitates a rigorous, quantitative approach to understanding their potential persistence. While anecdotal evidence and indirect signs might fuel speculation, it is through the application of ecological modeling that we can move beyond conjecture and towards a scientifically defensible assessment of a population's long-term viability. This subsection will explore the fundamental principles and methodologies employed in such modeling, focusing on how simulated population dynamics can illuminate the probabilities of survival for a species operating under challenging and often unknown environmental conditions. The core of this endeavor lies in translating observed or inferred ecological parameters into a mathematical framework that allows us to project future population trajectories.

At the heart of any population viability analysis (PVA) are key demographic rates: birth rates (natality) and death rates (mortality). For a hypothetical megafauna primate, establishing realistic values for these rates is the first crucial step. Birth rates can be influenced by factors such as age at first reproduction, litter size (or number of offspring per birth event), and the frequency of breeding. Mortality, conversely, is influenced by a complex interplay of factors including age-specific survival probabilities, predation, disease, and resource scarcity. In the context of a remote, potentially unstudied primate, estimating these parameters often involves drawing parallels from closely related species with known ecology, or making informed deductions based on the proposed life history traits of our target organism. For instance, if our hypothetical primate is assumed to be a large-bodied, slow-reproducing species—a common characteristic of many megafauna—then low birth rates and potentially extended periods of parental care would be incorporated into the models. Similarly, mortality rates might be presumed to be higher in juveniles due to their vulnerability, and potentially increase in older individuals due to senescence or reduced ability to compete for resources.

Beyond these intrinsic demographic rates, the concept of environmental carrying capacity (K) plays a pivotal role. Carrying capacity represents the maximum population size that a given habitat can sustainably support, based on the availability of essential resources such as food, water, and shelter. In a dynamic environment, K is not a static figure but rather fluctuates seasonally, annually, and in response to longer-term environmental changes like droughts or periods of abundance. Models often incorporate these fluctuations, either through stochasticity (random variation) in K or by explicitly defining seasonal patterns. For a large primate that likely requires significant amounts of specific vegetation or access to predictable water sources, the variability in K would be a critical determinant of population size fluctuations. A population that consistently exceeds K would experience increased mortality and reduced reproductive success, pushing its numbers back down. Conversely, periods of high resource availability could allow for temporary population increases, but also increase the risk of a severe crash if conditions subsequently deteriorate.

The integration of stochasticity is paramount in modeling the persistence of populations in the wild, particularly those in remote or poorly understood environments. Environmental stochasticity refers to unpredictable fluctuations in environmental conditions that can affect birth and death rates. These might include unpredictable weather patterns (e.g., a severe drought reducing food availability), disease outbreaks that sweep through a population, or the sudden appearance or disappearance of a key predator. Demographic stochasticity, on the other hand, relates to random variations in individual births and deaths, especially significant in small populations where the chance occurrence of a few extra deaths or births can have a disproportionate impact on population size. For a population of our hypothetical megafauna primates, especially if numbers are small, the chance of several individuals dying within a short period due to non-predictable events can significantly increase the risk of extinction. PVA models incorporate these forms of randomness by running simulations multiple times, each with slightly different random variations, to generate a range of possible population outcomes and their associated probabilities.

Furthermore, external threats, often referred to as catastrophic events or Allee effects, must be considered. Catastrophic events are sudden, drastic disturbances that can cause rapid population decline, such as large-scale wildfires, volcanic eruptions (if applicable to the region), or a devastating disease epizootic. Allee effects, conversely, describe situations where population growth rate declines at low population densities. This can occur because individuals have difficulty finding mates, locating prey, or cooperating in defense against predators. For a large primate that may rely on group coordination for foraging or protection, a significant reduction in population density could lead to a cascade of negative consequences, making it harder for the remaining individuals to survive and reproduce, even if resources are abundant. Modeling these factors requires careful consideration of the specific threats relevant to the hypothesized habitat and the primate's ecological niche.

The process of developing a PVA model typically involves several iterative steps. First, data are gathered or estimated for the key demographic parameters and environmental variables. This initial dataset is then used to build a baseline model. Simulations are run with this baseline, and the output—often presented as a graph of population size over time and a probability of extinction within a given timeframe—is scrutinized. If the initial parameters seem unrealistic or the results are too sensitive to minor changes, the process is repeated with refined estimates. Sensitivity analysis is a crucial part of this process, where modelers systematically alter individual parameters to determine which ones have the most significant impact on the outcome. This helps identify areas where more precise data is needed or where conservation efforts might be most effective.

For a species as potentially large and complex as our hypothetical megafauna primate, the spatial dynamics of populations are equally important. A single, isolated population is inherently more vulnerable than a network of interconnected populations. Metapopulation models, which explicitly account for the presence of multiple, discrete subpopulations linked by dispersal, are therefore often employed. These models can simulate scenarios where local populations go extinct but are subsequently recolonized by individuals from a surviving population. The rate and success of dispersal become critical factors in these models. If dispersal is limited, then extinction of one subpopulation significantly reduces the overall metapopulation's resilience. Conversely, robust dispersal can act as a buffer, rescuing declining populations and maintaining genetic diversity across the network.

The construction of such models is not without its challenges, especially when dealing with species and environments that are poorly understood. The "garbage in, garbage out" principle is particularly relevant. If the input parameters are based on weak assumptions or are heavily extrapolated from distantly related species, the model's predictions will be correspondingly uncertain. This is where the expertise of cryptozoologists and field biologists becomes indispensable. Their deep understanding of primate behavior, ecology, and the likely characteristics of remote habitats allows for the generation of more plausible parameter ranges and the identification of the most critical uncertainties. For example, knowledge of a primate's typical diet could inform assumptions about the carrying capacity of a particular habitat, while understanding its social structure might help in estimating the population thresholds at which Allee effects could become significant.

The utility of these modeling approaches extends beyond simply predicting extinction probabilities. They provide a framework for exploring hypothetical conservation interventions. For instance, a model could be used to assess the effectiveness of establishing wildlife corridors to facilitate dispersal between fragmented habitats, or to evaluate the impact of disease management strategies on population persistence. By simulating different management scenarios, researchers can identify the most cost-effective and impactful interventions, thereby guiding conservation efforts in potentially resource-limited situations. The very act of building a PVA model forces a detailed consideration of all known and suspected factors influencing a population's survival, thereby highlighting knowledge gaps and prioritizing future research.

Consider a hypothetical scenario for our megafauna primate. If initial estimates suggest a birth rate of 0.1 offspring per adult female per year and a maximum adult survival rate of 0.95, but then demographic stochasticity is introduced for a population of only 50 individuals, the models might reveal a significant chance of extinction within 100 years. Further refinement could involve incorporating environmental stochasticity, perhaps simulating a severe drought every 50 years that reduces the carrying capacity by 75% for two consecutive years. This would likely drastically increase the extinction probability. If, however, the model also includes the possibility of a small but consistent rate of immigration from a hypothetical neighboring population (e.g., 1-2 individuals per decade), the extinction risk might be substantially reduced, underscoring the critical importance of maintaining connectivity.

The fidelity of these models to reality is intrinsically linked to the quality and quantity of data available. In the absence of direct empirical data, researchers often rely on expert opinion, comparative analyses with well-studied analogues, and stochastic simulations to bridge these information gaps. The aim is not to achieve perfect prediction, but rather to generate robust probabilistic assessments that inform decision-making. For a species whose very existence is debated, the development of a PVA model can serve as a crucial tool in guiding future research and, if evidence of its existence strengthens, in prioritizing conservation actions. It moves the discussion from speculation to a structured, quantitative evaluation of a population's resilience in the face of ecological uncertainties. The outputs of such models, even if based on tentative parameters, provide a scientific rationale for allocating resources towards further investigation or, in the event of confirmed existence, immediate protective measures. They serve as a digital laboratory where hypotheses about survival can be tested against the unforgiving logic of mathematical representation, offering a glimpse into the potential future of a population whose present remains shrouded in mystery.

8

Case Studies: Bigfoot and Sasquatch

The Pacific Northwest of North America stands as the epicenter of anecdotal reports and scientific inquiry into the existence of Bigfoot, or Sasquatch. This vast, geographically diverse region, encompassing parts of Oregon, Washington, British Columbia, and even extending into Northern California and Idaho, presents a compelling suite of ecological conditions that theorists and researchers alike point to as ideal for a large, undiscovered primate. Understanding why this specific geographical area is so consistently associated with the phenomenon requires a deep dive into its environmental attributes—its dense, temperate rainforests, towering mountain ranges, extensive river systems, and the sheer immensity of its undeveloped wilderness.

The defining characteristic of the Pacific Northwest's terrestrial environment is its luxuriant, old-growth coniferous forest. These forests are dominated by towering trees such as Douglas fir (Pseudotsuga menziesii), Western hemlock (Tsuga heterophylla), Western red cedar (Thuja plicata), and Sitka spruce (Picea sitchensis) in coastal areas, transitioning to lodgepole pine (Pinus contorta) and various fir species at higher elevations. The dense canopy cover created by these giants is often so thick that it significantly limits light penetration to the forest floor, creating an environment that is both shaded and perpetually damp. This creates a complex, multi-layered habitat. The understory is typically rich with ferns, mosses, huckleberry bushes, and other shade-tolerant vegetation, providing a diverse food source and ample cover. For a large, terrestrial primate, such an environment offers a wealth of resources. The sheer biomass of vegetation could support a substantial herbivorous or omnivorous diet, while the dense foliage would provide excellent concealment from potential observers, whether human or animal. The complex structure of the forest, with fallen logs, dense thickets, and uneven terrain, also offers numerous places for shelter, nesting, or denning.

Beyond the immediate forest floor, the vertical dimension of these ecosystems is equally significant. The presence of large trees provides elevated pathways and vantage points, which could be utilized by a primate for foraging, travel, or observation. The extensive root systems and hollowed bases of ancient trees could also offer refuge or nesting sites. The sheer scale of these trees—some reaching heights of over 300 feet—means that a large primate could potentially move through the canopy with a degree of security and stealth unavailable in more open woodlands. This verticality is a critical component of the habitat's complexity and its potential to conceal a large, reclusive creature.

Intertwined with the dense forests are the region's formidable mountain ranges. The Cascade Range and the Coast Mountains, for instance, form vast, rugged interiors that are characterized by steep slopes, deep valleys, alpine meadows, and extensive networks of glaciers and snowfields at higher elevations. These mountainous areas often remain snow-covered for much of the year, effectively isolating vast tracts of land and creating natural barriers to human access. This remoteness is a crucial factor. The sheer difficulty of traversing these terrains means that large sections of the Pacific Northwest wilderness remain virtually unexplored by humans on a regular basis. Human presence is often concentrated along major river valleys, coastal areas, and established transportation corridors. The deep interior of these mountain ranges, with their challenging topography and harsh weather conditions, can thus serve as a vast sanctuary for any species seeking to avoid detection.

The presence of extensive river systems and their tributaries further shapes the ecological landscape. Rivers like the Columbia, Fraser, and Sacramento, along with countless smaller streams, carve through the region, providing vital water sources and creating riparian corridors that support different types of vegetation and wildlife. These riparian zones are often highly productive, attracting a variety of prey species and offering a rich foraging ground. For a large primate, these water sources would be essential, and the riverbanks might also serve as travel routes, albeit potentially more exposed ones. The valleys carved by these rivers often offer more gentle gradients and potentially more productive foraging areas compared to the steeper slopes of the surrounding mountains, creating pockets of high resource availability.

What truly sets the Pacific Northwest apart as a prime candidate for a Bigfoot habitat is the sheer scale of its wilderness. Large, contiguous tracts of undeveloped land are remarkably abundant in this region. Even today, vast areas remain under federal or provincial management as national forests, wilderness areas, and parks, which are managed to preserve their natural state and limit human development. This extensive wilderness provides the necessary space for a large animal population to exist without frequent encounters with humans. A species requiring significant territory, as a large primate likely would, needs these extensive, unbroken landscapes to thrive. The presence of multiple, overlapping ecological niches within these vast wilderness areas—from coastal temperate rainforests to high-elevation alpine zones—could support a diverse diet and a range of behavioral adaptations.

The historical context of human settlement in the Pacific Northwest also plays a role in why this region is so strongly associated with Bigfoot. While Indigenous peoples have inhabited these lands for millennia, European-American settlement, particularly the push for resource extraction like logging and mining, often focused on accessible areas. The more remote and rugged interiors were often bypassed or exploited less intensively. This historical pattern means that the deepest, most inaccessible parts of the wilderness may have remained relatively undisturbed for longer periods, providing a more stable environment for a reclusive species to persist. The logging industry, in particular, has created numerous old logging roads and trails that, while facilitating human access to some degree, also penetrate deep into otherwise inaccessible areas. These roads can be utilized by a large primate as travel corridors, and the clearings created by logging operations might offer access to different types of vegetation.

Furthermore, the biodiversity of the Pacific Northwest supports a complex food web, which is essential for any large mammal. The region is home to a rich array of wildlife, including large herbivores like Roosevelt elk (Cervus canadensis roosevelti) and mule deer (Odocoileus hemionus), which could serve as prey if the primate is an opportunistic carnivore or omnivore. A wide variety of smaller mammals, birds, fish, and a bounty of plant life—berries, roots, nuts, fungi—provide a diverse dietary palette. The presence of apex predators like cougars (Puma concolor) and black bears (Ursus americanus) suggests a robust ecosystem capable of supporting large-bodied animals. While these predators could pose a threat, they also indicate that the ecosystem can sustain a biomass that might include a large primate.

The concept of ecological niches is fundamental here. A hypothetical

Sasquatch would occupy a specific niche within this ecosystem. If it is primarily herbivorous, its niche would be defined by the availability of certain plant species, their seasonal abundance, and the primate's ability to digest them. If it is omnivorous or even carnivorous, its niche would involve foraging strategies for animal prey. The dense forest offers opportunities for ambush predation, while the open meadows might be used for foraging on vegetation or hunting smaller animals. The seasonal fluctuations in food availability—heavy berry crops in late summer, nut falls in autumn, and limited green forage in winter—would necessitate adaptive foraging strategies, potentially including seasonal migrations to areas with more abundant resources, or a capacity to subsist on less desirable food sources during lean times.

The perception of Bigfoot as a creature adept at evading detection is intrinsically linked to the habitat's characteristics. The dense undergrowth, the vastness of the terrain, and the frequent fog and precipitation common in the coastal Pacific Northwest all contribute to obscuring visibility and muffling sound. This environment is inherently conducive to stealth and concealment. A large animal could move through the forest with relative quiet, its passage masked by the rustling of leaves, the creaking of branches, and the pervasive sounds of the forest itself. The frequent rain and fog would further aid in its elusiveness, obscuring footprints and making visual confirmation exceptionally difficult.

The human element in this equation cannot be overlooked. The Pacific Northwest has a long history of logging, recreational activities like hiking and camping, and some degree of wilderness exploration. This has resulted in a higher number of reported sightings compared to many other regions, simply because more people are present in the habitat where Bigfoot is theorized to live. The very act of venturing into these wild areas, often on less-traveled trails or in search of solitude, increases the probability of encountering something unusual. Moreover, many of the reported encounters involve individuals who are experienced outdoorsmen, hunters, or trappers, people who are familiar with the natural world and unlikely to misidentify common wildlife. Their accounts, therefore, often carry a degree of weight and credibility.

The interconnectedness of the ecosystems within the Pacific Northwest is also relevant. The vast network of forests, mountains, and rivers creates a landscape where a large, mobile animal could potentially range over hundreds or even thousands of square miles. This expansive territory would allow a population to maintain itself without needing to concentrate in any single area, further increasing its chances of remaining undetected. Dispersal between different habitat types—for instance, moving from the coastal rainforests to the interior mountain ranges in search of different food sources or to avoid human activity—would be facilitated by these interconnected landscapes.

In essence, the Pacific Northwest represents a confluence of factors that make it an ideal hypothetical habitat for a creature like Bigfoot. It offers immense tracts of remote wilderness, complex and resource-rich forest ecosystems, challenging mountainous terrain that provides natural barriers, and a climate that aids in concealment. These elements combine to create an environment where a large, intelligent, and reclusive primate could potentially exist, and where occasional encounters with humans, while rare, might still occur, fueling the enduring legend. The very qualities that make this region so appealing to humans for recreation and resource utilization are the same qualities that would provide sanctuary for an elusive, unknown species.

Eyewitness accounts form the bedrock of the Bigfoot phenomenon, a tapestry woven from thousands of individual narratives spanning decades and vast geographical areas, primarily concentrated in the Pacific Northwest. These accounts, while anecdotal, offer a recurring description of a large, bipedal, ape-like creature. The consistency in these descriptions, despite the lack of any definitive physical specimen, is often cited by proponents as compelling evidence.

One of the earliest widely publicized accounts emerged in 1958 from Bluff Creek, California. A logger named Jerry Crew discovered large, human-like footprints near his construction site. These prints, measuring up to 15 inches long and 7 inches wide, were accompanied by reports of a large, hairy bipedal creature seen fleeing the area. The media quickly dubbed the creature "Bigfoot." While this incident was initially attributed to a hoax by some, the subsequent influx of similar reports in the region, particularly during the 1960s, cemented the Bigfoot legend in popular culture. The Bluff Creek prints, along with others that followed, often displayed characteristics that observers deemed inconsistent with known animal tracks or human deception. The stride length, the depth of the impressions, and the apparent weight distribution indicated a large, heavy creature.

Following the initial surge of interest, numerous individuals, often seasoned outdoorsmen, hunters, and loggers, have come forward with their own encounters. These individuals, familiar with the local fauna, frequently emphasize that what they saw was unlike any known animal. Stories often describe a creature of immense stature, typically between six and ten feet tall, covered in dark brown, black, or reddish-brown hair. The reports frequently mention a musky odor, a powerful build, and a distinctive gait. Some accounts describe the creature as having a conical head and a short neck, lacking a visible tail, and possessing large hands and feet. The eyes are often described as being dark and set deep within the skull.

The Patterson-Gimlin film, shot in 1967 in Willow Creek, California, remains one of the most iconic and debated pieces of visual evidence. The 16mm film, purportedly showing a female Bigfoot walking through the woods, has been subjected to intense scrutiny for decades. Proponents argue that the subject's musculature, gait, and proportion are incredibly difficult to replicate, especially with the technology available in the late 1960s. They point to the fluid, natural movement and the supposed anatomical details visible in the film as indicative of a real creature. Skeptics, conversely, have posited that the film depicts a person in an ape suit, citing perceived inconsistencies in movement and proportions, and the ease with which such a hoax could have been perpetrated at the time. The debate over its authenticity continues, with analyses ranging from biomechanical studies to forensic examinations of the film itself.

Beyond these famous cases, a multitude of less publicized eyewitness accounts contribute to the persistent belief in Bigfoot. These accounts often arise from solitary individuals or small groups engaged in activities that take them deep into remote wilderness areas: hiking, hunting, fishing, or simply camping. Many of these reports share common themes: a fleeting glimpse of a large, dark figure moving through dense forest, the sound of heavy footsteps or snapping branches, and an inexplicable sense of being watched. The emotional impact of these encounters is frequently described as profound, ranging from fear and awe to a deep-seated conviction that they have witnessed something extraordinary.

The consistency of descriptions across geographically disparate areas and time periods is a significant factor for believers. While the exact height and hair color might vary slightly, the overall morphology – large, bipedal, hairy – remains remarkably uniform. This perceived consistency, proponents argue, suggests an underlying reality rather than a collection of independent fabrications. They highlight how individuals who have never heard of Bigfoot or its legend have still described creatures that align with the established archetype.

However, the inherent subjectivity and potential for misidentification in eyewitness testimony present significant challenges to scientific validation. Humans are prone to misinterpreting ambiguous stimuli, especially under conditions of stress or excitement. The human brain can also be influenced by expectation and suggestion. In wilderness settings, where shadows, fog, and dense vegetation can play tricks on the eyes, and where familiar animals like bears or deer can appear strangely from certain angles or in poor light, misidentification is a plausible explanation for many sightings. The tendency for humans to pattern-recognize and impose familiar forms onto unfamiliar shapes further complicates the interpretation of eyewitness accounts.

The challenge for cryptozoologists and researchers is to sift through these subjective accounts to identify any patterns or details that might offer more objective clues. This often involves meticulously documenting sighting reports, noting the location, time of day, environmental conditions, the witness's background and familiarity with wildlife, and any specific details observed. Cross-referencing multiple reports from the same general area can sometimes reveal recurring themes or patterns that might warrant further investigation.

The issue of physical evidence is equally contentious. Beyond footprints, alleged evidence includes hair samples, scat, vocalizations, and photographic or video recordings, each presenting its own set of interpretive hurdles.

Footprints, as mentioned, are perhaps the most common type of physical evidence. Casts of large, bipedal footprints have been collected from numerous locations, particularly in the Pacific Northwest. These casts often show detailed features like dermal ridges, nail impressions, and specific anatomical characteristics that proponents argue are difficult to fake convincingly. The sheer number of these alleged prints, their consistency in size and morphology, and their presence in remote areas have led many to believe they represent genuine evidence of a large, unknown primate.

However, the scientific community remains skeptical. Footprints can be easily fabricated using carved wooden feet or other molds. The challenging terrain of the Pacific Northwest, with its soft earth, mud, and snow, is also conducive to the distortion of tracks, potentially making ordinary animal prints appear larger or more unusual than they actually are. Furthermore, the lack of any subsequent, verifiable trackways leading to or from these significant finds, or any clear evidence of the creature's passage (like broken branches at an unusual height or disturbed vegetation), often raises questions about the authenticity of isolated prints. The infamous "Fossil Footprints" of Yosemite, initially presented as evidence of ancient human giants, were later revealed to be a geological anomaly. While not directly related to Bigfoot, it highlights how natural processes can create intriguing formations that can be misinterpreted.

Hair samples attributed to Bigfoot have also surfaced sporadically. These samples are often found snagged on branches or embedded in tree bark. When analyzed, they have frequently been identified as belonging to known North American animals such as bears, deer, coyotes, or even domestic animals. In some instances, the hair has been human. Very rarely, analyses have yielded results that defy easy classification, but these have typically been insufficient to definitively identify a new species. The difficulty lies in the fact that without a known reference specimen, even an anomalous hair sample cannot be definitively linked to a creature like Bigfoot. Forensic geneticists require DNA from a known source to make a positive identification. The problem is circular: we need hair from Bigfoot to prove Bigfoot exists, but we can't prove the hair is from Bigfoot without knowing what Bigfoot hair looks like.

Vocalizations, often described as howls, screams, or grunts, are another layer of anecdotal evidence. Recordings of strange sounds in the wilderness have been attributed to Bigfoot. However, these recordings are notoriously difficult to analyze. The sounds could be any number of known animals, or even natural phenomena like wind whistling through rock formations. Without clear contextual information or the ability to definitively link the sound to a visible creature, these recordings remain ambiguous. The vastness of the wilderness means that many natural sounds are poorly understood and difficult to attribute to their source.

Photographic and video evidence, beyond the Patterson-Gimlin film, continues to emerge, often captured by trail cameras or by individuals with mobile devices. While some of these images and videos show ambiguous shapes or blurry figures in the background, they rarely provide clear, indisputable evidence. The quality of the footage is often poor, making definitive identification impossible. The proliferation of digital imaging and editing software also makes it easier than ever to create convincing-looking hoaxes. The scientific community generally requires high-quality, verifiable evidence, and much of the visual material presented for Bigfoot falls short of this standard.

The challenge in validating claims of Bigfoot evidence lies in the rigorous standards of scientific proof. Science demands reproducible results, verifiable data, and evidence that can withstand peer review. For a creature as elusive as Bigfoot is purported to be, gathering such evidence is incredibly difficult. If Bigfoot is a real, biological entity, it would likely be a large mammal with specific dietary needs, habitat requirements, and reproductive patterns. Such a creature would leave a more substantial ecological footprint than is currently demonstrable.

The scientific consensus remains that there is insufficient evidence to confirm the existence of Bigfoot. However, this does not necessarily mean that all reports and alleged evidence can be easily dismissed. The persistence of these accounts and the detailed nature of many eyewitness testimonies, coupled with the sheer volume of alleged evidence, continues to fuel research and debate. The critical task is to distinguish between genuine phenomena and misinterpretations, hoaxes, or wishful thinking. This requires a balanced approach that acknowledges the possibility of the unknown while maintaining a commitment to scientific rigor and skepticism. The allure of Bigfoot lies not only in the mystery of its potential existence but also in the ongoing quest to find irrefutable proof in a world increasingly explored and documented. The challenge, therefore, is to move beyond anecdote and speculation towards tangible, verifiable data that could, one day, either confirm or definitively refute the existence of this enigmatic creature.

The immense, ancient forests of the Cascade Range present a compelling, albeit challenging, stage for the hypothesized existence of a large, unclassified primate. To assess the ecological plausibility of such a creature, we must examine the fundamental requirements of any large mammal and the specific environmental conditions of this vast, temperate rainforest ecosystem. This involves a critical look at resource availability—food, water, and shelter—and the potential carrying capacity of these extensive wilderness areas, all considered against the backdrop of the lifestyle and morphology inferred from anecdotal data and eyewitness accounts.

The Cascade Mountains, stretching from Northern California through Oregon and Washington into British Columbia, are characterized by a rugged topography, volcanic peaks, and, most importantly for our discussion, extensive tracts of old-growth and secondary forests. These forests are dominated by coniferous trees such as Douglas fir, Western hemlock, Western red cedar, and various species of pine and spruce. Interspersed are riparian zones, subalpine meadows, and a complex understory comprising ferns, mosses, huckleberries, and other shrubs. This varied landscape is dissected by numerous rivers, streams, and glacial lakes, providing abundant fresh water. The sheer scale of these interconnected wilderness areas, extending over millions of hectares, is a crucial factor. Vast, roadless areas offer a degree of isolation that could, in theory, allow a large, reclusive animal to persist undetected.

The primary question regarding ecological plausibility hinges on the availability of sustenance. Eyewitness accounts and cryptozoological hypotheses generally describe Bigfoot as an omnivore, with a diet likely consisting of a broad spectrum of available resources. This would include significant quantities of plant matter: berries, roots, nuts, fungi, and various edible greens. The Pacific Northwest is rich in such resources. In the spring and summer, for instance, huckleberries, blueberries, salmonberries, and thimbleberries form a significant food source. Various species of roots and tubers would also be accessible, particularly for a creature with the presumed strength and digging capabilities often attributed to Bigfoot. The extensive fungal networks within these forests, especially in damp, decaying wood and forest floors, could also provide a vital and readily available food source throughout much of the year.

Beyond plant matter, a creature of Bigfoot's presumed size would require substantial protein. This could be derived from insects, grubs, small mammals, birds' eggs, and, crucially, carrion. However, the most significant protein source, and one frequently cited in anecdotal reports of Bigfoot's behavior, is fish. The rivers and streams of the Cascades are historically abundant with salmon and steelhead runs, particularly during the anadromous migrations. These fish provide a highly nutritious, calorie-dense food source. Anecdotal accounts often describe Bigfoot near rivers and streams, sometimes with fishing implements or evidence of fish consumption. If Bigfoot is an adept fisher, capable of catching fish from rivers or even tidal estuaries, this would significantly bolster its ability to sustain a large population. The seasonal availability of salmon runs could necessitate a strategy of food storage or a reliance on alternative protein sources during the off-season.

The energy expenditure of a large, bipedal primate is another critical consideration. A creature of the purported size and mass of Bigfoot would have considerable metabolic needs. Locomotion, particularly sustained movement across varied terrain, would demand significant caloric intake. The hypothesized arboreal or semi-arboreal tendencies in some descriptions, while perhaps less likely for a creature of immense size, would also contribute to energy expenditure. Therefore, the availability of calorie-dense foods is paramount. If the diet were primarily composed of low-calorie greens, supporting a large primate population would be far more challenging. The omnivorous, opportunistic diet, incorporating high-energy sources like fish, nuts, and stored fats from berries or roots, offers a more plausible scenario for meeting these metabolic demands.

Water, though abundant in the Cascades, is another essential resource. The dense network of rivers, streams, and lakes ensures that water is readily accessible across most of the habitat. Seasonal variations in precipitation and snowmelt influence water levels, but the underlying availability is rarely a limiting factor in this region.

Shelter and nesting sites are also vital. Bigfoot is described as largely terrestrial but capable of arboreal movement. In a temperate rainforest, ample shelter exists in the form of dense undergrowth, large tree cavities, overhangs, and rock formations. The construction of rudimentary shelters, perhaps from fallen branches and vegetation, is also within the realm of possibility for an intelligent primate. The availability of suitable nesting sites for reproduction would be important for population continuity. Large trees, particularly old-growth specimens with their expansive root systems and hollows, could provide secure locations.

The concept of carrying capacity is central to understanding the population dynamics of any species. Carrying capacity is the maximum population size of a biological species that can be sustained by that specific environment, given the food, habitat, water, and other necessities available. For a large primate like Bigfoot, with hypothesized low reproductive rates (typical of large mammals), the carrying capacity of even the vast Cascade wilderness would be a significant constraint. Estimates of carrying capacity for large mammals like bears or even cougars in similar environments provide a baseline for consideration. A population of large primates would require significant territory to forage and to avoid inter-group conflict.

If we consider Bigfoot as analogous to other large primate species, such as gorillas or orangutans, their population densities are generally very low. Gorillas, for instance, inhabit dense forests but occupy home ranges that are defended and where population densities are typically measured in individuals per square kilometer, rather than per hundred square kilometers. The sheer scale of the Cascade Range could accommodate a small, dispersed population, but likely not a large, thriving one. The hypothetical "family groups" or "tribes" sometimes suggested in anecdotal lore would require substantial, overlapping territories.

The ecological niche that Bigfoot would occupy is also a subject for consideration. In the Cascade ecosystem, apex predators like cougars and wolves, and large herbivores like elk and deer, already occupy well-defined roles. A large, omnivorous primate would need to find a niche that did not place it in direct, unsustainable competition with these established species. Its dietary flexibility—the ability to exploit plant matter, fish, insects, and carrion—allows it to occupy a broad niche, but it would still be competing for resources. The absence of a direct apex predator (other than perhaps cougars or wolves preying on young, though this is purely speculative) could theoretically allow a large primate to thrive, provided its food sources are sufficient.

One of the persistent arguments against the existence of Bigfoot is the lack of definitive physical evidence, such as bones, teeth, or clear skeletal remains, especially given the presumed mortality rates of any wild population. If a population of large primates exists, individuals would die from natural causes, predation, or accidents. Their remains should, over time, be discovered. The argument for their elusiveness, often posited by proponents, suggests that Bigfoot is exceptionally adept at avoiding detection and that its death rituals or dispersal of remains are also highly effective at remaining hidden. However, from a purely ecological standpoint, the accumulation of skeletal remains in any environment, even a damp forest, over decades or centuries, is a predictable consequence of population dynamics. The absence of such unequivocal fossil evidence, even in areas with suitable geological conditions for preservation, remains a significant hurdle to ecological plausibility.

Furthermore, the reproductive biology of a large primate would dictate population dynamics. Large mammals typically have long gestation periods, slow maturation rates for offspring, and often give birth to a single infant at a time. This means that populations grow very slowly, making them vulnerable to even minor disruptions. For a population to persist over centuries, the birth rate would need to exceed the death rate sufficiently to maintain viability, even with low numbers. This slow growth rate could explain why they remain so elusive, as recovering from a population dip would be a lengthy process.

The concept of migration and territoriality also plays a role. If Bigfoot is a solitary or loosely social animal, the territory required for a single individual would be substantial. If they live in family units, the combined territory would be even larger. The Cascade Range offers vast expanses, but the connectivity of these habitats is crucial. Whether Bigfoot populations are isolated pockets or part of a larger, interconnected metapopulation influences their long-term survival and genetic diversity. Fragmentation of habitat due to human development, such as logging roads and urban expansion, could pose a significant threat to any such population, limiting gene flow and increasing vulnerability.

The hypothetical "out-of-place" nature of Bigfoot, appearing in various ecosystems, raises further ecological questions. While the Pacific Northwest is the stronghold of reports, sightings occur elsewhere. For a large primate species to successfully colonize and persist in multiple, disparate environments, it would require remarkable adaptability and efficient dispersal mechanisms. Each new environment would present unique challenges in terms of food availability, climate, and competition. The success of such a species would depend on its behavioral plasticity and its ability to exploit available resources without becoming over-reliant on any single, potentially ephemeral, food source.

In summary, the Cascade Range offers a theoretically plausible environment for a large, unclassified primate, primarily due to its immense size, extensive forest cover, and abundant resources. The availability of varied food sources, particularly a flexible omnivorous diet that includes high-energy items like fish and nuts, along with ample water and shelter, provides the fundamental ecological requirements. However, the critical factors of carrying capacity, the need for consistent food sources throughout the year, the challenges posed by slow reproductive rates, and the conspicuous absence of definitive physical evidence, such as skeletal remains, present significant ecological hurdles to the confirmation of such a creature. While the wilderness of the Cascades is vast enough to potentially conceal a small, elusive population, the ecological footprint of a large mammal, even one as adept at concealment as Bigfoot is purported to be, would likely leave more discernible traces than have so far been scientifically verified. The argument for ecological plausibility remains a compelling one for proponents, resting on the sheer scale of the habitat, but it must be weighed against the rigorous demands of biological persistence and the scientific standard of verifiable evidence.

The proposition of an undiscovered primate species inhabiting the vast, remote wilderness areas of North America, particularly the Pacific

Northwest, inherently invites a robust and often skeptical response from the scientific community. While proponents of Bigfoot and Sasquatch existence often point to a substantial body of anecdotal evidence, including eyewitness testimonies, alleged footprints, and purported photographic or video documentation, these data points, for the most part, fail to meet the stringent criteria required for scientific validation. The chasm between anecdotal accounts and empirical evidence is wide, and it is within this divide that the scientific community's reservations primarily reside.

One of the most prevalent explanations for Bigfoot sightings centers on misidentification. The natural world, especially in the dense, often dimly lit forests of the Pacific Northwest, is replete with creatures that, under certain conditions, might be mistaken for something far more extraordinary. Large, dark-colored mammals such as American black bears (*Ursus americanus*) and even cougars (*Puma concolor*) are frequently cited as potential candidates for misidentification. Black bears, particularly when observed standing on their hind legs, can present a remarkably bipedal silhouette. Their size, fur coloration, and lumbering gait, especially at a distance or in poor lighting, could easily lead an observer, perhaps already primed by tales of Bigfoot, to interpret their appearance as that of the legendary hominid. The sheer ubiquity of bears in the very regions where Bigfoot is most often reported makes this a statistically probable explanation for many encounters.

Beyond bears, other known animals can contribute to the confusion. Large herbivores like elk (*Cervus canadensis*) or even moose (*Alces alces*), though quadrupedal, possess a significant size and mass that could, at a fleeting glance, be misconstrued. The movement of these animals through dense undergrowth, rustling leaves, and snapping twigs could create auditory cues that, combined with a glimpse of a large, dark shape, fuel the imagination. Even common animals, when seen in unusual circumstances or at a great distance, can become subjects of misinterpretation. The human eye, particularly when influenced by expectation or a desire to witness something unusual, is susceptible to what is known as pareidolia – the tendency to perceive a specific, often meaningful, image or sound in a random or ambiguous stimulus. This psychological phenomenon can lead individuals to see patterns, faces, or, in this context, primate-like figures where none objectively exist.

Furthermore, the terrain itself can play tricks on the observer. Shadows cast by trees, particularly during dawn and dusk, can morph familiar shapes into something monstrous. Fog, mist, and the undulating landscape of mountainous regions can distort distances and appearances, making it difficult to accurately assess size, shape, and movement. An individual experiencing these visual anomalies, coupled with the thrill or fear of being in a remote wilderness, might unconsciously embellish what they perceive, leading to a report that deviates significantly from the reality of the situation.

Another significant factor contributing to the skepticism is the undeniable presence of hoaxes. The allure of fame, notoriety, or simply the desire to play a prank has led individuals to deliberately create false evidence of Bigfoot. The history of cryptozoology is replete with examples of elaborate hoaxes designed to deceive the public and researchers alike. The Patterson-Gimlin film, arguably the most famous piece of purported Bigfoot evidence, has been subjected to intense scrutiny for decades, with many experts arguing that it depicts a person in an ape suit. While proponents maintain its authenticity, the technical challenges of creating such a film in 1967, coupled with subsequent analyses and testimony from individuals involved, have led a significant portion of the scientific community to view it with extreme skepticism.

The deliberate fabrication of footprints is another common method of hoaxing. Large, oddly shaped footprints have been found in mud, snow, and soft earth, often presented as irrefutable evidence of Bigfoot's passage. However, creating such tracks is relatively straightforward. Enthusiasts have demonstrated how simple wooden molds, or even oversized boots, can be used to impress misleading imprints into the ground. The consistency of certain alleged tracks, the presence of unusual gait patterns, and the context in which they are found have often raised red flags for investigators. Many alleged Bigfoot footprints bear hallmarks of deliberate creation, such as an unnatural symmetry, a lack of anatomical detail consistent with a living creature, or the presence of anomalies that seem designed to attract attention. The sheer number of reported footprints, often appearing in areas of high human traffic or in conjunction with other unsubstantiated claims, further fuels suspicion of fabrication.

The motivation behind hoaxes can vary. For some, it is a desire for financial gain through sensationalized stories or merchandise. For others, it is the thrill of deception itself, a form of performance art that plays on public fascination with the unknown. In some instances, hoaxes may even be perpetrated to discredit genuine research or to poke fun at the earnestness of Bigfoot enthusiasts. Regardless of the motive, the existence of deliberate deception casts a long shadow over all purported evidence, demanding an exceptionally high level of proof to distinguish genuine phenomena from manufactured ones.

Psychological factors also play a role in the interpretation of experiences and the perpetuation of belief. The power of suggestion is immense. Once an individual has been exposed to stories and images of Bigfoot, their perception can become biased. They may be more inclined to interpret ambiguous stimuli as evidence of the creature, consciously or unconsciously. This phenomenon is amplified in communities where Bigfoot lore is deeply ingrained, and where the presence of such a creature is considered a local phenomenon.

Confirmation bias is another significant psychological mechanism at play. Individuals who are predisposed to believe in Bigfoot are more likely to seek out, interpret, and remember information that confirms their existing beliefs, while disregarding or downplaying evidence that contradicts them. This can lead to a skewed perception of the available data, where isolated incidents or weak evidence are elevated to the status of proof. The internet and social media have, in some ways, exacerbated this effect by creating echo chambers where like-minded individuals can reinforce each other's beliefs without encountering significant dissenting viewpoints.

Furthermore, the human desire for mystery and the unknown is a powerful force. In an increasingly mapped and understood world, the idea of a large, undiscovered animal lurking in the wilderness offers a captivating sense of wonder. This yearning for the inexplicable can contribute to a willingness to accept evidence that might otherwise be dismissed. The romantic notion of a wild, untamed creature, a relic of a bygone era, resonates deeply with many, fostering a psychological environment where belief can flourish even in the absence of concrete proof.

The scientific community, by its very nature, operates under a rigorous framework of evidence-based reasoning and peer review. Extraordinary claims require extraordinary evidence. For a new species, especially a large mammal of such significant size and purported intelligence, to be accepted into the scientific canon, a substantial body of verifiable proof would be necessary. This typically includes:

1.

Physical Specimens: The most conclusive evidence would be the discovery of a specimen – a body, a carcass, or at least a complete skeletal structure. This would allow for detailed anatomical study, genetic analysis, and a definitive classification. The absence of such a specimen, despite decades of alleged sightings and purported evidence, is a major stumbling block. If a breeding population exists, individuals must die; their remains should, over time, be found. The vastness of the wilderness can explain elusiveness, but not the complete absence of mortality evidence.

2.

Type Specimens and Holotypes: For any new species, a "type specimen" or "holotype" is required – a specific, preserved individual specimen that serves as the definitive example of the species. This specimen is cataloged, studied, and deposited in a recognized scientific institution for future reference. Without such a specimen, classification remains impossible.

3.

Reproductive Evidence: Proof of reproduction, such as nests, juvenile individuals, or detailed observations of mating behavior, would be crucial. This would help establish the species' life cycle and ecological role.

4.

Genetic Material: DNA evidence from hair, scat, or tissue samples would be invaluable. However, many samples attributed to Bigfoot have been subjected to DNA analysis, and the overwhelming majority have been identified as belonging to known animals (bears, dogs, coyotes, deer, humans) or have been too degraded to yield conclusive results. Where unique genetic sequences have been claimed, these have often been disputed, based on contamination, flawed methodologies, or misinterpretation. The scientific standard for genetic evidence is high, requiring multiple, independent, and verifiable samples that consistently point to an unknown primate.

5.

Consistent and Verifiable Behavioral Observations: While anecdotal accounts are numerous, they often lack the detail, objectivity, and repeatability that scientific observation demands. Scientists require systematic, unbiased recordings of behavior, ideally over extended periods, to understand an organism's ecology, social structure, and interactions with its environment.

The scientific community's skepticism is not necessarily a denial of the *possibility* of an undiscovered species, but rather a refusal to accept unsubstantiated claims as fact. The burden of proof rests squarely on those who make the extraordinary claim. The lack of definitive, unambiguous, and repeatable evidence means that, for now, Bigfoot remains firmly in the realm of folklore and pseudoscience.

The challenge for any proponent seeking scientific acceptance is to move beyond anecdotal reports and sensationalized claims and to engage in rigorous, systematic investigation that adheres to established scientific methodologies. This would involve well-planned expeditions, the use of modern tracking and detection technologies (e.g., camera traps, acoustic monitoring, environmental DNA sampling), and a commitment to transparency and peer review. Until such evidence emerges, the scientific community will continue to explain Bigfoot reports through the lens of misidentification, hoaxes, and psychological phenomena, grounded in the well-established principles of biology, ecology, and psychology. The allure of the unknown is powerful, but scientific understanding is built on verifiable reality.

The persistent allure of Sasquatch and Bigfoot, despite the scientific community's prevailing skepticism, necessitates a structured approach for future investigative endeavors. To bridge the chasm between anecdotal accounts and empirical validation, a targeted strategy for further field research is not only warranted but crucial for advancing our understanding, whether that understanding ultimately confirms the existence of an unknown primate or solidifies existing explanations. This requires moving beyond generalized searches and focusing on specific, scientifically sound methodologies aimed at answering precise research questions. The primary objective of any future fieldwork should be to gather unambiguous, repeatable evidence that can withstand rigorous scientific scrutiny.

Central to this endeavor is the formulation of well-defined research questions. Instead of broadly asking "Does Bigfoot exist?", we must refine our inquiries to address specific ecological and biological puzzles. For instance, a key question could be: "What is the ecological niche, if any, that a large, bipedal primate might occupy in the Pacific Northwest, and what are the bio-signatures such a niche would produce?" This shifts the focus from a simple presence/absence debate to a more nuanced ecological investigation. Other critical questions might include: "What are the genetic markers that would definitively distinguish an unknown hominid from known primate and other mammalian DNA in the region?", "What are the specific vocalizations, if any, that distinguish this potential species from other wildlife, and how can these be reliably detected and differentiated?", and "What is the potential carrying capacity of the identified habitats for a large primate, and what evidence (e.g., predation patterns, resource competition) would support its presence?" Addressing these questions necessitates a multidisciplinary approach, integrating expertise from primatology, ecology, genetics, wildlife biology, and even behavioral psychology.

The methodologies employed in future field research must be as sophisticated as the challenge demands. Traditional methods, such as setting up basic camera traps or conducting general footprint surveys, have yielded limited definitive results, often falling prey to misidentification or deliberate fabrication. A more advanced strategy would involve a layered approach, combining remote sensing, targeted biological sampling, and advanced observational techniques.

Remote sensing technologies offer a non-intrusive way to survey vast and often inaccessible terrains. High-resolution satellite imagery, coupled with aerial drone surveys equipped with thermal and infrared cameras, could potentially detect anomalies in habitat use, unusual heat signatures, or large-scale disturbances that deviate from known animal or human activity patterns. For example, thermal imaging might reveal unusual heat signatures in areas where sightings are frequently reported, particularly during crepuscular or nocturnal hours when such a creature might be most active. Similarly, analyzing vegetation patterns for signs of unusual foraging or structural manipulation could provide indirect evidence. LiDAR (Light Detection and Ranging) technology, which can penetrate forest canopies, could be employed to map the landscape in minute detail, identifying potential shelter sites, travel corridors, or even subtle anthropogenic modifications to the environment that might be overlooked by conventional surveys.

Biological sampling remains paramount, but its execution needs to be more systematic and scientifically rigorous. The collection of potential hair, scat, or other biological material should be conducted with meticulous record-keeping, including precise GPS coordinates, date, time, and detailed environmental context. Crucially, protocols for sample preservation and handling must be established to prevent contamination and degradation, ensuring that any subsequent DNA analysis is reliable. Environmental DNA (eDNA) sampling offers a particularly promising avenue. By collecting water samples from streams and rivers, or soil and leaf litter samples from forest floors, researchers can detect DNA shed by organisms that have passed through the area. While eDNA has been successfully used to detect rare and elusive aquatic species, its application to terrestrial environments for detecting large mammals is still evolving but holds immense potential for identifying the presence of an unknown primate without direct observation or capture. If an unknown primate population exists, its DNA, however trace, should be detectable in these environmental samples.

A significant advancement would be the development of robust, standardized protocols for acoustic monitoring. Many anecdotal reports include descriptions of unusual vocalizations attributed to Bigfoot. Deploying networks of high-sensitivity, wide-spectrum microphones across known sighting hotspots and along potential travel corridors could capture a wider range of sounds than previously possible. Sophisticated audio analysis software, employing machine learning algorithms, could then be used to sift through vast amounts of audio data, identifying and isolating novel vocalizations that do not match known species. Comparing these unique sounds to known primate vocalizations from around the world could provide crucial clues about its taxonomic affiliation if indeed a new species is detected.

Behavioral studies, while challenging for an elusive creature, could also be enhanced. This might involve deploying more sophisticated motion-activated camera traps with extended battery life and higher resolution, capable of capturing more detailed behavioral sequences. The placement of these cameras should be strategic, focusing on areas identified through remote sensing or anecdotal reports as potential feeding grounds, water sources, or travel routes. Furthermore, researchers could focus on documenting the indirect signs of a large primate's presence. This includes examining tree breakage patterns, evidence of digging, unusual feeding sites, or waste deposits that are inconsistent with known local fauna. Detailed ethological analysis of these signs, comparing them to known primate behaviors, could offer further insights.

Ethical considerations are an integral part of any wildlife research, and studying a potentially rare and elusive species like Bigfoot adds layers of complexity. The paramount concern must be the welfare of any individual organisms that may exist. The objective should be to observe and document, not to capture or harm. This means avoiding methods that could lead to stress, injury, or disruption of natural behaviors. If a specimen is ever discovered, its recovery and study must be conducted under strict ethical guidelines, ideally involving collaboration with wildlife rehabilitation centers or established scientific institutions.

Moreover, any research involving potential human interaction, especially with individuals who have reported encounters, must be conducted with sensitivity and respect. While scientific rigor demands objectivity, the personal experiences of witnesses should be acknowledged. Establishing clear protocols for interviewing witnesses, ensuring anonymity when requested, and distinguishing between firsthand accounts and secondhand retellings is vital to avoid compounding misinterpretations.

The potential for hoaxes, as discussed previously, remains a significant hurdle. Future research must incorporate robust protocols for discrediting or accounting for potential hoaxes from the outset. This includes rigorous forensic analysis of any physical evidence, such as footprints (examining soil mechanics, hydration levels, and anatomical inconsistencies), hair samples (testing for microscopic anomalies or treatments inconsistent with natural shedding), and photographic or video evidence (analyzing frame rates, lighting, and digital artifacts). Establishing chain of custody for all collected evidence is non-negotiable.

Furthermore, a commitment to transparency and data sharing within the scientific community is essential. Any findings, whether they support or refute the existence of Bigfoot, should be published in peer-reviewed journals. This allows for independent verification, replication of studies, and critical evaluation by experts, which is the bedrock of scientific progress. Collaboration between cryptozoologists who possess deep knowledge of anecdotal evidence and mainstream biologists and ecologists who bring established scientific methodologies is also crucial. This cross-pollination of knowledge could lead to more informed research designs and a more balanced interpretation of results.

The vastness and remoteness of the potential habitats for such a creature present logistical challenges. Expeditions need to be carefully planned, considering the rugged terrain, unpredictable weather, and the need for self-sufficiency. Modern GPS technology, satellite communication, and advanced survival gear are indispensable. Furthermore, building relationships with local communities and indigenous peoples in areas with high sighting frequency can provide invaluable insights into local folklore, traditional knowledge of the land, and potentially even undocumented observations that have been passed down through generations. These local perspectives, when approached respectfully and critically, can help guide research efforts to areas of higher potential interest.

In conclusion, the pursuit of Sasquatch and Bigfoot, if it is to transition from the realm of folklore into scientific inquiry, requires a paradigm shift. This involves moving from passive observation and collection of anecdotal evidence to active, scientifically designed field research. By posing precise questions, employing advanced technological and biological methodologies, adhering to stringent ethical standards, and fostering transparency and collaboration, future research can systematically investigate the possibility of an unknown primate in North America. The ultimate goal is to gather definitive evidence, whatever its nature, that can contribute to our understanding of biodiversity and the natural world. This is not merely about confirming or denying the existence of a mythical creature, but about advancing scientific knowledge through rigorous investigation of the unknown. The challenge is immense, but the potential rewards, in terms of expanding our understanding of life on Earth, are equally profound. Such an undertaking would require significant funding, dedicated teams of researchers, and a sustained commitment over many years, but the scientific questions it seeks to answer are among the most captivating in natural history.

9

Case Studies: The Yeti of the Himalayas

The colossal mountain ranges of the Himalayas, a jagged scar across the roof of Asia, have long been the crucible of one of the world's most enduring enigmas: the Yeti. For centuries, these soaring peaks, perpetually cloaked in snow and ice, have harbored not only breathtaking beauty and formidable challenges for human endeavor but also the whispered legends of a reclusive, ape-like creature. These tales, woven into the very fabric of Himalayan cultures, speak of a being that moves with uncanny stealth through treacherous terrain, a creature more spirit than flesh to many who inhabit the lower reaches. To approach the question of the Yeti's existence from a biological and ecological standpoint requires us to first understand the environment that is purported to be its home. This is not merely a matter of identifying a habitat; it is about assessing the viability of such an environment for a large, endothermic primate, and exploring how the unique conditions of the Himalayas might shape, conceal, or even support such an organism.

The Himalayas present an ecological tapestry unlike any other on Earth. At the lower elevations, lush temperate forests gradually give way to the stark austerity of the alpine zone, and then to the frigid, windswept heights of the nival zone, where permanent snow and ice dominate. This dramatic altitudinal gradient creates a series of distinct ecosystems, each with its own specialized flora and fauna. The lower slopes, below approximately 2,000 meters, often support dense forests of oak, rhododendron, and pine, teeming with diverse life, including a rich array of primate species that are well-documented and understood. However, it is the higher altitudes, the domain of the Yeti lore, that pose the most intriguing questions regarding the potential presence of an unknown large mammal.

As one ascends, the vegetation thins dramatically. Between 2,000 and 3,500 meters, subalpine meadows dotted with hardy grasses, sedges, and resilient flowering plants emerge, interspersed with scattered dwarf rhododendrons and junipers. This is the traditional grazing ground for many of the region's ungulates, such as the bharal (Blue Sheep) and the Himalayan tahr, which themselves are adapted to steep slopes and high altitudes. Above this, typically beyond 3,500 meters, the landscape transforms into the harsh, barren terrain of the alpine desert. Here, life clings precariously to existence. Only the most tenacious plants, like saxifrages and edelweiss, can survive the extreme cold, low oxygen levels, and the relentless scouring of wind and snow. Further still, above the permanent snow line (generally around 5,000 meters), the environment becomes almost entirely devoid of macroscopic life, save for extremophile microorganisms, lichens, and the occasional insect or bird that ventures into these desolate realms.

The climatic conditions are equally formidable. Temperatures plummet far below freezing for much of the year, with significant diurnal and seasonal fluctuations. The air is thin, with considerably lower atmospheric pressure and oxygen content than at sea level, posing a significant physiological challenge for any mammal, particularly one of substantial size that would require a higher metabolic rate. Precipitation is predominantly in the form of snow, and avalanches are a constant, deadly hazard, reshaping the landscape and posing a perpetual threat to any inhabitants. The wind can be ferocious, capable of stripping flesh from bone, and visibility is often severely limited by fog, snowstorms, and blizzards. These are not conditions that would easily support a large, active primate population without extraordinary adaptations.

From an ecological perspective, the question arises: what could sustain a large primate in such an environment? The dietary requirements of an ape-sized creature would be substantial. If we consider the known ecological niches of large primates, such as gorillas or orangutans, their diets consist primarily of vegetation – leaves, fruits, shoots, and bark – supplemented with insects, and occasionally small vertebrates. In the high Himalayas, the availability of such food sources is highly seasonal and geographically restricted. During the brief summer months, the alpine meadows offer a burst of plant life, and fruits might be available in certain sheltered valleys. However, these resources would likely be insufficient to sustain a population through the long, harsh winters. A creature adapted to this environment might need to exploit a very broad range of food sources, including roots, tubers, and possibly even carrion, or perhaps exhibit extreme dietary flexibility.

The behavioral adaptations required would also be profound. A large primate would need strategies to cope with the extreme cold, such as developing a dense fur coat, accumulating significant body fat, and possessing a high metabolic rate. Many Himalayan mammals, like the yak and the snow leopard, have evolved thick insulation and specialized physiologies. The question is whether such adaptations could develop in a primate lineage within the time frame suggested by Yeti lore. Furthermore, a creature of significant size would require substantial territory to forage for sufficient food, especially if that food is dispersed and seasonal. This suggests that the Yeti, if it exists, would need to be incredibly elusive, ranging over vast, sparsely populated areas to avoid detection and minimize competition with other large predators and herbivores.

The terrain itself, while seemingly inhospitable, might offer refuge. The sheer verticality of the Himalayas, the vast networks of caves and overhangs, and the dense, often mist-shrouded forests at lower elevations could provide ideal hiding places. Anecdotal accounts frequently describe the Yeti as being able to navigate steep, snow-covered slopes with ease, a skill that would be essential for survival. Their purported ability to move silently through dense vegetation or across rocky scree would be a critical adaptation for avoiding predators (such as snow leopards at higher altitudes) and for hunting or foraging unnoticed.

It is within this context of extreme environment and limited resources that the cultural narratives of the Yeti take root. For the indigenous peoples of the Himalayas – the Sherpas, the Tibetans, the Nepalis – the mountains are not merely a geological feature but a sacred, often dangerous, realm imbued with spiritual significance. Legends of mountain spirits, elusive beasts, and hidden valleys are deeply ingrained in their worldview. The Yeti, or

Meh-teh (meaning "man-bear") as it is known in some regions, is often portrayed not as a simple animal but as something more primal, something that embodies the untamed spirit of the mountains.

These cultural narratives are not monolithic. Across different ethnic groups and geographical areas, the description and behavior of the Yeti can vary. Some accounts depict it as a terrifying, aggressive monster, while others describe a shy, reclusive creature that actively avoids humans. Many stories emphasize its intelligence, its ability to mimic sounds, and its deep connection to the natural world. The frequent association with footprints – often described as large, human-like, or sometimes bear-like with an unusual gait – has been a recurring theme throughout the history of Yeti lore, providing tangible, albeit ambiguous, "evidence" for its existence.

The earliest written accounts of a wild, man-like creature in the Himalayas can be traced back to Tibetan lamaic literature, though these are often allegorical or spiritual in nature. The Western fascination with the Yeti truly began in the early 20th century, fueled by the expeditions of mountaineers charting the world's highest peaks. In 1921, a British expedition to Mount Everest reported that local porters had found enormous footprints in the snow, which they attributed to a large, ape-like creature, the

metoh-kangmi – literally "man-bear snowman." This report, sensationalized by the press, gave birth to the Western term "Abominable Snowman."

Subsequent expeditions, driven by a mix of scientific curiosity and the allure of the unknown, continued to gather anecdotal evidence. Photographs of strange footprints, sightings reported by climbers and local villagers, and fragments of local folklore were pieced together, painting a picture of an elusive primate lurking in the remote valleys and desolate plateaus of the Himalayas. The scientific community, however, has largely remained skeptical, proposing that most sightings can be explained by misidentification of known animals (such as bears, langur monkeys, or even red pandas), unusual snow conditions creating distorted footprints, or outright hoaxes.

Despite the skepticism, the sheer volume and persistence of these accounts, spanning cultures and decades, warrant a serious examination of the ecological plausibility. Could a large primate species have evolved, or perhaps survived as a relict population, in the isolated and extreme environments of the Himalayas? Or are these legends simply the product of human imagination, fear, and the powerful suggestibility of an awe-inspiring landscape? The ecological conditions themselves offer both potential explanations for the creature's elusiveness and profound challenges to its existence. The sparse vegetation, extreme temperatures, and limited food availability at high altitudes present significant hurdles for any large mammal. Yet, the rugged terrain and the vast, unexplored wilderness could also provide the perfect sanctuary for an undiscovered species. The interplay between the biological requirements of a large primate and the unique environmental pressures of the Himalayas is the central question that cryptozoologists and ecologists alike must grapple with when considering the Yeti. It is a frontier where legend and biology meet, demanding rigorous inquiry into an environment that has, for millennia, guarded its secrets closely.

The enduring allure of the Yeti, often termed the "Abominable Snowman" in Western popular culture, is deeply entwined with a rich tapestry of historical accounts and indigenous lore. Far from being a recent Western invention, the concept of a large, ape-like or wild-man creature inhabiting the remote reaches of the Himalayas predates modern expeditions by centuries, existing as a vital component of the cultural and spiritual landscape for the peoples who call these mountains home. Understanding these historical narratives and the folklore of communities like the Sherpas is crucial to discerning the origins and persistence of the Yeti legend.

Early references to beings resembling the Yeti can be found in Tibetan lamaic literature, though these are often allegorical or framed within spiritual narratives. These texts speak of mountain spirits and elusive creatures that inhabit the high, inaccessible regions, embodying the wild and untamed essence of the Himalayas. While not explicitly scientific in their approach, these ancient stories demonstrate a long-standing cultural awareness of something unknown and powerful dwelling within the mountain wilderness. These early mentions reflect a worldview where the natural and supernatural realms were more fluid, and the awe-inspiring, often perilous, nature of the mountains was readily attributed to hidden forces and beings.

The Western world's introduction to the Yeti, and the subsequent explosion of interest, can be largely traced to the early 20th century and the era of ambitious mountaineering expeditions. In 1921, the British Mount Everest Reconnaissance Expedition, led by Lieutenant-Colonel Charles Howard-Bury, brought back reports that would fundamentally alter the perception of Himalayan mysteries. During their ascent, local porters accompanying the expedition discovered a series of unusually large footprints in the snow. These prints were unlike anything they had encountered before, leading them to attribute them to a creature they called the

metoh-kangmi, a term roughly translating to "man-bear snowman" or "wild man." This report, when disseminated through the press, was sensationalized and ultimately morphed into the more evocative "Abominable Snowman," a moniker that has since become synonymous with the creature. This early account, emphasizing the "snowman" aspect, immediately cast the Yeti as a creature of snow and ice, a fitting inhabitant for the extreme altitudes that mountaineers were beginning to explore. The details of these initial footprints were described as roughly humanoid in shape but significantly larger than any human foot, with the distinct impression of toes. The porters' explanation, rooted in their existing lore of mountain beings, provided a local context for what Westerners perceived as a startling discovery.

Following this initial report, subsequent expeditions to the region, driven by a potent mix of scientific curiosity, the thrill of exploration, and perhaps a touch of sensationalism, continued to gather anecdotal evidence. The accounts were varied and often contradictory, contributing to the enigma. Some climbers reported fleeting glimpses of large, dark figures moving with surprising agility across steep, snow-covered slopes. Others spoke of hearing strange, disembodied vocalizations echoing through the valleys – eerie cries or roars that seemed to emanate from the very rock and ice. These encounters, often brief and under challenging conditions of poor visibility, low light, and extreme altitude, were nonetheless compelling enough to fuel further speculation.

One of the most persistent pieces of "evidence" associated with the Yeti has been the discovery of anomalous footprints. Over the decades, numerous photographs and plaster casts of large, ape-like or crudely human-like tracks have been collected from remote areas. These footprints, often found in snow or soft earth, have been a focal point of Yeti investigations. While many have been attributed to known animals whose tracks might be distorted by melting or unusual gait, some have defied easy explanation. For instance, the famous photographs taken by mountaineer Eric Shipton in 1951 on the Menlung Glacier, near Everest, revealed a series of large, distinct prints that appeared to show an unusual gait and a five-toed impression, sparking renewed interest and debate. The prints were considerably larger than human feet and seemed to indicate a bipedal gait with widely splayed toes.

Beyond footprints, the lore surrounding the Yeti is also populated by alleged physical artifacts. Among the most intriguing and controversial have been purported Yeti scalps and handprints. Several monasteries in the Khumbu region of Nepal, particularly Tengboche Monastery, have historically claimed to possess Yeti scalps. These scalps, often leathery and matted with dark hair, were presented as tangible proof of the creature. However, scientific examination of these artifacts has consistently revealed them to be the remains of other known animals, most notably the Himalayan brown bear (

Ursus arctos isabellinus) or the Tibetan blue bear (*Ursus thibetanus*). The process of preparing these scalps would have involved local traditions, perhaps involving shaving or shaping the skin to resemble a more human-like or ape-like head, further obscuring their true origin. Similarly, alleged Yeti handprints, often found in mud or snow, have frequently been explained as misidentified prints of bears or other animals, or, in some cases, as deliberate fabrications.

The descriptions of the Yeti itself, as passed down through local lore and mountaineering accounts, are remarkably diverse. In Sherpa and Tibetan traditions, the creature is often referred to by names like

Meh-teh (meaning "man-bear"), *Dzu-teh* (meaning "bear"), or *Mi-teh* (meaning "man"). These names themselves suggest a connection to bears, which are known inhabitants of the Himalayas and whose tracks can sometimes be mistaken for those of a larger primate. However, many accounts emphasize a distinctly ape-like or hominid appearance. Descriptions frequently portray the Yeti as being covered in dark, shaggy hair, possessing broad shoulders, a short neck, and a relatively flat face with a prominent brow ridge. Its size is consistently described as imposing, often estimated to be between six and eight feet tall when standing erect.

The behavior attributed to the Yeti also varies significantly. Some narratives paint the creature as a fearsome, aggressive predator, a terrifying entity that poses a threat to humans and livestock. Tales of attacks on people or yaks, though rare and often lacking concrete corroboration, contribute to this image of a dangerous beast. Conversely, many other accounts, particularly those originating from local communities deeply connected to the mountain ecosystem, describe the Yeti as a shy, elusive, and even gentle creature that actively avoids human contact. These narratives often portray the Yeti as a guardian of the wild, a being that prefers solitude and retreats further into the wilderness when disturbed. Its intelligence is frequently remarked upon, with some stories suggesting it can mimic sounds, communicate with other Yetis, or even understand human intentions.

The reported sightings by Westerners, especially climbers who ventured into regions previously inaccessible, have played a pivotal role in shaping the popular image of the Yeti. Many renowned mountaineers, including those with scientific training, have reported encounters or discoveries that they could not readily explain. For example, mountaineer Reinhold Messner, one of the most accomplished climbers in history, dedicated significant time to investigating the Yeti phenomenon after his own experiences in the Himalayas. In his book "My Quest for the Yeti," Messner detailed encounters with local testimonies and his own observations, including a memorable instance of seeing what he believed to be a Yeti up close in the 1980s. While Messner ultimately concluded that the Yeti was likely a misidentified bear, his detailed accounts and dedication to the investigation highlight the persistent nature of these reports even among seasoned explorers.

The inconsistency in descriptions and artifacts is a hallmark of the Yeti phenomenon. The alleged scalps, handprints, and even purported Yeti fur samples that have been subjected to scientific analysis have invariably been identified as belonging to known species, most commonly bears, but also sometimes goats, monkeys, or even domestic animals. This recurring pattern of misidentification or fabrication raises significant questions about the reliability of some of the "evidence." However, it also highlights the cultural context in which these items were found and presented. In communities where the existence of the Yeti was a given, and where religious or spiritual traditions often involved animal parts, it is plausible that remains of known animals were interpreted through the lens of existing lore.

The Sherpa lore, in particular, provides a rich and nuanced perspective on the Yeti. For the Sherpa people, who have lived in close proximity to the highest peaks for generations, the mountains are imbued with spiritual significance. The Yeti, or

Meh-teh, is often viewed not merely as an animal but as a spirit of the mountains, a powerful entity that commands respect and caution. Sherpa elders often share stories passed down through oral tradition, detailing encounters and observations that blend naturalistic elements with supernatural ones. These narratives are not necessarily about finding a biological specimen but about understanding the place of this creature within their cosmology. The Sherpa often describe the Yeti as a solitary creature, a hermit of the peaks, that rarely reveals itself. Their understanding is often rooted in observation of its signs – its footprints, its calls, its presence in certain remote areas – rather than direct confrontation. These oral traditions are not static; they evolve and adapt, incorporating new observations and reinterpretations of phenomena.

One notable aspect of the historical accounts is the geographic clustering of reports. While Yeti sightings have been reported across a broad swathe of the Himalayas, certain regions, particularly the Khumbu region around Mount Everest and the Makalu-Barun National Park, have been hotspots for alleged encounters and footprint discoveries. This might be due to a combination of factors, including higher human activity in these accessible-yet-remote areas (due to expeditions and trekking), and the presence of established Sherpa communities who have long-standing traditions and knowledge of the mountain environment. The higher density of reports in these areas could also reflect the ecological suitability for certain large mammals that might be misidentified as Yeti, such as the Himalayan brown bear.

The persistence of Yeti lore is remarkable. Despite numerous expeditions and scientific investigations that have largely failed to produce conclusive evidence of an unknown primate, the legend endures. This persistence can be attributed to several factors: the vast, unexplored nature of the Himalayas, which leaves room for undiscovered species; the compelling nature of eyewitness accounts, even if they are later explained; the cultural significance of the Yeti within indigenous traditions; and the inherent human fascination with the unknown and the possibility of discovering something extraordinary. The historical trajectory of Yeti accounts, from ancient spiritual references to modern mountaineering reports and scientific investigations, reveals a complex interplay between cultural belief, environmental observation, and the human tendency to seek explanations for the mysterious. Each era and each group of investigators has interpreted the phenomenon through its own lens, contributing to the multifaceted and enduring enigma of the Yeti. The evolution of these accounts—from the initial

metoh-kangmi of the 1921 expedition to the more scientifically examined, yet still elusive, entity of later years—demonstrates not only the persistence of the legend but also the changing ways in which it has been perceived and investigated. The challenge for any cryptozoological inquiry lies in sifting through this rich, often contradictory, historical and cultural record to identify any kernels of verifiable truth, while acknowledging the powerful role of folklore and human imagination.

The vast and formidable Himalayan range presents an extreme environment, a crucible where life must contend with formidable challenges. For any large primate, let alone an unknown hominoid like the Yeti, to eke out an existence here would necessitate a remarkable suite of biological adaptations. The very notion of such a creature surviving in the alpine and sub-alpine zones, areas characterized by intense cold, thin air, and a scarcity of resources, compels us to examine the biological plausibility through the lens of ecological and evolutionary principles.

Firstly, the most immediate hurdle is the frigid climate. Temperatures in the Himalayas can plummet to well below freezing for much of the year, with significant snowfall often rendering the landscape impassable for extended periods. A large mammal inhabiting these altitudes would require substantial insulation. This could manifest as a dense, insulating layer of fur, significantly thicker and perhaps longer than that of primates found in temperate or tropical regions. The reported shaggy, dark hair of the Yeti in many accounts aligns with this requirement. Beyond passive insulation, such a creature would likely possess a higher basal metabolic rate to generate internal heat, or perhaps exhibit periods of torpor or hibernation during the harshest months, similar to many other high-altitude mammals that survive by conserving energy. Consider the adaptations seen in the snow leopard (

Panthera uncia), for instance. Its incredibly thick, dense fur provides exceptional insulation, and its large paws act like snowshoes, distributing its weight to traverse snowy terrain. While the Yeti is not a felid, the principle of specialized insulation is fundamental for survival in such an environment. Furthermore, the Yeti's hypothesized stature, often described as large and robust, could also contribute to a lower surface-area-to-volume ratio, a principle in physics that suggests larger bodies retain heat more efficiently than smaller ones. This Bergmann's Rule often applies to endotherms in colder climates, where populations of the same species tend to be larger in colder regions.

The availability of food is another critical factor. The alpine and subalpine zones offer a significantly more limited diet than lower elevations. Hardy grasses, sedges, lichens, and mosses would likely form the bulk of a herbivorous diet, supplemented by roots and tubers dug from the rocky soil when accessible. The Yeti's digestive system would need to be adept at extracting nutrients from these tough, fibrous plant materials. This might involve a specialized gut microbiome, similar to that found in ruminants, or a longer digestive tract to maximize nutrient absorption. Accounts of Yeti foraging for roots are not uncommon in local lore, suggesting a reliance on these more substantial, ground-level food sources. The hypothetical Yeti would also need to be opportunistic, capable of supplementing its diet with whatever animal protein becomes available. This could include insects, small rodents, birds, or even carrion. Some reports suggest the Yeti might prey on larger animals like wild sheep or goats, but this would require significant hunting prowess and energy expenditure in challenging terrain, making it a less likely primary food source compared to more readily available options. The ecological niche of a large, primate-like omnivore in such an environment is challenging to conceive, but not entirely outside the realm of possibility if it can effectively exploit a wide range of low-energy-density food sources and occasionally higher-energy ones.

To survive in an environment with scarce food and extreme weather, a large primate would also require access to suitable shelter. Caves, rock overhangs, and dense thickets of hardy shrubs would offer protection from the elements and predators. The ability to find and utilize such natural shelters would be paramount, especially during blizzards or periods of extreme cold. The rugged topography of the Himalayas, with its myriad of cliff faces, ravines, and glacial caves, could indeed provide ample opportunities for shelter. Imagine a creature that, like the orangutan or chimpanzee, can fashion rudimentary nests from branches and vegetation, but adapted to utilize rocky alcoves or even dig burrows, perhaps in areas with less permafrost. The ecological parallel here could be drawn with the Himalayan tahr (

Hemitragus jemlahicus) or the bharal (blue sheep, *Pseudois nayaur*), both of which are agile mountain ungulates that utilize steep, rocky terrain for both foraging and seeking refuge from predators and harsh weather. Their ability to navigate and survive in such precipitous environments demonstrates the potential for large mammals to adapt to the Himalayan landscape.

Furthermore, the reproductive strategies of any such creature would need to be efficient. The challenges of gestation, birth, and raising young in such a demanding environment would likely lead to slower reproductive rates, with longer intervals between births and extended parental care, similar to many large mammals in harsh climates. This would ensure that offspring are sufficiently developed and resilient before facing the full brunt of the mountain's challenges. The low population densities implied by such a reproductive strategy would also make detection by humans even more unlikely, contributing to the elusive nature of the Yeti.

The question of population size is also crucial. For a species to persist, especially a large-bodied one with potentially slow reproduction, a viable breeding population is necessary. This implies a significant geographic range and a sufficient number of individuals to maintain genetic diversity and avoid extinction. The vastness of the Himalayas could potentially support such a population, spread across remote valleys and high-altitude plateaus, avoiding human settlements. However, the isolation between these potential pockets of habitat could also lead to genetic bottlenecks and inbreeding if populations become too fragmented.

When considering the biological plausibility, it is also important to acknowledge the potential for misidentification. The Himalayan landscape is home to several large mammals whose tracks or brief sightings could be misinterpreted. The Himalayan brown bear (

Ursus arctos isabellinus) and the Tibetan blue bear (*Ursus thibetanus*) are significant contenders. Their footprints can be remarkably large and, when distorted by melting snow or soft ground, can bear a superficial resemblance to those of a bipedal hominoid. Furthermore, bears are known to stand on their hind legs, and their gait can sometimes appear unusual. The Himalayan wolf (*Canis lupus chanco*) or even large langur monkeys (*Semnopithecus entellus*) when moving in unusual circumstances, could also contribute to anecdotal reports. The thick fur of bears, combined with their bulk, could also lead to brief visual encounters being perceived as something larger and more ape-like, especially in low light conditions or at a distance.

The concept of ecological niches also plays a role. If a large primate were to exist, it would occupy a specific niche within the ecosystem. Its presence would inevitably have observable effects on its environment, from predator-prey dynamics to vegetation patterns. The absence of such observable impacts, or the consistent attribution of such signs to known species, weakens the argument for an unknown primate. However, the vastness and inaccessibility of much of the Himalayas mean that such impacts might simply have gone unnoticed or unrecorded by scientific observers.

The biological requirements for a large primate in the Himalayas are undeniably stringent, demanding specialized adaptations for cold tolerance, efficient foraging on sparse resources, and robust shelter-seeking behaviors. While the ecological challenges are immense, the history of life on Earth is replete with examples of organisms adapting to extreme environments. The question then becomes whether the specific adaptations required for a large primate to thrive in the Himalayas are biologically feasible and whether there is any indirect evidence supporting their existence. Drawing parallels with other high-altitude mammals, such as the yak (

Bos grunniens), snow leopard, and various species of mountain ungulates, highlights the capacity for life to flourish even in these seemingly inhospitable regions. These animals have evolved their own unique strategies for survival, from physiological adaptations to behavioral patterns. The hypothetical Yeti, if it exists or has existed, would have needed to carve out a similar, albeit primate-specific, niche within this complex and demanding ecosystem. The challenge for cryptozoologists and biologists alike is to find concrete evidence that transcends anecdotal accounts and misidentification, evidence that points unequivocally to a creature with the biological characteristics of a large, unknown primate adapted to the extreme conditions of the Himalayan alpine and sub-alpine zones. The continuing allure of the Yeti stems, in part, from this very biological puzzle: could such a creature, with all the necessary adaptations, truly exist, or has existed, in one of the world's most formidable landscapes? The biological plausibility remains a central, and as yet, unresolved facet of the enduring enigma.

The persistent allure of the Yeti, a cryptid as enigmatic as the snow-capped peaks it supposedly inhabits, has inevitably drawn the attention of the scientific community. For decades, expeditions, both large and small, have ventured into the remote reaches of the Himalayas with the explicit aim of uncovering definitive proof of its existence. These endeavors, driven by a mixture of scientific curiosity and the tantalizing prospect of a monumental discovery, have yielded a fascinating, albeit often frustrating, body of evidence. However, the very nature of the terrain, coupled with the elusive behavior attributed to the Yeti, has consistently conspired to make conclusive findings elusive, leaving behind a trail of ambiguous clues and unresolved questions.

One of the most enduring and frequently cited forms of evidence for the Yeti's existence comes in the form of footprints. Across the snow-laden slopes and rocky scree of the Himalayas, numerous individuals, from local Sherpa guides to seasoned mountaineers, have reported encountering large, human-like tracks. These prints, often described as being significantly larger than those of any known primate or bear, have become a cornerstone of Yeti lore. Early expeditions, such as those led by mountaineers like Eric Shipton in the 1950s, meticulously documented these tracks. Shipton's famous photograph of a large, bipedal footprint in the snow on the Menlung Glacier, taken in 1951, remains one of the most iconic images associated with the Yeti phenomenon. While Shipton himself was cautious, suggesting it might be a bear track distorted by melting, the image ignited public imagination and fueled further speculation. Subsequent expeditions have reported similar findings, with some prints exhibiting distinct features such as a prominent heel, a splayed forefoot, and even what appear to be claw marks, though the latter is often debated and inconsistent across reports.

The scientific analysis of these purported Yeti footprints has proven to be a complex and contentious issue. When samples of these tracks, or casts made from them, have been subjected to scrutiny, the interpretations have varied widely. Dr. Bernard Heuvelmans, a pioneer in cryptozoology, meticulously studied numerous Yeti footprint casts and photographs. He proposed that the prints were consistent with a large, bipedal primate, distinct from bears, and coined the term "anomalopithecus" for the creature. However, other zoologists and primatologists have offered more prosaic explanations. Many argue that the majority of these tracks can be attributed to known animals, particularly the Himalayan brown bear (*Ursus arctos isabellinus*) or the Tibetan blue bear (*Ursus thibetanus*). Bears are known to walk bipedally for short distances, and their large, padded paws, especially when leaving prints in soft snow or mud, can create distorted impressions. The melting and refreezing of snow can also dramatically alter the shape and size of an original track, elongating it, blurring distinct features, and sometimes creating the illusion of a bipedal gait. Moreover, the tracks of animals like the Himalayan wolf (*Canis lupus chanco*) or even large langur monkeys (*Semnopithecus entellus*), when viewed in unfavorable conditions or from a distance, could potentially be misidentified.

Beyond footprints, another significant category of alleged physical evidence comprises hair and fur samples. Over the years, numerous individuals have collected strands of hair and fur from areas where Yeti sightings have been reported or from alleged Yeti nesting sites. These samples, often described as coarse, dark, and unusually long, have been subjected to various scientific analyses. Early examinations, including those conducted by Dr. George Agogino in the 1960s, analyzed hairs sent to him by various expeditions. Some of these analyses, using microscopic examination, suggested that the hairs did not definitively match known primate species native to the region. This led to further speculation about an unknown primate.

However, more sophisticated genetic and morphological analyses in recent decades have largely debunked these early claims. In 2010, a team of researchers led by Dr. Bryan Sykes of Oxford University embarked on a comprehensive study of a large collection of purported Yeti hair samples, obtained from museums and private collections worldwide. These samples had been attributed to the Yeti based on anecdotal evidence and previous, less conclusive, examinations. Using advanced DNA sequencing techniques, Sykes' team analyzed mitochondrial DNA from nine of these specimens. The results were striking and largely disappointing for Yeti enthusiasts. The genetic analysis revealed that the vast majority of the samples, when they yielded usable DNA, were not from an unknown primate. Instead, they were identified as belonging to known animals, including the domestic dog (

Canis lupus familiaris), domestic horse (*Equus caballus*), domestic yak (*Bos grunniens*), and various species of bear, most notably the Himalayan brown bear and the Tibetan blue bear. One particularly intriguing sample, collected from Bhutan, showed a genetic match to a polar bear (*Ursus maritimus*) or its ancient ancestor, a finding that, while surprising in the Himalayas, points towards the possibility of misidentified bear remains or perhaps ancient migration patterns rather than an unknown ape.

The methodology and interpretation of these DNA studies, while rigorous, have not been entirely without their critics. Some cryptozoologists have argued that the samples analyzed may not have been representative of true Yeti evidence, or that the techniques used might not be sensitive enough to detect unique genetic markers if the creature is indeed a rare hominoid with a very small population. Nevertheless, the overwhelming consensus from these modern genetic studies is that the available physical evidence, particularly hair samples, does not support the existence of an unknown primate species in the Himalayas.

Another avenue of investigation has involved the collection of alleged Yeti scat or droppings. While scat can provide invaluable information about an animal's diet, health, and genetic makeup, Yeti scat samples have proven to be equally elusive and problematic. When found, these samples are often difficult to distinguish from the droppings of known animals like bears or large ungulates. Furthermore, the harsh environmental conditions of the Himalayas, with their extreme temperatures and high levels of UV radiation, can degrade biological material, making accurate genetic analysis challenging. The few scat samples that have been collected and analyzed have, much like the hair samples, typically been identified as belonging to known species.

The challenge of scientific investigation in the Himalayas extends beyond the ambiguity of the evidence itself. The sheer scale and inaccessibility of the mountain range present immense logistical hurdles. Vast swathes of the Himalayas remain virtually unexplored by modern scientific standards. Reaching these remote valleys, treacherous glaciers, and high-altitude plateaus requires significant resources, specialized equipment, and extensive planning. The weather can change with alarming speed, posing constant dangers to expeditions. The thin air at high altitudes can affect human physiology, limiting the duration and intensity of fieldwork. These environmental factors mean that any systematic scientific survey of potential Yeti habitats would be an undertaking of monumental proportions, far exceeding the scope and budget of most cryptozoological or even many standard biological research projects.

Furthermore, the socio-cultural landscape of the Himalayas adds another layer of complexity. The Yeti is deeply ingrained in the folklore and belief systems of the indigenous populations, particularly the Sherpa people. While this cultural significance is fascinating in its own right, it can also influence the reporting of sightings and the interpretation of evidence. Local communities may be reluctant to share information with outsiders, or conversely, they may feel pressure to conform to expectations by reporting encounters. The line between genuine observation, folklore, and the desire to believe can become blurred. While these cultural factors do not invalidate potential evidence, they underscore the need for careful, objective data collection and interpretation, free from preconceived notions.

The scientific expeditions that have attempted to find concrete evidence have often been characterized by a range of approaches. Some have been focused solely on cryptozoological pursuits, actively searching for the Yeti. Others have been broader ecological surveys or mountaineering expeditions that happened upon potential clues. For instance, the 1954 expedition funded by the Daily Mail, led by Sir John Hunt, famously searched for the Yeti, collecting footprint casts and hair samples. While they ultimately found no conclusive proof, their efforts galvanized public interest. More recently, teams have employed modern technologies such as camera traps and satellite imagery, hoping to capture definitive visual evidence or identify unusual patterns of movement. However, the effectiveness of camera traps in such vast and rugged terrain is limited by battery life, the need for specific placement, and the sheer difficulty of covering enough ground.

The persistent lack of unambiguous, scientifically verifiable evidence is, for many, the most compelling argument against the Yeti's existence. In the realm of biology, for a large, complex organism to persist, there must be a viable breeding population. This implies a significant number of individuals spread across a sufficiently large geographic range. Such a population would, over time, leave behind a more substantial and consistent trail of evidence: bones, unambiguous DNA traces in droppings or remains, clear visual sightings by trained observers, and observable impacts on the ecosystem. The fact that, despite numerous expeditions and countless anecdotal reports, no such definitive evidence has materialized is a significant challenge for the Yeti hypothesis. The scientific method relies on falsifiability and reproducibility. To date, the Yeti remains frustratingly elusive to these fundamental principles. The evidence gathered, while intriguing to some, has consistently fallen short of scientific standards, often being attributable to known species, misinterpretations, or environmental artifacts. The story of scientific investigations into the Yeti is, therefore, a narrative of well-intentioned efforts grappling with an exceptionally challenging environment and a phenomenon that consistently eludes definitive capture, leaving the legend to thrive in the realm of the possible, rather than the proven.

The Yeti, often referred to as the "Abominable Snowman," transcends mere cryptozoological curiosity; it is a potent cultural icon deeply woven into the fabric of Himalayan folklore and the global imagination. Its presence in local traditions and its enduring appeal in popular culture speak volumes about our fascination with the unknown, the wild, and the untamed aspects of our world. For the indigenous peoples of the Himalayas, particularly the Sherpa and other ethnic groups inhabiting the higher altitudes, the Yeti is not simply a creature of myth but a tangible, if rarely seen, entity with a significant role in their spiritual and cultural landscapes.

In the traditional belief systems of these mountain communities, the Yeti is often perceived as a powerful, solitary being that inhabits the remote, inaccessible regions of the mountains. It is seen as a guardian of the wilderness, a spirit of the high peaks, and at times, a formidable force to be reckoned with. Stories about the Yeti are passed down through generations, serving as cautionary tales, explanations for unexplained phenomena, and expressions of respect for the immense power of the natural environment. These narratives often describe the Yeti as a creature of immense strength, capable of great feats of endurance and possessing an acute awareness of its territory. Local shamans and spiritual leaders may incorporate the Yeti into their rituals, viewing it as an important element of the mountain ecosystem, a being that commands respect and demands a cautious approach from humans venturing into its domain. The perceived duality of the Yeti—its potential danger and its sacred connection to the mountains—makes it a complex figure, reflecting the inherent respect and apprehension that mountain dwellers have for their formidable environment. The reverence shown to the Yeti is intrinsically linked to the reverence for the mountains themselves, which are often considered sacred places, inhabited by spirits and deities.

The physical characteristics attributed to the Yeti in these local accounts, while varying in detail, consistently point to a large, ape-like creature, covered in thick, shaggy hair, with a powerful build and capable of walking upright. These descriptions, often passed down orally, have formed the bedrock of the Yeti legend, predating modern scientific inquiry and Western fascination. The deep roots of these beliefs mean that for many locals, the existence of the Yeti is not a matter of debate but a matter of lived cultural reality, informed by ancestral knowledge and spiritual understanding. This deep cultural embedment makes it challenging for external scientific perspectives to entirely detach the phenomenon from its rich cultural context.

Beyond the Himalayas, the Yeti has captured the global imagination, evolving into a pervasive figure in popular culture. From adventure novels and films to cartoons and consumer products, the "Abominable Snowman" has become a widely recognized symbol of mystery and the wild. This popularization, however, has often led to a simplification and sometimes sensationalization of the legend. The nuanced cultural significance in its native regions can be overshadowed by a more generalized, often anthropomorphic, portrayal in Western media. This is not to say that the Yeti's presence in popular culture is entirely superficial; it taps into a fundamental human desire to believe in the existence of undiscovered wonders and to explore the edges of our known world. The Yeti embodies the allure of the unexplored, the last frontier on Earth that might still harbor secrets.

The enduring appeal of the Yeti as a cultural icon can be understood through several lenses. Firstly, it taps into our innate fascination with the unknown. In an era where much of the planet has been mapped and explored, the idea of a large, intelligent creature lurking in the remote, inaccessible corners of the world offers a potent antidote to the perceived predictability of modern life. It represents the possibility of discovery, the thrill of the hunt for something extraordinary. Secondly, the Yeti resonates with our connection to the natural world and our anxieties about its preservation. In a time of increasing environmental degradation and species extinction, the Yeti can be seen as a symbol of the wild, an ambassador for the untouched wilderness that we are in danger of losing. Its elusiveness can be interpreted as a testament to its successful adaptation to a harsh environment, a creature perfectly suited to its remote habitat, perhaps even a victim of human encroachment and the shrinking of wild spaces.

Furthermore, the Yeti serves as a powerful metaphor for the limits of human knowledge and the vastness of the unknown. Despite our technological advancements and extensive scientific exploration, the Himalayas, with their extreme altitudes and treacherous terrain, continue to present formidable challenges. The Yeti's continued elusiveness, even after numerous expeditions and investigations, highlights how much we still have to learn about our planet's ecosystems, especially those in extreme environments. It prompts us to consider what might still exist beyond the reach of our current understanding and observational capabilities. The Yeti, in this sense, becomes a symbol of scientific humility, a reminder that the world may hold wonders that defy easy explanation or categorization.

The narrative of the Yeti is also intertwined with the history of exploration and adventure. The early expeditions into the Himalayas, often undertaken by intrepid explorers and mountaineers seeking fame, fortune, or scientific discovery, brought the Yeti to international attention. These expeditions, fueled by a sense of romantic adventure and the allure of the exotic, often returned with tales that, while lacking definitive proof, added layers to the legend. The very act of searching for the Yeti in such a challenging environment became part of the adventure, and the stories generated by these quests further solidified its place in the popular imagination. Mountaineers, in particular, have played a significant role in propagating Yeti lore. Many seasoned climbers have reported fleeting glimpses, strange sounds, or unusual tracks during their ascents of the world's highest peaks. These accounts, coming from individuals accustomed to observing wildlife and navigating the mountains, carry a certain weight, even if they remain anecdotal.

The Yeti's transformation into a global icon is also a testament to the power of storytelling and the human capacity for belief. Once the idea of a mysterious mountain ape was introduced to the wider world, it resonated with existing archetypes of wild men, monsters, and hidden civilizations found in folklore across many cultures. This resonance allowed the Yeti to find fertile ground in the collective consciousness, becoming a readily adopted symbol of the uncanny and the undiscovered. Its image, often a composite of descriptions and artistic interpretations, has become almost as recognizable as its name.

However, this cultural ubiquity also presents a significant challenge for scientific inquiry. The deep-seated cultural belief in the Yeti, coupled with its widespread presence in popular media, can create an environment where wishful thinking and confirmation bias can influence the interpretation of evidence. For scientists seeking objective data, it can be difficult to disentangle genuine observations from ingrained beliefs or manufactured narratives. The very passion and enthusiasm that the Yeti inspires can, ironically, become an obstacle to dispassionate scientific assessment.

The enduring enigma of the Yeti also prompts deeper reflection on the nature of scientific exploration, particularly in extreme and remote environments. The Himalayas represent one of the most challenging frontiers for biological research on Earth. The sheer scale of the region, its extreme climatic conditions, the limited accessibility of vast areas, and the logistical difficulties associated with conducting sustained fieldwork all contribute to significant gaps in our knowledge. Even with modern technology, thoroughly surveying such an immense and rugged landscape for elusive fauna is an undertaking of immense difficulty and cost. The Yeti, therefore, might not just be a product of misidentification or folklore; it could also be a consequence of the sheer limitations of our ability to comprehensively study these vast, wild places.

The lack of definitive evidence, while frustrating for cryptozoologists, is also, for many scientists, the most compelling argument against the Yeti's existence as a distinct biological entity. For a large mammal to persist, especially a primate species that would likely be social or at least territorial, a viable breeding population is required. Such a population would necessitate a significant geographic range and, over time, would leave behind a consistent and discernible ecological footprint. This would include more frequent and unambiguous physical evidence such as skeletal remains, more readily identifiable droppings for genetic analysis, and a higher probability of direct sightings by trained observers. The absence of such robust evidence, despite decades of focused searching and a general increase in human activity in the Himalayas, weighs heavily against the hypothesis of an unknown primate species.

In essence, the Yeti stands at the intersection of myth and potential reality, a creature that embodies the untamed spirit of the Himalayas and the persistent human quest for discovery. Its existence as a cultural icon is undeniable, deeply embedded in local traditions and enthusiastically embraced by global popular culture. As a scientific enigma, however, it continues to elude definitive proof, serving as a potent reminder of the vastness of our planet and the enduring mysteries that lie within its wildest, most inaccessible corners. The Yeti phenomenon thus offers a unique case study, not just in cryptozoology, but in the complex interplay between culture, belief, and the scientific endeavor to understand the natural world. It highlights how our perception of nature is shaped not only by empirical observation but also by our deepest desires, our most cherished stories, and the profound sense of awe that the most formidable landscapes on Earth can inspire. The continued fascination with the Yeti underscores a fundamental human truth: we are drawn to the unknown, and the possibility of undiscovered wonders fuels our curiosity and shapes our engagement with the world around us.

10

Other Global Megafauna Primate Reports

The Almas, a term whispered in the remote, rugged landscapes of Central Asia, conjures images of a creature remarkably similar to the Yeti of the Himalayas, yet distinct in its geographical origin and the specific cultural tapestries that enfold its legend. While the Yeti roams the snow-capped peaks of the greater Himalayan range, the Almas is primarily associated with the formidable mountain systems of Mongolia, Siberia, and parts of Kazakhstan – regions characterized by vast steppes, dense taiga forests, and towering, often arid, mountain ranges like the Altai. These are environments that, by their very nature, foster isolation, nurture unique ecosystems, and, consequently, can serve as sanctuaries for elusive life forms, whether they be known or, as the lore suggests, unknown.

The nomenclature itself offers a glimpse into the creature's perceived essence. "Almas" is thought to derive from an ancient Turkic or Mongolian word that translates to "wild man" or "man of the wilderness." This linguistic connection immediately situates the Almas within a global archetype, a recurring theme in human folklore that speaks to the perceived existence of wild, untamed humanoids inhabiting the fringes of civilization. These beings, whether benevolent or terrifying, represent the untamed aspect of nature, a primal force that exists beyond the boundaries of human society and its constructed order. The Almas, in this context, is not merely an animal but a being that occupies a liminal space between the human and the wild, a concept that has resonated with inhabitants of these regions for centuries, if not millennia.

Historical accounts of the Almas are as varied as the landscapes it is said to inhabit. Indigenous peoples, including the nomadic herders of Mongolia and the indigenous groups of Siberia, have long spoken of encounters with such beings. These testimonies, often passed down through oral traditions, describe a creature that is bipedal, powerfully built, and covered in dense, reddish-brown or grey hair. Unlike the more ape-like descriptions often associated with the Yeti, some Almas accounts suggest a more human-like facial structure, albeit rugged and weathered, with deep-set eyes and a pronounced brow ridge. The stature is consistently reported as being tall, often exceeding that of an average human, with broad shoulders and long limbs, indicative of adaptation to traversing difficult, mountainous terrain.

One of the recurring themes in Almas lore is its apparent intelligence and ability to evade human detection. The creatures are described as being incredibly shy and wary, often observed from a distance or glimpsed fleetingly. They are said to possess an acute awareness of their surroundings and a remarkable capacity for camouflage, melting into the shadows of dense forests or the stark, rocky outcrops of the mountainsides. This elusiveness has, of course, contributed significantly to the difficulty in obtaining concrete evidence, reinforcing the creature's legendary status.

Early investigations into Almas reports were often hampered by the vastness of the territory and the rudimentary nature of scientific tools available. However, the persistent nature of these accounts, spanning across different ethnic groups and geographical pockets within Central

Asia, began to attract the attention of more systematic researchers, particularly during the Soviet era. Scientists, often driven by a combination of genuine scientific curiosity and a desire to catalog the full biodiversity of the vast Soviet empire, embarked on expeditions to collect testimonies and search for physical evidence.

One of the most significant periods of investigation into the Almas occurred in the mid-20th century. Soviet ethnographers and cryptozoologists, such as Boris Porshnev, dedicated considerable effort to documenting Almas sightings and collecting oral traditions. Porshnev, a controversial but influential figure in Soviet paleoanthropology, was a fervent believer in the existence of undiscovered hominids, including the Almas. He meticulously gathered accounts from local populations, cross-referencing descriptions and attempting to identify patterns that might suggest a common source or a single species. His work, while often criticized for its subjective interpretations and lack of definitive proof, played a crucial role in bringing the Almas into a more formalized study, moving it beyond the realm of pure folklore and into the nascent field of cryptozoology.

Porshnev and others analyzed numerous individual testimonies, many of which shared striking similarities. For instance, the descriptions of the Almas's footprints often noted their size and shape, suggesting a large, human-like foot with a distinct arch, different from that of bears or other known animals. There were also reports of the creature leaving behind scat, or droppings, which, if analyzed, could potentially offer genetic clues. However, obtaining such samples proved exceedingly difficult, with the reported droppings often being collected by locals who were either unaware of their scientific significance or who were unwilling to part with potentially rare and sacred artifacts.

The ecological context of the Almas's purported habitat is crucial to understanding the plausibility of such a creature's existence. The Altai Mountains, a vast and remote mountain range forming part of the border between Russia, Mongolia, China, and Kazakhstan, are characterized by a rugged topography, with peaks reaching over 4,500 meters. This region boasts a diverse range of ecosystems, from alpine meadows and coniferous forests (taiga) to arid steppes and high-altitude deserts. Such varied and extensive terrain could indeed support a large, elusive primate species, provided there is a sufficient food supply and a breeding population capable of sustaining itself over generations.

The Altai Mountains, in particular, are known for their rich biodiversity, including large mammals such as ibex, argali sheep, wild boar, bears, and various species of deer. The presence of a robust prey base, combined with the dense forest cover and rugged, inaccessible areas, presents a theoretically viable habitat for a large, undiscovered primate. Furthermore, the region's remoteness and limited human population density in many areas would have provided an ecological niche for such a creature to evolve and persist, largely undetected by the modern world. The existence of species like the snow leopard, also known for its elusiveness and mastery of camouflage in challenging environments, demonstrates that large, rare mammals can indeed thrive in these remote mountain ecosystems.

Despite the compelling ecological arguments and the wealth of anecdotal evidence, concrete scientific proof of the Almas has remained elusive. No living specimen has ever been captured, and no undisputed physical remains, such as bones or teeth, have been recovered. The scientific community, by and large, remains skeptical, attributing Almas sightings to misidentifications of known animals (such as bears or primates in zoos, whose enclosures might be described by locals), folklore, or outright fabrication.

However, the persistent nature of the reports, particularly those from individuals with no prior knowledge of the Almas legend, continues to fuel interest. Some researchers point to genetic studies of hair samples or scat that have yielded anomalous results, suggesting the presence of an unknown primate. These findings, however, are often inconclusive, with the DNA potentially belonging to a known but rare subspecies, or being too degraded for definitive identification.

One of the most intriguing aspects of the Almas phenomenon is its potential connection to Neanderthals or other extinct hominin species. Some theories propose that the Almas could be a surviving population of Neanderthals or a closely related hominin that managed to persist in isolated pockets of Central Asia long after their supposed extinction elsewhere. This hypothesis gains some traction from the description of the Almas as being more primitive or ape-like in its features than

Homo sapiens, with a robust build and a pronounced brow ridge. The idea of a relict hominin population is a captivating one, suggesting that our understanding of human evolutionary history might be incomplete.

The Altai region has also been a focal point for investigations into other cryptids, most notably "the Altai Dragon," a creature sometimes described as a large, serpent-like reptile. While seemingly disparate, the existence of multiple local legends about unusual creatures in a region known for its remoteness and unique ecological niches highlights the potential for undiscovered or poorly understood fauna to exist in such environments.

The scientific approach to the Almas, as with other cryptozoological subjects, involves a rigorous application of the scientific method. This requires verifiable evidence that can be independently studied and replicated. For the Almas, this would ideally involve unambiguous photographic or video evidence, clear biological samples (hair, scat, or bone) that yield unique genetic sequences attributable to an unknown primate, or, in the most definitive scenario, the discovery of a living specimen.

The challenges of conducting such research in the remote and vast territories where the Almas is reported are immense. Logistical difficulties, extreme weather conditions, and the sheer scale of the search area make comprehensive surveys incredibly difficult and expensive. Moreover, gaining the trust and cooperation of local communities, who may view the Almas with a mixture of fear, respect, and a reluctance to share their traditions with outsiders, is another significant hurdle.

Despite these challenges, the allure of the Almas persists. It represents a tantalizing possibility – the chance that a large, intelligent hominid could still exist in the wild, unknown to modern science. This possibility speaks to a deep-seated human desire to believe that our planet still holds secrets, that there are still frontiers to explore, and that the wild holds wonders yet to be discovered. The Almas, like its Himalayan cousin the Yeti, embodies this enduring mystery, a testament to the power of legend and the enduring fascination with the unknown inhabitants of our world's wildest places. The ongoing, albeit often sporadic, interest in the Almas underscores the fact that while science demands proof, the human spirit often thrives on the possibility, on the whisper of the wild man in the remote mountains of Central Asia.

The vast continent of Australia, renowned for its unique marsupial fauna and ancient landscapes, also harbors its own enduring tales of elusive, large-bodied hominoid creatures. Among these, the Yowie stands as the most prominent and persistent of Australian cryptid legends, a figure deeply woven into both the indigenous Dreamtime and the modern tapestry of anecdotal sightings. Unlike the snow-dwelling Yeti or the Central Asian Almas, the Yowie is intrinsically linked to the continent's diverse and often temperate wilderness, from the dense, humid rainforests of the east coast to the rugged bushlands and mountain ranges further inland. Its narrative, spanning millennia of Aboriginal oral tradition and contemporary reports, paints a picture of a powerful, intelligent, and remarkably shy being that has eluded definitive scientific discovery.

The origins of the Yowie legend are deeply rooted in the rich and complex belief systems of Aboriginal Australians. For tens of thousands of years, Indigenous peoples have inhabited this land, developing an intricate understanding of its flora, fauna, and the spiritual forces that permeate it. Within these traditions, numerous totemic beings and spirits populate the landscape, and among them are creatures that bear a striking resemblance to the modern descriptions of the Yowie. These ancestral stories often speak of powerful, wild beings that dwell in the remote, untamed corners of the continent, far from human settlements. While direct, consistent nomenclature across all Aboriginal language groups is challenging due to the continent's vast linguistic diversity, the essence of a large, hairy, bipedal entity is a recurring theme. These beings were often described as guardians of sacred sites, possessing immense strength and an intimate knowledge of the land. They were entities to be respected, feared, and, above all, avoided. The Aboriginal accounts provided an early framework for understanding a creature that occupied a liminal space between the human and the animal, a spirit of the wild that predated and existed alongside human presence. The integration of these ancient narratives with more recent eyewitness accounts is crucial to appreciating the Yowie's enduring presence in the Australian psyche.

The modern era has seen a surge in reported Yowie sightings, particularly from the 20th century onwards. These accounts, while varied in their specifics, often share a remarkable consistency in describing the creature's physical characteristics. Eyewitnesses frequently report a tall, powerfully built primate-like being, standing anywhere from six to ten feet in height. Its body is typically described as being covered in thick, matted hair, ranging in color from dark brown and black to reddish-brown or even grey. The proportions are often noted as being robust, with broad shoulders and long arms, suggesting a creature adapted for climbing or traversing dense terrain. The face is usually described as ape-like or primitive, with a receding forehead, a pronounced brow ridge, and deep-set eyes. Some reports also mention a distinct, musky odor emanating from the creature, a detail often associated with large mammals.

Geographical distribution is another key element in the Yowie narrative. While sightings have been reported across various parts of Australia, there is a notable concentration of reports emanating from the eastern coastal regions, particularly New South Wales and Queensland. These areas are characterized by extensive temperate and tropical rainforests, rugged mountain ranges such as the Blue Mountains and the Great Dividing Range, and vast tracts of bushland – environments that provide ample cover, food sources, and seclusion for a large, undiscovered animal. The dense, ancient rainforests, with their towering trees, tangled undergrowth, and challenging terrain, represent an ideal habitat for a creature that could remain hidden from human observation. These ecosystems are rich in biodiversity, supporting a range of mammals, birds, and reptiles, and potentially offering a sufficient food base for a large omnivorous or herbivorous primate.

The ecological context for a creature like the Yowie within Australia is complex but not entirely implausible. Australia's unique evolutionary history, marked by its long isolation as a continent, has led to the development of a specialized fauna. While the continent is famous for its marsupials, it also supports a number of native placental mammals, including bats, rodents, and the dingo. However, the absence of any indigenous great apes or large primates is a significant factor that fuels scientific skepticism. The prevailing scientific consensus is that the Yowie represents a misidentification of known animals, folklore, or deliberate hoaxes. Known animals that could potentially be mistaken for a Yowie include large marsupials like the eastern grey kangaroo when seen at a distance or in unusual circumstances, or even bears if they were introduced to areas, though bears are not native to Australia. However, the consistent descriptions of a bipedal, ape-like creature with specific facial features and gait often distinguish Yowie reports from sightings of kangaroos or other native fauna.

The question of how such a large primate could have remained undiscovered in Australia, particularly given the continent's relatively sparse human population in many areas, is a central challenge for Yowie proponents. One hypothesis suggests that the Yowie might be a relict population of an ancient hominoid species that either evolved independently in Australia or arrived on the continent during periods of lower sea levels when land bridges connected it to Southeast Asia. However, fossil evidence for such large primates in Australia is notably absent, which is a significant hurdle for this theory. Alternatively, some speculate that the Yowie could be a highly adapted, unknown species of ape or hominid that evolved in isolation.

The ecological niche that a Yowie might occupy is another area of consideration. If it is primarily a herbivore or omnivore, it would need access to a plentiful supply of fruits, berries, leaves, and potentially insects or small animals. The rainforests and woodlands of eastern Australia offer a diverse range of such food sources. The creature's arboreal or semi-arboreal nature, as suggested by some descriptions, would allow it to exploit resources within the forest canopy, further contributing to its elusiveness. Its powerful build and agility would also be advantageous for navigating the dense undergrowth and climbing trees to forage or escape danger.

One of the most significant challenges in substantiating Yowie reports is the nature of the evidence. Much of the information available consists of anecdotal eyewitness accounts. While compelling, these accounts are subjective and prone to misinterpretation, especially when encounters are brief, occur in low light conditions, or when the witness is under duress. The lack of definitive physical evidence, such as undisputed footprints, hair samples that can be genetically analyzed to reveal an unknown primate, or skeletal remains, remains a considerable obstacle for scientific acceptance.

However, there have been instances where purported Yowie evidence has been presented. Footprint casts are among the most common types of evidence. These impressions are often described as being large, human-like in shape, but with distinct differences, such as a greater toe spread or deeper heel impression. While some casts are dismissed as misinterpretations of bear or kangaroo tracks, or even deliberate hoaxes, others possess characteristics that proponents argue are consistent with an unknown bipedal primate. Hair samples have also been collected from various locations associated with sightings. Analysis of these samples has sometimes yielded results that are inconclusive, identifying them as belonging to known Australian marsupials or simply as unknown animal hair. In a few rare cases, genetic analysis has suggested the presence of DNA that does not match any known species, though these results are often questioned due to potential contamination or the degradation of the samples.

The scientific community, by and large, remains highly skeptical of the Yowie's existence. The lack of conclusive physical evidence, combined with the ecological improbability of a large, undiscovered primate species existing in Australia without leaving a more substantial fossil or genetic record, are major points of contention. The prevailing view is that Yowie sightings are more likely attributable to misidentifications of known animals, particularly large marsupials, bears in zoos that may have been described to locals, or the psychological effects of encountering unfamiliar wildlife in remote environments. The power of suggestion and the influence of folklore also play a significant role in perpetuating such legends.

Despite the scientific skepticism, the Yowie phenomenon continues to capture the public imagination and inspire ongoing investigations. A dedicated community of researchers and enthusiasts actively documents sightings, collects evidence, and conducts expeditions into the areas where Yowie reports are most frequent. These efforts, while often operating outside the mainstream scientific establishment, contribute to the ongoing discourse surrounding the possibility of undiscovered megafauna. They highlight the persistent allure of the unknown and the desire to believe that our planet, even in its well-explored regions, might still harbor hidden wonders.

The ecological context of Yowie sightings, particularly within the rainforests of Queensland and New South Wales, is critical. These regions are characterized by a high degree of endemism, meaning many species are found nowhere else on Earth. They are also home to a number of large animals, including kangaroos, wallabies, koalas, and dingoes, as well as large reptiles like crocodiles. The dense vegetation and rugged terrain provide excellent camouflage, and the sheer size and remoteness of some of these areas mean that a reclusive animal could potentially exist for extended periods without being observed. The Blue Mountains, for instance, a vast wilderness area near Sydney, has been a hotspot for Yowie sightings for decades, with numerous reports detailing encounters with tall, hairy bipeds in its dense eucalyptus forests and deep gullies. The Dandenong Ranges in Victoria, and pockets of rainforest in Tasmania, have also seen their share of Yowie lore.

The psychological and cultural aspects of the Yowie legend are also significant. The creature taps into primal fears of the unknown and the wilderness, representing a wild, untamed force that exists beyond the boundaries of human control. For Indigenous Australians, the Yowie, in its various ancestral forms, was often a part of the natural and spiritual order, a creature to be respected and understood within the framework of their worldview. For contemporary Australians, the Yowie has become a kind of national cryptid, a mysterious inhabitant of their own backyard, embodying the wild, untamed spirit of the Australian continent. The narratives surrounding the Yowie, whether rooted in ancient lore or modern anecdotes, serve as a reminder that even in a technologically advanced world, there are still frontiers of mystery that beckon human curiosity and a lingering possibility that the wild still holds secrets yet to be unveiled. The persistence of the Yowie legend, across millennia and diverse cultural contexts, underscores the enduring human fascination with the unexplained and the potent allure of the great unknown lurking within our planet's wildest places.

The humid, subtropical wetlands and dense forests of the southeastern United States present a distinct ecological theater for potential undiscovered megafauna, and within this unique environment, tales of the Skunk Ape have long held sway. Unlike the snow-capped peaks of the Himalayas or the vast Australian bush, the Skunk Ape is intrinsically tied to the water-logged cypress swamps, dense pine forests, and tangled palmetto scrub that characterize states such as Florida, Alabama, Georgia, and South Carolina. These are lands where human encroachment often meets untamed wilderness, creating a rich tapestry of anecdotal encounters and lingering mysteries.

The habitat itself is a crucial element in understanding the Skunk Ape phenomenon. The Everglades in Florida, with its labyrinthine network of sloughs, sawgrass marshes, and mangrove forests, is perhaps the most iconic locale associated with these sightings. This vast, intricate ecosystem is a haven for a rich diversity of wildlife, including alligators, panthers, manatees, and numerous bird species. The dense vegetation, often waterlogged, provides excellent cover, and the sheer scale of the area makes it plausible, in theory, for a large, elusive creature to evade detection. Similarly, the Okefenokee Swamp, straddling the Georgia-Florida border, is another prime candidate for such an inhabitant. Its dense cypress stands, cloaked in Spanish moss, its dark, tannin-stained waters, and its remote, marshy interior create an atmosphere of primeval wilderness, a perfect sanctuary for a creature that prefers to remain unseen. Beyond these prominent wetlands, reports also emanate from the pine-heavy forests and coastal regions of states like Alabama, Mississippi, and South Carolina, suggesting a distribution that is not solely confined to the deepest swamps but extends into adjacent forested areas.

The ecological requirements for a large, primate-like creature in this region would be complex. A primate of substantial size would necessitate a diet rich in calories and nutrients. The southeastern wetlands and forests offer a varied larder: fruits from various trees and bushes, nuts, berries, roots, insects, and potentially small vertebrates such as fish, frogs, turtles, and small mammals. The abundance of water would also be a significant advantage, providing a constant source of hydration and supporting a diverse aquatic food web. The dense canopy of the cypress and pine forests would offer opportunities for arboreal foraging and shelter, while the undergrowth and swampy terrain could provide camouflage and refuge. The ability to navigate through shallow water, dense vegetation, and potentially climb trees would be essential adaptations for survival in this environment. The characteristic musky odor, often associated with the Skunk Ape, is a common trait among many large mammals, particularly those that rely on scent marking for communication or territorial defense, or as a defense mechanism.

The anecdotal evidence for the Skunk Ape is as varied as the landscape it is said to inhabit. Reports often describe a creature standing between five and eight feet tall, covered in dark, reddish-brown or black hair, with broad shoulders and a powerful build. Unlike more ape-like descriptions, some accounts suggest a more hominoid appearance, with a distinct head, sloping shoulders, and a stooped posture. The "skunk" moniker derives from the foul odor that witnesses frequently report, often described as a potent combination of decaying matter, ammonia, and the pungent smell of a skunk. These encounters are typically fleeting, occurring at dusk or dawn, often in remote, wooded areas or along the fringes of swamps.

One of the most frequently cited categories of evidence involves footprint casts. These impressions, often found in muddy areas or along the soft earth of forest trails, are typically described as being large, often exceeding 12 inches in length, and resembling a human footprint but with distinct differences. Some casts exhibit a more splayed toe arrangement, a deeper heel impression, or a distinct arch, leading proponents to argue that they are inconsistent with known animal tracks or human footwear. While many such casts are readily explained as misidentifications of bear, alligator, or even distorted human footprints, a persistent subset of these casts possesses characteristics that defy easy classification. The Florida panther, for instance, is a powerful, elusive predator of these regions, and its tracks are well-documented; however, Skunk Ape footprints are generally described as being significantly larger and possessing a more primate-like morphology.

Hair samples are another common, though often inconclusive, piece of evidence. In areas where sightings have been reported, unusual strands of hair are sometimes discovered snagged on branches or found near purported tracks. These samples, when subjected to analysis, have frequently been identified as belonging to known animals such as deer, raccoons, or even domestic animals that may have wandered into the area. However, a small number of hair samples have defied conventional identification, showing characteristics that do not match any known North American fauna. These results, while intriguing, are often complicated by issues of contamination, sample degradation over time, and the difficulty of definitively ruling out extremely rare or regionally endemic variations of known species.

Eyewitness accounts form the bedrock of the Skunk Ape narrative. These are often dramatic and unsettling, with individuals reporting fleeting glimpses of the creature moving through dense undergrowth, crossing roads, or observing them from the edge of the woods. Some reports detail encounters where the creature seemed to exhibit a degree of intelligence or curiosity, while others describe it as fearful and intent on escaping notice. The emotional impact of these encounters is often profound, with witnesses describing feelings of terror, awe, and disbelief. The consistency in the descriptions of its physical appearance, gait, and the characteristic odor across numerous independent reports, spanning decades and different geographical locations within the southeastern United States, lends a certain weight to the collective testimony.

The ecological plausibility of a large, undiscovered primate in the southeastern United States is a subject of considerable debate among scientists. The primary argument against the existence of such a creature centers on the lack of definitive fossil evidence and the relative scarcity of its reported habitat compared to more widespread primate species. Fossil hominoids are not native to North America, and the evolutionary history of the continent does not suggest the development of large, apelike primates. The prevailing scientific consensus often attributes Skunk Ape sightings to misidentifications of known animals, most commonly black bears, which are abundant in the region and can, at a distance or under unusual circumstances, appear bipedal. The black bear's ability to stand on its hind legs, its dark fur, and its size can lead to confusion, particularly in low light conditions or when the observer is startled. Other potential misidentifications include large raccoons, wild hogs, or even unusual behavior by humans in remote areas.

However, proponents of the Skunk Ape's existence point to several factors that complicate these simple explanations. They argue that the consistent descriptions of a creature with distinctly primate-like features, such as a receding forehead, a pronounced brow ridge, and a specific gait, often differ significantly from bear sightings. The musky odor is also a detail frequently mentioned in Skunk Ape reports but less commonly associated with bears in a manner that would cause such alarm. Furthermore, the specific ecological niches occupied by Skunk Apes in some reports—deep swamps, areas with specific plant life—are not always the preferred habitat of black bears, which tend to favor more forested areas.

The question of how a creature of this size could remain undetected in a region that, while vast in parts, is also heavily populated and frequently traversed by humans, is another significant challenge. The southeastern United States, despite its extensive wilderness areas, has a long history of human habitation and exploration. The argument is often made that if a population of large primates existed, there would be a greater likelihood of finding more concrete evidence, such as skeletal remains, more frequent and unambiguous tracks, or even documented predation on livestock or humans.

Despite the scientific skepticism, the Skunk Ape phenomenon persists, fueled by a dedicated community of cryptozoologists and enthusiasts who continue to document sightings, collect evidence, and explore the wild regions where the creature is said to roam. The allure of the Skunk Ape lies in its embodiment of the untamed wilderness that still exists within the populated southeastern landscape. It represents a primal mystery, a reminder that even in well-traveled territories, there may still be corners where the unknown lurks, a testament to the enduring fascination with what might be hidden just beyond the edge of our current understanding. The dense, often difficult-to-traverse terrain of the swamps and forests, coupled with the creature's purported nocturnal habits and keen sense of smell, provides a plausible ecological scenario for a creature that has managed to elude definitive scientific discovery. The very environment that is its supposed home is also its greatest ally in remaining hidden.

The global tapestry of cryptozoological reports, when examined through the lens of primate-like creatures, reveals a fascinating, albeit often speculative, pattern of distribution and recurring characteristics. While many of these accounts remain firmly in the realm of folklore and misidentification, a persistent undercurrent of shared descriptions across vastly different geographical regions suggests an intriguing, if unproven, phenomenon. This subsection aims to explore these global patterns, drawing connections between disparate reports and seeking common threads that might illuminate our understanding of these elusive beings, moving beyond the localized narratives of creatures like the Skunk Ape.

Across the globe, tales of large, unknown primates emerge from environments as varied as the dense, mist-shrouded rainforests of Asia and South America to the rugged, less explored mountainous regions of North America and even parts of Africa. A common denominator in many of these sightings is the description of a creature that falls somewhere between a known ape and a human, often possessing immense physical strength and a bipedal or semi-bipedal gait. The "Yeti" of the Himalayas, the "Bigfoot" or "Sasquatch" of North American forests, and the "Orang Pendek" of Sumatra, though differing in specific details, share a fundamental resemblance in their purported morphology and elusive nature. These creatures are frequently depicted as being covered in dark hair—ranging from reddish-brown to black or grayish—and exhibiting a robust build, suggesting adaptations for a life that demands physical prowess in challenging terrains.

The habitat preferences of these alleged cryptid primates are remarkably consistent with ecological niches that could potentially support large, undiscovered mammals. Reports overwhelmingly place them in remote, densely vegetated areas, far from significant human populations. This includes vast tracts of primary forest, rugged mountain ranges, and deep, unexplored valleys. The logic here is ecologically sound: such environments offer ample food resources, cover for concealment, and the isolation necessary for a breeding population to remain undetected. For instance, the Amazon rainforest, a region teeming with biodiversity and vast areas of unexplored territory, has been the source of numerous reports of ape-like beings, sometimes described as having more human-like proportions than true apes. Similarly, the ancient forests of the Pacific Northwest of North America, with their towering trees and deep ravines, provide an ideal environment for a large, arboreal or semi-arboreal creature to thrive. The appeal of these environments lies in their ability to mask the presence of even a substantial animal, offering both sustenance and sanctuary.

One of the most striking commonalities across these reports is the recurring mention of a distinctive odor, often described as foul, musky, or pungent. While the Skunk Ape's name directly references this characteristic, similar olfactory descriptions are associated with Bigfoot and other alleged hairy hominoids. This persistent detail is significant because it aligns with biological realities. Many large mammals use scent marking for territorial communication, defense, or to deter predators. The presence of a strong, unusual odor would be a logical adaptation or byproduct for a creature that relies on stealth and evasion, potentially using scent to warn off intruders or as a byproduct of its metabolic processes. The consistency of this detail, even among individuals who have had no prior knowledge of other reports, lends it a degree of credibility within the anecdotal evidence.

The descriptions of locomotion also reveal a shared pattern. While often reported as walking bipedally, a significant number of sightings suggest a knuckle-walking or at least a posture that deviates from typical human gait. This hints at a primate anatomy that might be a mosaic of ape-like and hominid-like features, or perhaps adaptations for a specific mode of movement suited to their environment. The power and speed attributed to these creatures, often moving with surprising agility through dense undergrowth or across uneven terrain, further supports the idea of a creature well-adapted to its wild habitat.

Furthermore, the geographical distribution itself suggests an interesting biogeographical puzzle. Reports of large, unknown primates are not concentrated in a single evolutionary hotspot but appear independently in areas with plausible ecological conditions for large mammals. The divergence in environments—from the humid tropics of Southeast Asia to the temperate rainforests of North America and the high altitudes of the Himalayas—implies that if these creatures are real, they may represent a case of convergent evolution or, more speculatively, surviving relic populations of ancient hominoids. The presence of fossil hominoids in Asia, for example, adds a layer of paleoanthropological intrigue to the reports of creatures like the Gigantopithecus, a massive ape that once roamed parts of Asia. While Gigantopithecus is widely believed to be extinct, some researchers propose that smaller, more agile descendants could have survived in remote areas.

The evidence, though largely anecdotal, often includes traces such as footprints, hair samples, and broken branches or disturbed vegetation. The "Bigfoot" footprint, in particular, is an iconic image within cryptozoology—a large, five-toed or sometimes splayed-toed impression that proponents argue cannot be attributed to known animals or human footwear. While scientific consensus often attributes these to misidentification, erosion, or deliberate hoaxes, the sheer volume and geographical spread of such reports are difficult to dismiss entirely. Similarly, hair samples recovered from various locations have, in some instances, defied conventional identification, presenting unique morphological characteristics or DNA sequences that do not align with known species. These findings, while often inconclusive due to contamination or methodological issues, fuel the ongoing investigation.

The concept of isolated populations, perhaps remnants of an ancient lineage that once had a wider distribution, is a recurring theme in discussions of cryptid primates. In a world where large predators and prey have often been driven to extinction or confined to remote reserves, the idea of a highly intelligent, elusive primate surviving in the vast, undeveloped wilderness areas is, for some, a plausible scenario. The Amazon, Borneo, and the remote wildernesses of North America and Asia represent some of the last great frontiers for undiscovered large fauna.

The ecological plausibility of such creatures also hinges on diet. Large primates require significant caloric intake. The fruits, nuts, roots, and potential for scavenging or hunting small prey available in these diverse habitats could, in theory, sustain a population of large, unknown primates. The specific dietary adaptations would, of course, vary with the environment. A tropical rainforest primate might subsist on a wide array of fruits and insects, while a creature in a more temperate or mountainous region might rely more on roots, berries, and opportunistic predation.

One fascinating aspect is the potential for different "species" or variations of these unknown primates to exist in different parts of the world, yet still share common traits. This could be explained by a shared ancestral lineage that diversified as populations became isolated, or by convergent evolution driven by similar environmental pressures. For instance, the differences between the robust, mountain-dwelling Yeti and the forest-dwelling Bigfoot might reflect adaptations to distinct ecological niches, while their underlying primate characteristics remain. The Orang Pendek, described as a smaller, more bipedal creature from the Sumatran rainforest, further broadens this spectrum, suggesting a range of primate forms adapted to different forest strata and microhabitats.

The challenges in studying these creatures are immense. They inhabit remote, often dangerous terrains, possess acute senses of awareness, and are presumably intelligent enough to avoid human detection. The lack of definitive physical evidence—a body, clear photographic or video proof, or a verifiable DNA sample—remains the primary hurdle for scientific acceptance. However, the persistence of these reports, spanning continents and centuries, continues to inspire a dedicated community of researchers and enthusiasts who believe that somewhere in the world's remaining wilderness, a truly remarkable primate discovery awaits. The recurring themes in their descriptions—the bipedal gait, the hairy covering, the elusive nature, and the distinctive odor—weave a compelling, albeit unproven, narrative that transcends geographical boundaries, hinting at a global phenomenon that continues to elude our scientific grasp.

The enduring mystery of these global cryptid primate reports compels us to consider the vastness of our planet's unexplored regions and the limitations of our current biological knowledge. While skepticism is warranted and the scientific method demands rigorous proof, the sheer volume and consistency of anecdotal evidence from disparate corners of the world present a puzzle that cryptozoologists and some ecologists find too intriguing to dismiss outright. The potential implications of their existence—ranging from the discovery of a living hominoid lineage to a better understanding of primate evolution and biogeography—remain a tantalizing prospect, fueling continued exploration and debate.

The very notion of a large, unknown primate species existing in remote corners of the globe presents a profound challenge to our established taxonomic systems. Current classifications, meticulously built over centuries of scientific inquiry and fossil evidence, delineate a clear evolutionary path for primates. The discovery of a new megafauna primate would necessitate a significant re-evaluation of this framework, forcing us to question the completeness of our understanding of primate evolution and distribution. It would imply that our current models, which largely rely on fossil records and genetic analysis of known species, have overlooked a substantial branch of primate diversity, potentially a relic lineage that has persisted in isolation. This would not merely be the addition of a new entry in a zoological catalog; it would signify a fundamental shift in our perspective on primate history and the biogeographical forces that have shaped it.

Consider the implications for the classification of hominoids. Our understanding of the hominoid family tree, which includes humans, great apes (gorillas, orangutans, chimpanzees, and bonobos), and their extinct ancestors, is based on a robust body of evidence. The discovery of a creature that exhibits traits suggestive of both ape-like and human-like characteristics, yet does not fit neatly into any existing genus or species, would create a taxonomic conundrum. Would it be placed within the Hominidae family, prompting a debate about its precise evolutionary relationship to other hominids? Or would it necessitate the creation of an entirely new taxonomic rank, perhaps a new subfamily or even superfamily, to accommodate its unique evolutionary trajectory? The possibility of such a creature being a living descendant of a group previously thought to be extinct, like

Gigantopithecus, further complicates matters. If confirmed, this would not only redefine our understanding of primate evolution but also suggest that our paleontological record, vast as it is, is far from complete. The criteria for species delimitation, often based on reproductive isolation and morphological distinctness, would be tested to their limits. If these unknown primates were indeed a distinct species, what would be their closest living relatives, and how would their genetic divergence be explained? These are not merely academic questions; they delve into the very core of how we define life and organize our knowledge of it.

The gaps in our knowledge of primate evolution and distribution are significant. The fossil record, by its very nature, is fragmented. Many regions of the world with potential primate evolutionary history remain unexplored or inaccessible due to challenging terrain, political instability, or the sheer expense of paleontological expeditions. This leaves considerable room for surprise. The idea of a surviving megafauna primate species challenges the prevailing assumption that all large primate lineages have been discovered and documented. It suggests that the "empty" spaces on our evolutionary maps might not be so empty after all, and that pockets of evolutionary innovation or stasis could exist in areas we have historically overlooked. The distribution of known primate fossils, particularly those of large-bodied species, often hints at broader historical ranges than currently observed. If these unknown primates are real, their existence would imply that these historical distributions may have persisted far longer and in more geographically dispersed areas than previously theorized.

The biogeographical implications are equally profound. The current distribution of primates is well-documented, largely confined to tropical and subtropical regions of Africa, Asia, and the Americas. The emergence of a large primate in, for instance, the temperate forests of North America or the high altitudes of the Himalayas would defy our current biogeographical models. Such a discovery would force us to reconsider the environmental parameters that we believe are necessary for primate survival and diversification. Were our assumptions about their ecological requirements too narrow? Could large primates adapt to a wider range of climatic conditions and dietary resources than we currently acknowledge? The presence of cryptid primate reports from such diverse environments suggests a remarkable adaptability, perhaps indicating an evolutionary history that involved adaptation to a wider array of ecological niches than those occupied by their known relatives. This could imply a much older and more complex evolutionary narrative for primates than is currently understood, with lineages branching off and adapting to distinct environmental pressures over vast stretches of time.

Furthermore, the potential existence of such creatures challenges our understanding of extinction events. Mass extinction events, such as the one that wiped out the non-avian dinosaurs, have profoundly shaped the course of life on Earth, often opening up ecological niches for surviving groups to diversify. If a large primate lineage has managed to survive for millions of years, perhaps through periods of significant environmental change, it speaks to an exceptional evolutionary resilience. It would imply that certain ecological refugia, areas that remained relatively stable through periods of upheaval, may have harbored these elusive species. Identifying such refugia would become a critical focus for future research, both in understanding past biodiversity and in guiding conservation efforts for potentially undiscovered species. The history of life is replete with examples of organisms surviving against all odds, and a large primate could be another testament to this enduring capacity for survival.

The very definition of "megafauna" in the context of primates would also be recalibrated. While we readily classify elephants, rhinoceroses, and large bovids as megafauna, applying this term to primates typically conjures images of extinct giants like

Gigantopithecus. If a living species of large, ape-like creature is confirmed, it would necessitate a redefinition of what constitutes megafauna within the primate order. This would then open the door to reconsidering other, less substantiated reports of large, unknown mammals across various taxa. It would signal that our current inventory of the planet's largest creatures is potentially incomplete. The impact on evolutionary biology would be immense. It could provide crucial insights into the evolutionary pressures that drive gigantism in primates, the factors that contribute to their long-term survival, and the ecological roles they might play in their respective ecosystems. For example, understanding the dietary habits and reproductive strategies of a newly discovered megafauna primate could shed light on the dynamics of ancient ecosystems where such creatures were potentially more common.

The implications extend to the very process of scientific discovery. The history of zoology is filled with examples of species that were once considered mythical or legendary, only to be scientifically described years later. The okapi, for instance, was known to indigenous peoples of the Congo for centuries before it was formally documented by Western science in the early 20th century. The discovery of a large, unknown primate would undoubtedly follow a similar pattern, initially relying on anecdotal evidence and local accounts before rigorous scientific investigation could yield definitive proof. This process itself highlights the limitations of relying solely on traditional scientific methodologies, which can sometimes be slow to acknowledge or incorporate indigenous knowledge and eyewitness testimony. A new megafauna primate would underscore the importance of bridging the gap between folklore, local expertise, and empirical scientific research. It would necessitate a more open-minded and interdisciplinary approach to biological exploration, one that is willing to consider unconventional evidence and pursue leads that might otherwise be dismissed. The scientific community would need to develop new protocols for investigating and validating such extraordinary claims, ensuring that genuine discoveries are not overlooked due to premature skepticism or adherence to rigid methodological paradigms.

Moreover, the genetic and physiological characteristics of such a creature would be of paramount interest. How would its genetic makeup compare to known primate species? Would it represent a distinct evolutionary lineage with a deep divergence, or would it show closer affinities to existing ape or hominid groups, suggesting a more recent evolutionary split or perhaps even interbreeding events in the distant past (though the latter is highly speculative)? Its physiology, adapted to its specific environment, could reveal novel biological mechanisms for locomotion, thermoregulation, or nutrient extraction. For instance, a primate living in a cold, mountainous region would likely possess different physiological adaptations than one inhabiting a humid tropical forest, and studying these differences would enrich our understanding of primate adaptability. The metabolic rates, dietary specializations, and even the sensory capabilities of such a creature could offer groundbreaking insights into the diversity of primate life.

The question of why such a large creature has remained hidden also demands consideration. The prevailing scientific consensus is that all extant primate species have been identified. The persistence of reports of unknown large primates therefore suggests either a profound level of stealth and intelligence on the part of these creatures, or that our methods of surveying biodiversity in remote areas are insufficient. These animals would likely occupy niches that are difficult for humans to access or survey regularly. Their behavior would have to be exceptionally secretive, perhaps nocturnal, and highly adept at avoiding detection. Understanding the ecological factors that enable such secrecy would be as important as confirming their existence. This could involve studying their preferred habitats, their social structures, and their predator-prey relationships, all of which would contribute to a more comprehensive understanding of the ecological dynamics of the regions they inhabit.

The challenge of studying them would be immense, requiring innovative techniques in tracking, observation, and sample collection, likely employing advanced technologies such as remote sensing, camera traps, and non-invasive genetic sampling.

In conclusion, the hypothetical discovery of a large, unknown primate species would represent a paradigm shift in taxonomy and evolutionary biology. It would highlight the vastness of what we do not know about primate evolution and distribution, challenge our current understanding of biogeography and extinction, and necessitate a re-evaluation of the criteria used to define species and megafauna. Such a discovery would underscore the importance of interdisciplinary research, the integration of local knowledge with scientific inquiry, and the continuous refinement of our methods for exploring and understanding the natural world. It would be a testament to the enduring mysteries of life on Earth and a powerful reminder that our scientific knowledge, while advancing, still has significant frontiers to conquer. The very possibility fuels the imagination and underscores the vital importance of continued exploration and conservation of the planet's remaining wild places.

11

Evidence Beyond Anecdotes: Track and Trace

The meticulous study of animal tracks, a discipline known as ichnology, offers a powerful lens through which to peer into the lives of creatures that might otherwise remain unseen. While often associated with paleontology and the study of dinosaur footprints, the principles of forensic ichnology are equally applicable to contemporary wildlife biology and, crucially, to the investigation of potential undiscovered primate species. When confronted with evidence of an unknown creature, particularly in the form of its tracks, a systematic analysis can yield a wealth of information, transforming a fleeting impression in mud or snow into a detailed portrait of the animal that left it behind.

The fundamental premise of footprint analysis is that every step taken by an animal leaves behind a unique signature. This signature is not merely an imprint of the foot itself but a complex interplay of biomechanics, substrate characteristics, and environmental conditions. A skilled ichnologist can decipher this signature by examining a range of factors. The first and most apparent is the morphology of the track – its shape, size, and the presence of distinct features such as toe pads, claws, or heel impressions. For primates, this is of particular interest. The five-fingered hand and foot structure, the presence or absence of an opposable thumb or hallux, and the relative lengths of the digits are all critical morphological markers that can help distinguish one primate species from another. Even subtle differences in the webbing between toes or the robustness of the metatarsals can provide clues. In the context of an unknown primate, detailed measurements of these features, compared against the known diversity of primate foot morphology, could begin to narrow down the possibilities or, more intriguingly, suggest a form entirely outside our current classifications.

Beyond the static imprint, the dynamic aspects of locomotion are revealed through the arrangement of multiple tracks, forming a trackway. Gait analysis is a cornerstone of ichnology. The pattern in which tracks are laid down – whether they are in a straight line, a curve, or a more complex pattern indicating a change in direction or speed – provides insights into how the animal moves. For instance, a consistent diagonal gait, where the hind foot lands near the forefoot on the opposite side, is typical of many quadrupedal mammals. However, primates, with their capacity for bipedalism, brachiation, and a variety of arboreal and terrestrial gaits, present a more complex scenario. The presence of alternating hind limb prints, or a more plantigrade (flat-footed) gait pattern, as opposed to digitigrade (toe-walking), could indicate a primate. The spacing between consecutive prints, known as the stride length, offers information about the animal's height and leg length. A longer stride typically suggests a larger animal with longer limbs. Conversely, a shorter stride might indicate a smaller individual or a slower gait.

The depth of a footprint is another crucial piece of data. The substrate – be it soft mud, sand, snow, or even damp earth – plays a significant role in how deeply an animal's foot sinks. By understanding the properties of the substrate (its moisture content, compaction, and grain size), an experienced tracker can estimate the pressure exerted by the animal's foot, and by extension, its weight. A heavier animal will leave a deeper impression than a lighter one of similar foot size. This is particularly relevant when considering the possibility of a large, undiscovered primate. If footprints are found to be exceptionally deep in a substrate that is only moderately yielding, it strongly suggests a significant body mass, potentially far exceeding that of any known primate species. Forensic techniques involving casting and cross-sectioning of footprints can provide even more precise data on depth profiles and the distribution of pressure across the foot.

Furthermore, the impression of the entire foot, or just portions of it, can reveal details about the animal's posture and locomotion. For example, the presence of distinct toe impressions alongside a heel mark suggests a plantigrade gait, common in bears and apes. If only toe marks are evident, it points towards a digitigrade or semi-digitigrade posture. The angle at which the foot is placed relative to the direction of travel, known as the angle of turnout or toe-out, can also be informative. While many animals exhibit a degree of toe-out for stability, significant variations can hint at specific adaptations or anatomical features. In the case of a potential new primate, a lack of significant toe-out, or even a slight toe-in, might be observed, especially if the creature exhibits a more upright posture or a gait that is less adapted to uneven terrain.

The analysis extends to the negative evidence within the trackway. The absence of claw marks, for instance, is a significant indicator. Most canids and felids possess claws that are often imprinted in their tracks, though they can be retracted in some species. Many primates, however, lack significant claws, possessing nails instead. The presence of impressions of nails, or the distinct lack of claw marks, can therefore serve as a crucial diagnostic feature. Similarly, the pattern of wear on the foot can provide information about the animal's age and activity. Older animals may show signs of arthritis or injuries that alter their gait and the appearance of their tracks.

When investigating reports of undiscovered primates, the process begins with careful documentation. This involves not just measuring and photographing tracks, but also noting the surrounding environment, the substrate, and the context in which the tracks were found. Ideally, multiple tracks forming a trackway will be present, allowing for gait analysis. Casting of the tracks, using materials like plaster or dental stone, is essential for preserving the three-dimensional details of the impression and allowing for more detailed laboratory analysis. These casts can be studied for subtle anatomical features, such as the presence of an arch in the foot, the spacing between the toes, and the shape of the heel pad.

The comparative aspect of ichnology is vital. Once data is collected from an unknown track, it is compared against a database of known animal tracks. For primate investigations, this database would include the tracks of all known primate species, from the smallest prosimians to the great apes. However, the challenge with an unknown primate is that it might not fit neatly into existing categories. This is where the expertise of primatologists and paleoanthropologists becomes invaluable. They can help interpret unusual features and consider evolutionary possibilities. For example, if tracks exhibit a clear five-toed structure with a hallux (big toe) that appears capable of grasping, this strongly suggests a primate. The size of the impression would then indicate whether it belongs to a known species or potentially a larger, undiscovered one.

Consider the hypothetical discovery of footprints attributed to

Gigantopithecus, a prehistoric ape whose fossilized jawbones and teeth suggest an animal of immense size, perhaps up to 3 meters tall and weighing as much as 500 kg. If such a creature were to exist today, its footprints would likely be monumental, significantly larger than those of any living ape. The stride length would be correspondingly enormous. The depth of the impression in a substrate like wet soil would be considerable, reflecting its massive weight. Furthermore, its gait would be a subject of intense debate. While often depicted as quadrupedal, some researchers have speculated that such a large ape might have adopted a bipedal or knuckle-walking posture for at least some of its locomotion. Footprints could provide critical evidence for these hypotheses. For instance, if the tracks consistently show impressions of a large, flat foot with a pronounced heel, and perhaps evidence of knuckle-striking in front of the foot impression, it could support a more human-like or knuckle-walking gait rather than a purely quadrupedal one. The presence of a grasping hallux might be less pronounced if the animal was primarily terrestrial and less arboreal than its ape relatives.

The scientific rigor applied to footprint analysis is comparable to that used in forensic science. Factors that can lead to misidentification are carefully considered. These include:

Substrate deformation: Not all impressions are true tracks. Rocks can shift, roots can exert pressure, and erosion can create misleading shapes. A skilled tracker distinguishes between genuine biological impressions and natural geological or botanical phenomena.

Melanism and Albinism: While these conditions affect coat color, they can indirectly influence track morphology if associated with skeletal abnormalities, though this is rare.

Injury and Disease: Animals with injuries, diseases, or deformities can produce abnormal tracks. These must be recognized and ideally distinguished from the typical track of the species.

Environmental Factors: Weathering, water flow, and subsequent animal activity can obscure or alter tracks, making interpretation challenging. A single, isolated track is far less informative than a clear trackway.

Tread Patterns of Known Animals: In some instances, unusually large or distorted tracks of known animals can be mistaken for something more exotic. For example, the hind footprint of a large bear can sometimes appear surprisingly primate-like due to its shape and the presence of toe pads. However, bears typically have distinct claw marks, and their hind feet are proportionally longer and narrower than their forefeet, leading to a different trackway pattern.

When analyzing tracks of potential unknown primates, the focus shifts to specific primate characteristics. The presence of a distinct arch in the foot, as opposed to a relatively flat foot like that of a canid, is a key feature. The relative lengths of the digits are also critical. In many primates, the second toe (or hallux in the case of the foot) is shorter than the other digits, but still clearly defined and often capable of grasping. The distance between the toe pads and the heel pad, and the overall proportion of the foot, are measured.

The concept of a "primate track index" could be developed, similar to indices used in other areas of forensic science, to quantify certain ratios of foot measurements that are characteristic of different primate groups. For example, the ratio of the length of the hallux to the length of the second toe, or the ratio of the heel pad length to the overall foot length, could be used to differentiate between ape-like tracks and those of other primates or mammals.

In the pursuit of cryptid primates, the challenge is often the scarcity of evidence. Many reports consist of single, isolated footprints that are difficult to interpret definitively. This is where persistent, systematic searching in areas where such sightings are reported becomes crucial.

The goal is to find a trackway that allows for gait analysis, or multiple clear impressions that reveal consistent morphological details. The use of advanced technology, such as motion-activated cameras strategically placed near water sources or known animal trails, can potentially capture images of the animals themselves, or at least provide irrefutable photographic evidence of their tracks.

Furthermore, the study of fossil primate footprints, though rare, can provide comparative data. While the direct study of fossil footprints of extinct ape species like

Gigantopithecus is limited by the scarcity of such evidence, the footprints of other extinct hominoids and hominins offer valuable insights into their locomotion and foot morphology. For instance, the Laetoli footprints, dating back 3.6 million years, provide compelling evidence of bipedal locomotion in early hominins, showing a clear heel strike and a non-divergent hallux, similar to modern human footprints. Comparing potential modern unknown primate tracks to these ancient hominin prints could reveal evolutionary affinities or divergences.

The scientific value of footprint analysis extends beyond mere identification. It can reveal details about behavior, population dynamics, and ecological roles. A series of tracks might show territorial marking, migratory patterns, or interactions between individuals. The presence of juveniles alongside adults in a trackway would suggest a family group or breeding population. The location and distribution of tracks can indicate habitat preferences, foraging areas, and pathways used by the animals. For a large, elusive primate, understanding these aspects through track analysis could provide the first tangible evidence of its ecological niche and its potential impact on the environment.

In conclusion, the science of footprint analysis, or forensic ichnology, offers a robust and objective methodology for investigating claims of unknown primate species. By systematically examining the morphology, gait, depth, and context of tracks, and by applying principles of comparative anatomy and biomechanics, it is possible to glean a remarkable amount of information about the creature that made them. While the interpretation of tracks is not without its challenges, the potential to gather evidence that moves beyond anecdote and into the realm of scientific data makes ichnology an indispensable tool in the quest to understand the full spectrum of primate biodiversity, whether known or yet to be discovered. The imprint left in the earth, however transient, can speak volumes to those who possess the knowledge and patience to listen.

The allure of the unknown, particularly concerning the potential existence of large, undiscovered primates, often manifests most dramatically in the form of anomalous footprints. These impressions, discovered in remote locales and frequently described as being significantly larger and more human-like than those of any known great ape, form a cornerstone of anecdotal evidence for cryptid primates. However, the scientific rigor demanded by cryptozoology, much like in wildlife biology or forensic science, necessitates a critical and systematic examination of such findings, moving beyond mere description to rigorous interpretation. This section delves into the methodology of evaluating these anomalous footprints, dissecting their characteristics against known biological and physical principles, and exploring the common pitfalls and potential for misinterpretation that often accompany them.

When confronted with an alleged footprint of a large, unknown primate, the initial step is a meticulous documentation process. This extends beyond simple photography and measurement; it involves understanding the substrate in which the print is found. Is it soft mud, damp earth, sand, snow, or a combination of materials? The consistency and moisture content of the substrate are paramount, as they dictate the fidelity and depth of the impression. A well-preserved print in a yielding substrate will capture fine details, such as the texture of the skin, the shape of the toe pads, and the presence or absence of a distinct heel. Conversely, a print in dry, hard-packed earth might only provide a vague outline, easily subject to misinterpretation.

The morphology of the print is then scrutinized. For a primate, certain anatomical features are expected. A pentadactyl (five-fingered/toed) structure is a fundamental characteristic. The presence of a divergent hallux (opposable big toe) is a strong indicator of primate ancestry, particularly for arboreal species, although its degree of divergence can vary significantly. The relative lengths of the digits are also critical. In many ape species, the second toe is often shorter than the first or third, but its presence and form are significant. The heel pad, if clearly impressed, provides information about the animal's weight distribution and gait. A broad, well-defined heel pad is typical of apes, suggesting a more plantigrade (flat-footed) stance compared to the digitigrade (toe-walking) stance of many other mammals.

However, anomalous footprints often present features that defy straightforward categorization. For instance, footprints attributed to the legendary Bigfoot or Sasquatch are frequently described as being enormous, oblong, and possessing five toes, with the first toe sometimes depicted as being somewhat abducted or opposable. The stride length accompanying these prints is often reported as being exceptionally long, suggesting a bipedal creature of considerable height. The challenge lies in comparing these descriptions with the known variation within primate foot morphology. While the great apes (gorillas, orangutans, chimpanzees, bonobos) exhibit variations in their foot structure and gait, none of their typical tracks approach the reported dimensions or morphology of these alleged cryptid prints.

One of the most common explanations for anomalous primate-like footprints is misidentification of tracks made by known animals. Large bears, particularly their hind footprints, can sometimes be mistaken for primate tracks, especially if viewed in poor light or a partially obscured substrate. Bears are plantigrade, and their hind feet have a relatively broad heel pad. However, bear tracks typically exhibit distinct claw marks, which are usually absent in primate tracks (though primates may have nail impressions). Furthermore, the arrangement of the toes in bears, while superficially primate-like, often differs in relative size and spacing. The powerful claws of a bear digging into the substrate can also create impressions that might be misinterpreted.

Another frequent candidate for misidentification is the track of a large canid, such as a wolf or a coyote. While canids are digitigrade, their tracks can sometimes appear more rounded in certain substrates. However, the typical elongated shape of a canid track, the prominent claw marks, and the often symmetrical arrangement of the two central toes are usually distinguishing features. It is also important to consider the possibility of tracks made by hooved animals in certain unusual circumstances, though these are less likely to be confused with primate prints.

The biomechanical feasibility of a footprint is also a crucial consideration. The size and shape of a footprint are directly related to the skeletal structure of the foot and the forces applied during locomotion. For a primate of immense size, as often described in cryptid reports, the skeletal structure would need to be proportionally robust to support its weight and withstand the stresses of movement. If a footprint suggests a certain foot structure, for instance, an extremely elongated foot with disproportionately long toes, it raises questions about the functional anatomy of the creature. Could such a foot effectively support the weight of a massive animal? Would it be efficient for locomotion? Biomechanical modeling can help assess these possibilities. For example, if a footprint appears to show an exceptionally high arch combined with a narrow heel, it might be inconsistent with the anatomy of a large, heavy-bodied primate.

The depth of a footprint is another critical indicator, directly correlating with the animal's weight and the substrate's resistance. Anomalous footprints are often described as being exceptionally deep, suggesting a creature of immense mass. However, interpreting depth requires careful consideration of the substrate. A print in very soft, waterlogged mud will naturally be deeper than a print of the same animal in dry, firm soil. Conversely, a relatively shallow impression in soft mud might indicate a lighter animal or one that did not fully place its weight into the substrate. If a footprint is deep in a moderately firm substrate, it strongly suggests significant weight.

The concept of a "primate footprint index" could be a valuable tool in cryptid investigations, similar to indices used in forensic science to classify and compare tracks. This would involve calculating ratios of key measurements, such as the ratio of the length of the hallux to the length of the second toe, the ratio of the heel pad width to the overall foot width, or the ratio of the stride length to the estimated leg length. By comparing these indices from anomalous prints to those derived from known primate species, it might be possible to identify unusual deviations or even potentially novel morphological patterns. For example, an unusually long and slender foot with widely spaced toes might suggest a primate adapted for gripping branches, but if the stride length and depth suggest a terrestrial creature of significant weight, this combination would be biomechanically puzzling.

The phenomenon of "mud packing" or "false tracks" is a significant factor in the misinterpretation of evidence. In certain substrates, particularly soft mud or snow, water movement, falling debris, or even the pressure from a larger animal's body passing over the area can create impressions that superficially resemble footprints. A careful investigator will look for corroborating evidence, such as a consistent trackway showing a natural gait pattern, and will differentiate between isolated, ambiguous impressions and clear, biological prints. The presence of multiple, clear prints forming a trackway, revealing a consistent gait and foot morphology, is far more compelling than a single, isolated impression.

Hoaxes are, of course, a constant consideration when evaluating anecdotal evidence, especially in the realm of cryptozoology. Anomalous footprints are prime candidates for deliberate fabrication. Methods of creating convincing fake footprints can range from simple carving of logs or large stones to more sophisticated techniques involving casting and embedding them in the substrate. The discovery of footprints in highly accessible areas, near human habitation, or in conjunction with other unusual phenomena can raise suspicions of a hoax. However, it is crucial for investigators to avoid assuming a hoax simply because the evidence is unusual. Extraordinary claims require extraordinary evidence, but a premature dismissal of possibilities can also hinder scientific inquiry. A critical approach involves looking for inconsistencies in the alleged evidence, such as unnatural patterns, signs of artificial creation, or a lack of corroborating biological signs (e.g., no scat, hair, or feeding signs in the vicinity).

In cases where multiple tracks are found, analysis of the trackway provides additional insights. The gait pattern – how the feet are placed relative to each other – can reveal a great deal about the animal's locomotion. A consistent alternating pattern of left and right hind footprints is typical of bipedal locomotion. The stride length (distance between consecutive prints of the same foot) and the step length (distance between consecutive prints of alternating feet) can be used to estimate the animal's height and leg length. For a hypothetical undiscovered primate, a stride length suggesting a creature of 3 meters or more in height would require a correspondingly massive limb structure. The consistency of this stride length over a series of tracks is important; erratic stride lengths could indicate an injured animal or a less efficient gait.

The concept of "primate gait signatures" is an area of ongoing research. While quadrupedal mammals often have distinct gaits, primates, with their diverse adaptations for arboreal and terrestrial locomotion, exhibit a wider range of movement patterns. Gorillas, for instance, are knuckle-walkers, and their tracks often show impressions of their knuckles in front of the footprint. Orangutans are primarily arboreal and their terrestrial locomotion can be quite deliberate and careful. Chimpanzees and bonobos are more agile and can move bipedally for short distances. An undiscovered primate might exhibit a unique gait that leaves a distinctive trackway. For example, a primate with a highly efficient bipedal gait might leave prints with a clear heel strike and a relatively straight trackway, similar to humans. However, if the print morphology suggests a more ape-like foot structure, a consistent and efficient bipedal gait would present a biomechanical puzzle.

The issue of trace evidence in conjunction with footprints is also critical. The discovery of hair samples, scat (feces), or feeding signs in the vicinity of anomalous footprints could provide further evidence to support the existence of an unknown primate. DNA analysis of hair or scat samples could potentially identify a known or unknown primate species. The nature of any feeding signs – for example, stripped bark from trees at a height inaccessible to known animals, or the remains of specific prey – could also offer clues about the creature's diet and ecological niche. The absence of such corroborating evidence, especially over multiple discoveries, can weaken the case for a genuine biological entity.

When assessing anomalous footprints, it is essential to consult with experts in primatology, anatomy, and ichnology. A primatologist can provide context on the known diversity of primate foot morphology, locomotion, and evolutionary potential. An anatomist can assess the biomechanical plausibility of the observed print characteristics and hypothesize about the underlying skeletal structure. An ichnologist can provide expertise in track identification, interpretation, and the recognition of false tracks and natural impressions. Collaboration between these disciplines is crucial for a comprehensive and objective evaluation.

The scientific community generally requires clear, repeatable, and verifiable evidence before accepting the existence of a new species. For cryptid primates, this typically means finding not just a few ambiguous footprints, but a consistent pattern of evidence, ideally including multiple trackways, photographic or video evidence of the creature itself, or biological specimens like hair or scat that can be conclusively identified as belonging to an unknown primate. The remarkable rarity of such conclusive evidence for alleged undiscovered primates, despite numerous reports of anomalous footprints over decades, is a significant factor in why these claims remain in the realm of speculation rather than established scientific fact.

In summary, the interpretation of anomalous footprints attributed to undiscovered primates is a complex endeavor that demands a rigorous scientific approach. It involves meticulous documentation, detailed morphological and biomechanical analysis, comparison with known species, consideration of substrate effects, and an awareness of potential misidentifications and hoaxes. While the tantalizing possibility of discovering a new primate species fuels interest in these unusual tracks, scientific credibility hinges on moving beyond anecdotal accounts to verifiable, objective evidence. The footprints, while often the first and most compelling piece of evidence, are just the starting point in a long and arduous process of scientific investigation.

The allure of the unknown, particularly concerning the potential existence of large, undiscovered primates, often manifests most dramatically in the form of anomalous footprints. These impressions, discovered in remote locales and frequently described as being significantly larger and more human-like than those of any known great ape, form a cornerstone of anecdotal evidence for cryptid primates. However, the scientific rigor demanded by cryptozoology, much like in wildlife biology or forensic science, necessitates a critical and systematic examination of such findings, moving beyond mere description to rigorous interpretation. This section delves into the methodology of evaluating these anomalous footprints, dissecting their characteristics against known biological and physical principles, and exploring the common pitfalls and potential for misinterpretation that often accompany them.

When confronted with an alleged footprint of a large, unknown primate, the initial step is a meticulous documentation process. This extends beyond simple photography and measurement; it involves understanding the substrate in which the print is found. Is it soft mud, damp earth, sand, snow, or a combination of materials? The consistency and moisture content of the substrate are paramount, as they dictate the fidelity and depth of the impression. A well-preserved print in a yielding substrate will capture fine details, such as the texture of the skin, the shape of the toe pads, and the presence or absence of a distinct heel. Conversely, a print in dry, hard-packed earth might only provide a vague outline, easily subject to misinterpretation.

The morphology of the print is then scrutinized. For a primate, certain anatomical features are expected. A pentadactyl (five-fingered/toed) structure is a fundamental characteristic. The presence of a divergent hallux (opposable big toe) is a strong indicator of primate ancestry, particularly for arboreal species, although its degree of divergence can vary significantly. The relative lengths of the digits are also critical. In many ape species, the second toe is often shorter than the first or third, but its presence and form are significant. The heel pad, if clearly impressed, provides information about the animal's weight distribution and gait. A broad, well-defined heel pad is typical of apes, suggesting a more plantigrade (flat-footed) stance compared to the digitigrade (toe-walking) stance of many other mammals.

However, anomalous footprints often present features that defy straightforward categorization. For instance, footprints attributed to the legendary Bigfoot or Sasquatch are frequently described as being enormous, oblong, and possessing five toes, with the first toe sometimes depicted as being somewhat abducted or opposable. The stride length accompanying these prints is often reported as being exceptionally long, suggesting a bipedal creature of considerable height. The challenge lies in comparing these descriptions with the known variation within primate foot morphology. While the great apes (gorillas, orangutans, chimpanzees, bonobos) exhibit variations in their foot structure and gait, none of their typical tracks approach the reported dimensions or morphology of these alleged cryptid prints.

One of the most common explanations for anomalous primate-like footprints is misidentification of tracks made by known animals. Large bears, particularly their hind footprints, can sometimes be mistaken for primate tracks, especially if viewed in poor light or a partially obscured substrate. Bears are plantigrade, and their hind feet have a relatively broad heel pad. However, bear tracks typically exhibit distinct claw marks, which are usually absent in primate tracks (though primates may have nail impressions). Furthermore, the arrangement of the toes in bears, while superficially primate-like, often differs in relative size and spacing. The powerful claws of a bear digging into the substrate can also create impressions that might be misinterpreted.

Another frequent candidate for misidentification is the track of a large canid, such as a wolf or a coyote. While canids are digitigrade, their tracks can sometimes appear more rounded in certain substrates. However, the typical elongated shape of a canid track, the prominent claw marks, and the often symmetrical arrangement of the two central toes are usually distinguishing features. It is also important to consider the possibility of tracks made by hooved animals in certain unusual circumstances, though these are less likely to be confused with primate prints.

The biomechanical feasibility of a footprint is also a crucial consideration. The size and shape of a footprint are directly related to the skeletal structure of the foot and the forces applied during locomotion. For a primate of immense size, as often described in cryptid reports, the skeletal structure would need to be proportionally robust to support its weight and withstand the stresses of movement. If a footprint suggests a certain foot structure, for instance, an extremely elongated foot with disproportionately long toes, it raises questions about the functional anatomy of the creature. Could such a foot effectively support the weight of a massive animal? Would it be efficient for locomotion? Biomechanical modeling can help assess these possibilities. For example, if a footprint appears to show an exceptionally high arch combined with a narrow heel, it might be inconsistent with the anatomy of a large, heavy-bodied primate.

The depth of a footprint is another critical indicator, directly correlating with the animal's weight and the substrate's resistance. Anomalous footprints are often described as being exceptionally deep, suggesting a creature of immense mass. However, interpreting depth requires careful consideration of the substrate. A print in very soft, waterlogged mud will naturally be deeper than a print of the same animal in dry, firm soil. Conversely, a relatively shallow impression in soft mud might indicate a lighter animal or one that did not fully place its weight into the substrate. If a footprint is deep in a moderately firm substrate, it strongly suggests significant weight.

The concept of a "primate footprint index" could be a valuable tool in cryptid investigations, similar to indices used in forensic science to classify and compare tracks. This would involve calculating ratios of key measurements, such as the ratio of the length of the hallux to the length of the second toe, the ratio of the heel pad width to the overall foot width, or the ratio of the stride length to the estimated leg length. By comparing these indices from anomalous prints to those derived from known primate species, it might be possible to identify unusual deviations or even potentially novel morphological patterns. For example, an unusually long and slender foot with widely spaced toes might suggest a primate adapted for gripping branches, but if the stride length and depth suggest a terrestrial creature of significant weight, this combination would be biomechanically puzzling.

The phenomenon of "mud packing" or "false tracks" is a significant factor in the misinterpretation of evidence. In certain substrates, particularly soft mud or snow, water movement, falling debris, or even the pressure from a larger animal's body passing over the area can create impressions that superficially resemble footprints. A careful investigator will look for corroborating evidence, such as a consistent trackway showing a natural gait pattern, and will differentiate between isolated, ambiguous impressions and clear, biological prints. The presence of multiple, clear prints forming a trackway, revealing a consistent gait and foot morphology, is far more compelling than a single, isolated impression.

Hoaxes are, of course, a constant consideration when evaluating anecdotal evidence, especially in the realm of cryptozoology. Anomalous footprints are prime candidates for deliberate fabrication. Methods of creating convincing fake footprints can range from simple carving of logs or large stones to more sophisticated techniques involving casting and embedding them in the substrate. The discovery of footprints in highly accessible areas, near human habitation, or in conjunction with other unusual phenomena can raise suspicions of a hoax. However, it is crucial for investigators to avoid assuming a hoax simply because the evidence is unusual. Extraordinary claims require extraordinary evidence, but a premature dismissal of possibilities can also hinder scientific inquiry. A critical approach involves looking for inconsistencies in the alleged evidence, such as unnatural patterns, signs of artificial creation, or a lack of corroborating biological signs (e.g., no scat, hair, or feeding signs in the vicinity).

In cases where multiple tracks are found, analysis of the trackway provides additional insights. The gait pattern – how the feet are placed relative to each other – can reveal a great deal about the animal's locomotion. A consistent alternating pattern of left and right hind footprints is typical of bipedal locomotion. The stride length (distance between consecutive prints of the same foot) and the step length (distance between consecutive prints of alternating feet) can be used to estimate the animal's height and leg length. For a hypothetical undiscovered primate, a stride length suggesting a creature of 3 meters or more in height would require a correspondingly massive limb structure. The consistency of this stride length over a series of tracks is important; erratic stride lengths could indicate an injured animal or a less efficient gait.

The concept of "primate gait signatures" is an area of ongoing research. While quadrupedal mammals often have distinct gaits, primates, with their diverse adaptations for arboreal and terrestrial locomotion, exhibit a wider range of movement patterns. Gorillas, for instance, are knuckle-walkers, and their tracks often show impressions of their knuckles in front of the footprint. Orangutans are primarily arboreal and their terrestrial locomotion can be quite deliberate and careful. Chimpanzees and bonobos are more agile and can move bipedally for short distances. An undiscovered primate might exhibit a unique gait that leaves a distinctive trackway. For example, a primate with a highly efficient bipedal gait might leave prints with a clear heel strike and a relatively straight trackway, similar to humans. However, if the print morphology suggests a more ape-like foot structure, a consistent and efficient bipedal gait would present a biomechanical puzzle.

The issue of trace evidence in conjunction with footprints is also critical. The discovery of hair samples, scat (feces), or saliva in the vicinity of anomalous footprints could provide further evidence to support the existence of an unknown primate. However, the analysis of such biological samples presents its own unique set of challenges, particularly in the context of cryptozoological investigations. Unlike established wildlife surveys where reference samples from known species are readily available, identifying the source of an unknown hair or scat can be an exceptionally difficult task, especially if the alleged creature is truly novel.

The primary challenge lies in the potential for contamination. Biological samples are fragile and can easily be mixed with DNA from other animals, insects, or even the environment itself. For instance, a hair found snagged on a branch might have picked up DNA from numerous other creatures that have passed through the area. Similarly, scat samples can contain fragments of plant material or prey that have their own DNA signatures, complicating the analysis. Forensic DNA techniques, while incredibly powerful, are highly sensitive to such contamination, and an improperly collected or handled sample can render the results inconclusive or misleading.

Despite these difficulties, the potential payoff from analyzing trace evidence is immense. The advent of advanced DNA sequencing technologies, particularly mitochondrial DNA (mtDNA) and nuclear DNA (nDNA) analysis, offers a pathway to identifying the species of origin. Mitochondrial DNA is particularly useful because it is present in multiple copies per cell and is more resilient to degradation than nuclear DNA, making it ideal for analyzing older or degraded samples. Comparing the obtained mtDNA sequences to vast genetic databases like GenBank can often identify the source species. If the sequence does not match any known species, it could be strong evidence for a novel organism.

Nuclear DNA analysis, while more susceptible to degradation, provides a richer source of information, including sex determination and potentially even individual identification if sufficient genetic markers are recovered. In a cryptozoological context, obtaining nuclear DNA from a hair shaft, for example, could confirm if the sample belongs to a primate and potentially provide enough genetic material to distinguish it from known primates. The process typically involves extracting DNA from the sample, amplifying specific genetic regions using polymerase chain reaction (PCR), and then sequencing these amplified regions.

The analysis of primate scat, or feces, can also yield valuable information. Beyond DNA, the examination of undigested materials within the scat can reveal dietary habits. For instance, the presence of specific seeds, fruit remnants, or animal remains can shed light on the creature's ecological niche and foraging strategies. For a large primate, one might expect to find evidence of significant plant matter consumption, potentially including tough fibrous materials, or evidence of opportunistic scavenging or predation. However, distinguishing the scat of a large, unknown primate from that of other large mammals, such as bears or large canids, can be challenging based on macroscopic examination alone. Microscopic analysis and DNA testing become crucial for definitive identification.

The discovery of saliva, perhaps from a bite mark on a tree or a food source, presents another avenue for investigation. Saliva contains epithelial cells from the mouth and tongue, which are rich in DNA. Analysis of this DNA could potentially identify the species and even an individual if sufficient genetic material is recovered. However, finding such samples in situ is rare, and like other biological samples, contamination is a significant concern.

A notable success story in species discovery through trace evidence, albeit in a more conventional wildlife biology context, highlights the potential. The discovery of the Saola (Pseudoryx nghetinhensis), an elusive ungulate in Vietnam, was significantly aided by the analysis of hair and scat samples found in the region. Although these were not primate samples, they demonstrate how genetic analysis of biological traces can lead to the identification of previously unknown species. Similarly, the occasional identification of new species of rodents, insects, or marine life has been driven by genetic analysis of collected specimens or environmental DNA (eDNA) samples.

In the context of potential undiscovered primates, imagine finding a hair sample snagged on a thorny bush in a remote mountainous region. Initial microscopic examination might reveal characteristics atypical of known local fauna. If the hair is sufficiently preserved, DNA extraction could proceed. If the mtDNA sequence obtained does not match any primate in the global database, it would be an extraordinary finding, warranting extensive further investigation, including attempts to secure more samples and potentially even observe the creature itself. The nuclear DNA, if recovered, could then be used to build a more complete genetic profile.

However, it is crucial to temper optimism with realism. The vast majority of hair and scat samples collected in the wild belong to known species. Distinguishing the sample of a truly novel primate from that of a known primate with unusual morphology or from a sample that has degraded or been contaminated is a significant challenge. For instance, a bear's hair might be misinterpreted by an untrained eye due to its size and texture, and while DNA analysis would definitively identify it as ursine, the initial misinterpretation could lead to false hopes.

Furthermore, the cost and complexity of advanced genetic analysis mean that it is not always feasible to test every ambiguous biological sample found. Resources are often directed towards the most promising leads. In cryptozoology, this means samples found in conjunction with other compelling evidence, such as consistent trackways or credible eyewitness accounts, are more likely to garner the attention and funding for detailed genetic analysis.

The scientific community's threshold for accepting the existence of a new species is high. A single anomalous DNA sequence, while intriguing, would not be sufficient. It would require replication, confirmation from independent laboratories, and ideally, the discovery of a physical specimen – a body, a live capture, or at least clear photographic or video evidence of the animal itself. The genetic data would serve as a powerful supporting piece of evidence, guiding further research and increasing the likelihood of eventual discovery.

Therefore, while the prospect of identifying an unknown primate through hair, scat, or saliva analysis is scientifically exhilarating, it remains a long shot, contingent on overcoming substantial logistical and analytical hurdles. The collection and preservation of uncontaminated samples in areas where these creatures are rumored to exist are paramount. The methodology must be as rigorous as any forensic investigation, ensuring that the data obtained is reliable and interpretable. Without this meticulous approach, even the most promising biological trace might yield ambiguous results, leaving the mystery of the undiscovered primate, for now, in the realm of speculation. The scientific method, in its pursuit of verifiable truths, demands that we move beyond suggestive anecdotes and towards irrefutable biological evidence.

The interpretation of anomalous primate trackways extends far beyond static impressions in the substrate. A truly scientific approach necessitates a dynamic analysis of locomotion, examining the biomechanical signatures left behind. This involves dissecting the rhythm and mechanics of movement as revealed by the sequence of prints, offering insights that a single footprint alone cannot provide. Understanding these dynamics is crucial for differentiating potential cryptid primates from known fauna and for assessing the biological feasibility of such creatures.

When a series of prints is discovered, the investigator must move from evaluating individual impressions to analyzing the entire trackway. The stride length – the distance between successive placements of the same foot – is a primary indicator of an animal's size and locomotion efficiency. For a hypothetical large primate, an unusually long stride would suggest a creature of considerable height and limb length. However, simply measuring the longest stride is insufficient. A consistent stride length over several meters, or even tens of meters, is far more indicative of a natural gait. Erratic stride lengths can suggest an injured animal, an awkward gait, or even a series of unrelated prints. Furthermore, the stride length must be considered in relation to the width of the trackway – the lateral distance between the left and right foot placements. A narrow trackway with a long stride is typical of efficient bipedal locomotion, where the body's center of gravity is kept directly over the supporting foot. A wider trackway, especially with a shorter stride, might indicate a more quadrupedal or knuckle-walking gait, where greater lateral stability is required.

The heel strike and toe-off phases of locomotion leave distinct impressions that can reveal much about the mechanics of movement. In human bipedalism, a distinct heel strike is common, followed by a rolling motion through the foot to a powerful toe-off. For apes, the pattern differs. Gorillas, as knuckle-walkers, often leave impressions of their weight-bearing knuckles in front of their footprints, and their terrestrial gait involves a different pattern of pressure distribution across the foot. Orangutans, with their highly adaptable feet designed for arboreal life, exhibit a more deliberate, flat-footed terrestrial gait, often with a clear imprint of the entire sole and digits. If an anomalous trackway shows consistent, deep heel impressions followed by the splaying of toes at the point of push-off, it might suggest a more human-like bipedalism. Conversely, if the prints are consistently deeper at the front, with minimal heel impression, it could indicate a digitigrade or semi-digitigrade stance, which is not typical of known great apes.

The relative positioning of the toes within the footprint is also dynamic. While the static morphology of the toes is important, their placement during locomotion reveals function. For example, a primate that uses its feet for grasping branches would likely exhibit a more splayed toe pattern, even when terrestrial, to maintain stability or potentially for propulsion. If an alleged large primate trackway shows five distinct toe impressions in every print, with the first toe consistently positioned in a way that suggests opposability or a significant grasping function, this could be a biomechanical clue. However, it is also important to consider the possibility of substrate deformation. In soft mud, toes can splay outwards due to the pressure of the foot pushing into the material, creating an illusion of greater divergence than is anatomically present.

The concept of a "primate gait signature" is complex due to the diverse locomotor adaptations within the order. However, certain patterns are more common. For instance, terrestrial quadrupedalism often involves a diagonal gait, where the hind foot on one side is placed close to or ahead of the forefoot on the opposite side, leading to a relatively straight trackway with alternating left and right prints. Bipedalism, whether obligate like humans or facultative like chimpanzees, typically results in a more direct, alternating placement of hind feet. A trackway showing a consistent pattern of left hind, right hind, left hind, etc., with a minimal amount of twisting or lateral deviation, strongly suggests bipedal locomotion. The angle of the feet relative to the direction of travel is also informative. Human feet typically point forward or slightly outward. Ape feet, even when used for bipedalism, often show a more outward angle, reflecting their evolutionary history as arboreal climbers. An anomalous trackway with feet consistently pointed outwards at a significant angle, combined with a long stride, might suggest a primate with a more apelike foot structure attempting efficient bipedal movement, which could present biomechanical challenges.

The depth of the impressions within a trackway offers a crucial insight into the distribution of weight and the forces applied during locomotion. While a single deep print might be caused by an animal pausing or shifting its weight, a consistent pattern of depth across a trackway reveals the dynamic forces. For a heavy creature moving bipedally, the deepest impressions would likely be where the heel strikes and where the toes push off. If the substrate is uniform, consistent depth variations across a series of prints can indicate the points of greatest pressure during the gait cycle. For example, a consistent deep impression of the heel followed by a less defined, shallower impression of the forefoot might suggest a different form of bipedalism than that of a human.

Furthermore, the analysis of a trackway can sometimes reveal evidence of how the animal navigates obstacles or changes its gait. For instance, if the trackway encounters a small log or a rise in the terrain, the animal might briefly transition to a quadrupedal gait or adjust its stride length and foot placement. Observing such transitions in an anomalous trackway can provide valuable clues about the creature's agility and adaptability. A sudden change from a long, striding bipedal gait to a shorter, wider, more shuffling gait could indicate difficulty or a change in the terrain.

The biomechanics of a powerful push-off are also evident in the substrate. A strong toe-off in bipedal locomotion leaves impressions of the toes digging into the ground, particularly the ball of the foot and the tips of the toes. The degree to which the toes are flexed or extended at the point of push-off can also be discernible. If an anomalous trackway shows deep impressions of the toes, especially the hallux, digging into the substrate in a manner consistent with powerful propulsion, it supports the idea of an active, dynamic locomotion.

In some instances, the substrate itself can provide a "negative impression" that reveals information about the animal's structure. If a creature steps in very soft mud and then lifts its foot, the surrounding mud may collapse inward, creating a deeper impression than the actual foot. However, the areas that were in direct contact with the foot may remain relatively undisturbed or have different textures. Advanced imaging techniques, such as lidar scanning, can sometimes capture subtle variations in depth and texture that are not apparent to the naked eye, offering a more detailed three-dimensional reconstruction of the foot's interaction with the substrate.

The interplay between the morphology of the foot, as suggested by individual prints, and the dynamics of the trackway is where truly compelling evidence can emerge. For example, if prints suggest a foot that is relatively broad and flat, with a non-opposable hallux, yet the trackway demonstrates a long, efficient stride with clear heel strikes and toe-offs, this presents a biomechanical anomaly. It would challenge our understanding of primate locomotion. Similarly, if the prints suggest an ape-like foot with a highly flexible mid-tarsal joint and significant toe abduction, but the trackway shows a consistent, straight line of progression with minimal lateral deviation, this also raises questions. Such inconsistencies might point towards a hoax, a misidentification, or, most intriguingly, a primate with a unique evolutionary trajectory in its locomotion.

The presence of claw marks, while typically associated with non-primates, can occasionally be observed in primate tracks under specific conditions. For instance, if a primate is moving on a very steep incline or a slippery surface, its nails might scrape against the substrate. However, consistent and prominent claw marks in a trackway described as primate-like would be a significant point of contention, strongly suggesting a misidentification with a canid or felid. Conversely, the absence of any such marks, even in substrates where they might be expected, further supports a primate origin.

The concept of "primate foot flexibility indices," analogous to those used for static prints, can also be applied dynamically. By analyzing how the toes splay or compress during the gait cycle, one might infer the flexibility of the metatarsals and phalanges. A primate adapted for arboreal life might show greater flexibility and splaying of the toes during propulsion, even on terrestrial surfaces, compared to a primate that has evolved for more efficient terrestrial bipedalism.

When evaluating anomalous trackways, it is also crucial to consider the environmental context. The type of substrate, the presence of water, wind, or subsequent animal activity can all alter or obscure track evidence. A trackway found in a rapidly flowing stream, for instance, would be highly unlikely to preserve the fine details of a gait. Similarly, a trackway discovered in an area prone to rockfall or significant plant debris would be difficult to interpret definitively. The most compelling evidence often comes from trackways found in stable, undisturbed substrates that have preserved the impressions clearly over a significant distance.

The investigation of cryptid primates, much like any field biology or forensic investigation, relies on the principle of corroboration. A single anomalous footprint, or even a short series of prints, can be explained by numerous factors, including misidentification, substrate deformation, or deliberate fabrication. However, a long, consistent trackway exhibiting a unique gait pattern, coupled with prints that display internally consistent and biomechanically plausible morphology for a large primate, begins to build a more robust case. If this trackway is also found in conjunction with other forms of trace evidence, such as hair or scat, and is located in an area where such a creature might reasonably exist, the cumulative evidence becomes increasingly difficult to dismiss.

The scientific method demands that hypotheses are tested rigorously. In the case of anomalous primate trackways, the hypothesis might be that they represent the locomotion of an unknown primate species. Testing this hypothesis involves comparing the observed dynamics and morphology of the trackway against the known biomechanical capabilities and locomotion patterns of all known primate species, as well as other large mammals that might be confused with them. If the trackway exhibits characteristics that are consistently outside the range of variation for known species, and if it appears biomechanically feasible for a large primate, then it becomes a significant piece of evidence warranting further investigation. The analysis of gait dynamics is not merely an academic exercise; it is a powerful tool for distinguishing the signals of genuine biological phenomena from the noise of misinterpretation and deception.

The allure of anomalous trackways, whether they hint at an undiscovered primate or a creature of even greater mystery, is undeniable. However, as any seasoned field biologist or cryptozoologist will attest, fascination must be tempered with a rigorous adherence to scientific principles. The very nature of trace evidence, particularly when dealing with impressions in substrates like mud, sand, or snow, presents inherent limitations that cannot be ignored. These impressions are ephemeral, susceptible to environmental degradation, and can be remarkably ambiguous. A single footprint, or even a short series of prints, can be subject to a multitude of interpretations, many of which do not involve an unknown biological entity.

One of the most significant challenges lies in the variability of substrates. The same creature will leave vastly different impressions depending on whether it walks on firm, dry earth, soft, waterlogged mud, loose gravel, or powdery snow. In soft mud, for instance, a footprint can be significantly enlarged by slumping and erosion, creating an impression far larger than the actual foot. The edges can blur, details of dermal ridges can be lost, and the depth can be exaggerated. Conversely, on hard-packed surfaces, impressions might be shallow and indistinct, revealing little more than the basic outline of the foot. This variability means that direct comparisons between trackways found in different environments can be problematic, and assumptions about an animal's size or gait based solely on print dimensions can be dangerously misleading. The alleged "enlargement effect" in soft substrates is a well-documented phenomenon, and it must be a primary consideration when evaluating any purported evidence of large, unknown creatures. A trackway that appears exceptionally large might simply be the result of the substrate's properties rather than the size of the animal that created it.

Furthermore, the issue of contamination is a constant concern in fieldwork. Trackways can be partially or wholly obscured by the passage of other animals, the accumulation of natural debris such as leaves and twigs, or even human interference. If a trackway is discovered in a heavily trafficked area, either by known fauna or by humans, it becomes exceptionally difficult to isolate the impressions made by the hypothesized unknown creature. Multiple overlapping prints can create a confusing jumble, making it impossible to discern a coherent gait pattern. Even seemingly pristine environments can be subject to subtle contamination. A light dusting of snow might preserve a track for a time, but a subsequent snowfall can either obscure it or create a new layer of impressions on top, rendering the original evidence uninterpretable.

The question of intentional fabrication, or hoaxes, also looms large. Throughout history, numerous individuals have created fake footprints or trackways to deceive others, for amusement, notoriety, or even financial gain. The famous "Hoxne footprints" of the 1950s, initially presented as evidence of a large, unknown ape-like creature in East Anglia, were eventually revealed to be elaborate hoaxes created by a local farmer. Such deceptions often mimic genuine tracks but contain subtle inconsistencies or an unnerving perfection that can, upon closer inspection, betray their artificial origin. The sheer effort and ingenuity that can go into creating a convincing hoax cannot be underestimated. This necessitates a skeptical mindset and a thorough examination of any evidence for signs of manipulation, such as unnaturally uniform print shapes, suspiciously clean edges where debris might be expected, or a trackway that appears too perfectly formed to be natural.

Beyond the physical limitations of the trace evidence itself, there are broader scientific imperatives that must be met. The cornerstone of scientific inquiry is the demand for verifiable, reproducible evidence. While anecdotal accounts and intriguing photographs or videos can spark interest and suggest avenues for further investigation, they rarely constitute proof on their own. For a claim of an unknown species to be accepted by the scientific community, the evidence must be of a nature that can be examined, tested, and ideally, independently verified by multiple researchers.

This brings us to the crucial requirement for unambiguous identification markers. In the study of known fauna, definitive identification often relies on a combination of morphological characteristics, behavioral observations, and, increasingly, molecular evidence. For a cryptid, the evidence needs to be equally robust. For instance, if we are to consider the possibility of an undiscovered primate, the trace evidence needs to exhibit characteristics that are consistently and unequivocally primate-like, and yet demonstrably different from any known primate. This means that the morphology of the prints, when analyzed in conjunction with the gait dynamics, should align with primate biomechanics while simultaneously defying easy categorization within existing species.

DNA analysis has revolutionized many fields of biology, and it holds immense potential for cryptozoological investigations. Obtaining a clear, uncontaminated DNA profile from a hair, scat sample, or even a bloodied substrate associated with an anomalous trackway would be a monumental step forward. However, even DNA evidence is not without its challenges. Contamination from other animals, environmental DNA, or even human handlers during sample collection can lead to misleading results. Furthermore, a DNA profile, if it were to be novel, would still need to be corroborated with other forms of evidence. A unique DNA sequence is suggestive, but without associated physical evidence – a specimen, clear photographs of the creature, or a consistent trackway – it remains an isolated anomaly. The scientific community requires not just a novel genetic signature but a compelling biological context to accompany it.

Similarly, photographic and video documentation, while powerful tools, can be easily manipulated or misinterpreted. The ubiquity of digital editing software means that images and videos can be convincingly faked. Even without deliberate manipulation, the quality of the documentation can be a limiting factor. Poor lighting, low resolution, distance, and the presence of obscuring elements can all render visual evidence inconclusive. An alleged blurry video of a large primate moving through dense foliage might be exciting, but it is unlikely to satisfy the standards of scientific proof. Clear, well-lit, high-resolution images or videos, ideally captured from multiple angles and showing the creature's distinct morphological features and behavior, are necessary to move beyond speculation. The ability to positively identify specific anatomical traits, such as the structure of the hands and feet, the facial features, or the proportions of the limbs, is paramount.

The concept of "reproducibility" is also critical. In scientific experiments, findings are considered robust when they can be replicated by independent researchers. In cryptozoology, direct replication of the initial discovery of an anomalous trackway is often impossible, as the conditions that led to its formation are unique. However, the underlying principles of evidence collection and analysis must be reproducible. This means that the methods used to document, measure, and analyze the trace evidence should be clearly described and follow established scientific protocols, allowing other researchers to understand how the conclusions were reached and to potentially apply similar methodologies to future discoveries. The absence of reproducible methodologies or a lack of transparency in data collection can cast doubt on the validity of any claims.

Ultimately, the path from intriguing trace evidence to scientifically accepted proof of an unknown species is a long and arduous one, fraught with challenges. It requires not only the discovery of compelling evidence but also the rigorous application of scientific principles to its interpretation. The limitations of trace evidence – its ephemerality, susceptibility to environmental factors, and ambiguity – mean that extraordinary claims require truly extraordinary evidence. This evidence must go beyond anecdotal accounts and superficial impressions, demanding verifiable, reproducible markers such as clear genetic material, unambiguous visual documentation, and trackways that exhibit consistent and biomechanically plausible characteristics that cannot be explained by known species or natural phenomena. Until such irrefutable proof is presented, the existence of cryptid primates, and other unknown creatures, will remain in the realm of speculation, awaiting the discovery that can bridge the gap between intriguing possibility and scientific certainty.

12

Evidence Beyond Anecdotes: Audio and Visual

The study of primate vocalizations offers a rich avenue for understanding communication within existing species and provides a crucial baseline for evaluating anomalous sonic evidence. Our closest living relatives, the great apes, exhibit a sophisticated array of vocalizations that serve a multitude of purposes, from maintaining social bonds to signaling danger and coordinating group activities. Understanding this known repertoire is paramount before we can begin to interpret the purported vocalizations of unknown primate species, often referred to as cryptid primates.

Within the family Hominidae, which includes chimpanzees, bonobos, gorillas, and orangutans, vocal communication is highly developed. Chimpanzees, for instance, possess a diverse vocal repertoire that includes hoots, pant-hoots, barks, grunts, and screams. These calls are not random utterances but are context-dependent and convey specific meanings. Pant-hoots, often performed in unison by multiple individuals, are a prominent example, used to reinforce group identity, announce presence, and sometimes to establish territorial boundaries. The intensity, duration, and modulation of these hoots can communicate different emotional states, ranging from excitement to aggression. A short, sharp hoot might signal alarm, while a long, drawn-out series could indicate general excitement or anticipation.

Gorillas, particularly the dominant silverbacks, utilize a range of vocalizations including grunts, roars, hoots, and chest-beating vocalizations. Their deep chest beats are not purely visual displays; they are accompanied by resonant vocalizations that can carry considerable distances, serving to intimidate rivals and to communicate with their social group. Specific grunts are used for routine social interactions, indicating contentment, mild displeasure, or the need for attention. Screams, on the other hand, are typically associated with distress, fear, or aggression, and can be incredibly potent warnings to both conspecifics and potential predators.

Orangutans, though generally more solitary than other great apes, also employ vocalizations. Males, in particular, produce long, resonant calls known as "long calls," which can last for several minutes and serve to advertise their presence, attract females, and deter rival males. These calls are characterized by a series of pulses and an accompanying exhaled “holler” that can be heard for several kilometers through dense forest. Females and juveniles also vocalize, but their calls are generally softer and shorter, used for maintaining contact within their dispersed social network.

Bonobos, often called the "make-love-not-war" apes, have a vocal repertoire that shares similarities with chimpanzees, including hoots, screams, and barks. However, their social structure, which is more female-dominated and relies heavily on social bonding through sexual behavior, may influence the nuances and frequency of certain vocalizations. Their alarm calls, for instance, are crucial for warning others of danger, and their "play panting" is a clear indicator of amicable intent during social interactions.

The study of these known vocalizations relies on a variety of methodologies. Field researchers employ specialized recording equipment, often including parabolic microphones and digital recorders capable of capturing a wide range of frequencies and nuances. Spectrographic analysis then allows for the detailed examination of these calls, breaking them down into their constituent frequencies, amplitudes, and temporal patterns. This objective analysis reveals patterns and variations that might not be immediately apparent to the human ear and provides a quantifiable basis for comparing different calls and vocalizing individuals. By cataloging the types of calls, the contexts in which they are used, and the responses they elicit, scientists build a comprehensive understanding of primate communication. This body of knowledge is the foundation upon which any claim of anomalous primate vocalizations must be assessed.

When we turn our attention to the realm of cryptozoology and the alleged vocalizations of undiscovered megafauna primates, a pattern of purported sounds emerges from various eyewitness accounts and folklore. These descriptions often involve vocalizations that are perceived as distinct from those of known animals, and frequently described as being more powerful, unusual, or unsettling. Common descriptors include deep howls, eerie whistles, high-pitched screams, and guttural roars, often attributed to creatures like Bigfoot, the Yeti, or other ape-like entities.

Consider the alleged vocalizations attributed to Bigfoot, often referred to as Sasquatch in North American folklore. Reports frequently describe a deep, resonant howl, sometimes described as a "whoop" or a "scream," that can carry for remarkable distances through dense forests.

These sounds are often reported during the night or in remote wilderness areas, and are frequently associated with sightings or the discovery of unusual tracks. The sheer power and volume attributed to these sounds suggest an animal of considerable size and, potentially, a specialized vocal apparatus capable of producing such resonant frequencies. Some researchers propose that these howls could function similarly to the long calls of orangutans or the territorial vocalizations of other large mammals, serving to demarcate territory, attract mates, or warn off rivals. The sheer variability in descriptions—ranging from a mournful howl to a piercing scream—might reflect different emotional states, ages, or sexes of the creature, or perhaps different functions for the vocalizations themselves.

In the Himalayan regions, accounts of the Yeti often include descriptions of its vocalizations. While less frequently reported than visual sightings or tracks, some folklore and eyewitness testimony mention eerie screams, shouts, or low growls emanating from the high-altitude wilderness. These sounds are often described as being distinctly non-human and unnerving, contributing to the mystique and fear associated with the creature. The stark, often barren, and acoustically resonant environments of the Himalayas could amplify and carry such vocalizations, making them particularly striking to those who hear them. The potential function in such an environment might involve communication over vast distances, alerting others to the presence of potential threats or signaling the location of prey.

Further afield, reports of ape-like creatures in other parts of the world, such as the Skunk Ape of the southeastern United States or the Yowie of Australia, also occasionally include descriptions of vocalizations. These are often described as aggressive roars, guttural growls, or high-pitched shrieks, again emphasizing a departure from the known primate vocal spectrum. The "skunk" aspect of the Skunk Ape, for instance, might suggest a vocalization accompanied by a strong odor, though this is purely speculative and likely a conflation of sensory inputs.

Analyzing these alleged megafauna primate vocalizations from a biological and ecological perspective presents significant challenges, primarily due to the lack of concrete, verifiable recordings. Most accounts are anecdotal, relying on the subjective interpretation of sounds heard by individuals. However, we can draw inferences by comparing these descriptions to the known vocal capabilities of large primates and other large mammals.

The deep, resonant howls attributed to creatures like Bigfoot could, in theory, be produced by a large-bodied primate. The size of the vocal folds, the dimensions of the resonant cavities (larynx, pharynx, oral cavity), and the ability to control airflow are all critical factors in sound production. Large primates, such as gorillas, possess impressive vocal capabilities, and an even larger, unknown hominoid might possess the anatomical structures to produce sounds of greater volume and lower frequency. The phenomenon of "infrasound," sounds below the range of human hearing, is also relevant. While not directly audible, infrasound can be felt and can travel long distances, potentially influencing animal behavior. However, direct reports of infrasonic emissions from cryptids are rare, and most descriptions focus on audible phenomena.

Eerie whistles, on the other hand, are less commonly associated with typical primate vocalizations. Whistles are often produced by forcing air through a narrow aperture, and while some primates use high-pitched calls, the specific modulation of a whistle is not a hallmark of great ape communication. It's possible that such descriptions might arise from misinterpretations of wind through natural formations, the calls of other animals (like birds or bats), or perhaps even a specialized form of primate communication that has not been observed in known species. However, in certain contexts, such as signaling over very long distances in open or mountainous terrain, a high-frequency, directional whistle could be an effective means of communication.

The high-pitched screams and shrieks reported are more akin to the distress calls of known primates, such as chimpanzees or gorillas when threatened. However, the sheer intensity and sustained nature of some reported screams might suggest a greater capacity for vocal projection, perhaps linked to their hypothetical size and physiological adaptations. Such screams could serve as a potent alarm, a defensive threat display, or even an expression of extreme pain or agitation.

A crucial aspect to consider is the function of these vocalizations within a hypothetical primate ecological niche. If these creatures are indeed large, arboreal or semi-arboreal primates, their vocalizations would likely serve purposes analogous to those of known apes and monkeys:

1. **Territoriality and Spacing:** Long-distance calls, like the purported howls, could be essential for maintaining territories, especially in dense forest environments where visual contact is limited. This would prevent unnecessary aggression between individuals and ensure adequate resource distribution.
2. **Social Cohesion and Mate Attraction:** Vocalizations can play a role in maintaining social bonds within groups (if they live in groups) and in attracting mates over considerable distances. The variations in calls could signal reproductive readiness, social status, or availability.
3. **Alarm and Predator Warning:** As with all prey animals, vocalizations would be critical for alerting conspecifics to the presence of

danger. The type of alarm call might vary depending on the nature of the threat, with different calls signaling aerial predators versus terrestrial ones.

4. **Interspecific Communication:** While less common among primates, it's conceivable that certain vocalizations might be directed at or perceived by other species, though this is highly speculative.

The challenges in verifying these reported vocalizations are immense. Firstly, obtaining clear, unambiguous audio recordings of these alleged creatures is incredibly difficult. The environments in which they are reported to vocalize are often remote, acoustically complex, and prone to masking from other natural sounds (wind, water, other animals). Secondly, even if a recording is obtained, attributing it definitively to an unknown primate requires ruling out all known sources. Many of the sounds described—howls, screams, whistles—can be produced by a wide range of known animals, including canids (wolves, coyotes), felids (cougars), birds of prey, and even non-biological sources like wind interacting with natural formations.

Moreover, the human element of interpretation cannot be overlooked. Fear, expectation, and the inherent human tendency to anthropomorphize can lead to the misinterpretation of ambiguous sounds. A distant coyote howl heard under stressful conditions in a remote wilderness might easily be perceived as something more sinister and larger, especially if the listener is predisposed to believe in such creatures. The "phantom" or "mystery" sound phenomenon is well-documented, where unusual or inexplicable noises are often attributed to unknown causes simply because their origin is not immediately apparent.

For cryptozoological audio evidence to gain traction within the scientific community, it would need to meet a stringent set of criteria. This would involve high-quality, clear recordings that capture a significant portion of the vocalization and its context. Crucially, these recordings would need to be analyzed by bioacousticians who specialize in animal sounds, employing sophisticated spectrographic analysis to identify unique acoustic features. These features would then need to be compared against extensive databases of known animal vocalizations. The goal would be to identify characteristics that are demonstrably distinct from all known species and that align with the hypothesized biology of an undiscovered primate.

Furthermore, consistent patterns in the vocalizations would be important. If multiple independent recordings from different locations and by different individuals capture remarkably similar acoustic signatures that cannot be attributed to known animals, this would strengthen the case. Likewise, if these vocalizations are consistently associated with other forms of evidence, such as trackways, sightings, or biological samples (hair, scat), it would lend further credibility.

The role of technology in this endeavor is expanding. Advances in acoustic monitoring technology, such as the deployment of remote sensing devices equipped with highly sensitive microphones and long-duration recording capabilities, could potentially capture elusive vocalizations. These devices can be placed in areas with a history of reported sightings or anomalous sounds, allowing for continuous, passive monitoring over extended periods. Analyzing the vast amounts of data generated by such systems would, however, require sophisticated algorithms capable of filtering out ambient noise and identifying target vocalizations.

In conclusion, while the vocalizations attributed to alleged megafauna primates remain largely in the realm of anecdote and speculation, their study offers a compelling intersection of cryptozoology, biology, and acoustics. The known vocal repertoires of great apes provide a vital biological framework for understanding the potential communication strategies of undiscovered primate relatives. The reported howls, whistles, and screams, while needing rigorous scientific scrutiny, hint at the possibility of complex acoustic signaling within the ecological niches these creatures are believed to inhabit. Ultimately, bridging the gap between folklore and scientific acceptance will require the capture of unambiguous, verifiable audio evidence that stands up to thorough bioacoustic analysis and can be demonstrably linked to an unknown biological entity, moving beyond the realm of subjective interpretation and into the domain of objective, reproducible scientific data.

The spectral analysis of purported Bigfoot and Yeti vocalizations represents a frontier where cryptozoology grapples with the rigorous methodologies of bioacoustics. While anecdotal accounts abound, often painting vivid sonic landscapes of the unknown, the challenge lies in translating these descriptions into quantifiable data that can withstand scientific scrutiny. This endeavor requires not only the collection of audio evidence but also its objective interpretation, differentiating it from the cacophony of known terrestrial sounds.

One of the most frequently reported vocalizations attributed to Bigfoot is the "scream," a sound that consistently evokes strong emotional responses in those who claim to have heard it. These are not the excited yelps of a fox or the distress calls of a domestic animal. Instead, witnesses often describe a piercing, sustained shriek that can carry for impressive distances, even through dense forest canopies. The sheer volume and unnerving quality of these screams have led some to speculate about the physiological adaptations required for such vocal projection. If a creature the size of Bigfoot were capable of producing such sounds, it would necessitate a robust vocal apparatus, potentially involving enlarged laryngeal structures and a powerful respiratory system. Spectrographic analysis of such a scream, were a verifiable recording to be obtained, would likely reveal a broad frequency spectrum, with a significant emphasis on higher frequencies associated with piercing quality, and a notable amplitude, indicating considerable power. The duration and modulation of the scream would also be critical, providing insights into the emotional state of the vocalizer—whether it signifies extreme fear, pain, territorial aggression, or perhaps even a warning call designed to propagate rapidly through a challenging acoustic environment. When compared to the documented screams of known primates, such as the alarm calls of chimpanzees or the distress vocalizations of gorillas, a hypothetical Bigfoot scream would likely differ in its raw power and possibly its pitch range, assuming a larger body size correlates with lower fundamental frequencies but the capacity for higher-pitched, intensely projected sounds. However, it is imperative to consider the potential for misidentification. Many known animals can produce startling screams. The shrieks of a cougar, the agitated calls of owls, or even the sound of wind whistling through specific geological formations can be profoundly unsettling and, in a primal, isolated setting, might be misinterpreted by a fearful or expectant listener. The key difference, from a bioacoustic perspective, would lie in the distinct acoustic signature—the unique patterns of frequencies, harmonics, and temporal structures that characterize a specific species.

Another category of alleged Bigfoot sounds is the "howl," often described as deep, resonant, and carrying a mournful or eerie quality. This differs from the sharp, urgent nature of a scream and is more akin to the territorial calls of large predators or the long-distance communication of social mammals. Descriptions often evoke sounds that are unlike a wolf's howl, possessing a more guttural, drawn-out, and sometimes modulated character. Some accounts suggest a "whoop" or a "wail," implying a complex vocalization that might involve variations in pitch and timbre. From a bioacoustic standpoint, such a howl would be characterized by a strong fundamental frequency, likely in the lower range due to the hypothesized size of the creature, with overtones that contribute to its resonance and carrying capacity. The spectrographic signature would likely show distinct harmonic structures and potentially a fluctuating pitch contour, suggesting deliberate modulation rather than a simple, steady tone. The function of such a howl could be manifold: establishing territorial boundaries, communicating with conspecifics over vast distances, attracting mates, or even as a form of intimidation. If these creatures are indeed solitary or live in small, dispersed family units, long-distance vocalizations would be crucial for maintaining social connections and managing territory. Comparing this to known primate vocalizations, the long calls of orangutans, though distinct in their "holler" component, share the function of advertising presence and territory over distance. However, the deep, resonant howl described for Bigfoot suggests a vocalization that might leverage different acoustic principles, perhaps more akin to the vocalizations of large cervids or canids, but with a primate-derived vocal tract. The challenge here, again, is differentiation. The howl of a coyote, especially when amplified by canyon walls or dense forest, can sound larger and more menacing than it is. Similarly, certain bird calls or even the sounds of collapsing trees or falling rocks in remote areas can create auditory illusions. A verifiable spectrographic analysis of a Bigfoot howl would need to exhibit a unique acoustic fingerprint, one that could not be replicated by known fauna or environmental noise.

The Yeti, inhabitants of the towering Himalayas, are also associated with a range of vocalizations, though perhaps less frequently documented than those attributed to Bigfoot. Accounts from this region often describe eerie screams, shouts, and low growls emanating from the high-altitude wilderness. The stark, often acoustically resonant environment of the Himalayas could amplify and carry such sounds, making them particularly striking. The "screams" associated with the Yeti are sometimes described as higher-pitched and more frantic than those attributed to Bigfoot, possibly reflecting adaptations to a different ecological niche or different predator-prey dynamics. The mountainous terrain might favor vocalizations that are directional and can travel over long distances, perhaps utilizing echoes and reverberations to their advantage. Spectrographic analysis would be essential to determine if these screams possess unique spectral characteristics that distinguish them from the calls of known mountain animals, such as snow leopards, bharals (blue sheep), or even certain species of birds adapted to high altitudes. The "shouts" reported could be interpreted as shorter, sharper vocal bursts, potentially used for short-range communication or as alarm signals. The low growls attributed to the Yeti are more consistent with the defensive or aggressive vocalizations of large mammals. These would likely exhibit a strong emphasis on low frequencies, with a rough, broadband spectral quality indicative of vocal cord tension and turbulent airflow. The context of such growls would be critical—were they defensive, territorial, or predatory? Analyzing the spectrographic signature could reveal patterns indicative of arousal, aggression, or fear. When considering the Yeti's environment, one must also account for the unique acoustic properties of thin, cold air and the way sound propagates across vast, open, and often barren landscapes. Sounds can travel surprisingly far in such conditions, and wind can create a multitude of deceptive noises. The possibility of misidentification of known animals is again significant. The distressed cries of a lost yak, the calls of a Himalayan monal pheasant, or the roar of a rare leopard could all be misinterpreted in the vastness and isolation of the Himalayas, particularly when filtered through the lens of expectation and folklore.

Beyond these commonly cited examples, there are other reported vocalizations that add to the complexity of the cryptid primate sonic puzzle. These include reports of whistles, clicks, and even guttural roars.

High-pitched whistles are particularly intriguing, as they are not a typical feature of great ape vocal communication. While some primates use high-frequency calls, the modulated whistle as a deliberate signal is less common. If an unknown primate were to utilize whistles, it might suggest an adaptation for long-distance communication in open terrain, or perhaps a specialized form of signaling that exploits directional sound propagation. Spectrographic analysis of a whistle would reveal a narrow bandwidth with a relatively pure tone, the pitch of which could be modulated over time. The function could range from contact calls to alarm signals. However, the likelihood of misinterpreting wind noise, the calls of specific birds (like whistles of certain thrushes or eagles), or even mechanical sounds in remote human settlements must be carefully considered.

Guttural roars, on the other hand, align more closely with the vocal capabilities of large, powerful animals. These sounds would be characterized by a low fundamental frequency, a significant degree of spectral noise, and a broad frequency range, indicating the involvement of the entire vocal tract and potentially the use of flapping vocal folds. Such roars could serve as powerful territorial displays or intimidation tactics, similar to those of lions or bears. Their spectrographic profile would show a rich, complex spectrum indicative of significant vocal power and a wide range of harmonic content. The sheer volume and intensity implied by "roars" suggest an animal of considerable size and physiological capacity.

The process of verifying these alleged vocalizations requires a systematic approach, moving beyond mere description to rigorous analysis. The first and most crucial step is the acquisition of clear, unambiguous audio recordings. This is a monumental challenge in itself. Such creatures are by definition elusive, inhabiting remote and often inaccessible regions. The environments in which they are reported to vocalize are frequently characterized by significant background noise: wind, flowing water, animal activity, and sometimes even distant human sounds. Effective recording requires sophisticated equipment capable of capturing a wide dynamic range and frequency response, often employing directional or parabolic microphones to isolate the target sound from its surroundings. Furthermore, the recording needs to be contextualized. Knowing the time of day, the weather conditions, the surrounding environment, and the presence of any other potential sound sources is vital for accurate interpretation.

Once a recording is obtained, the critical phase of bioacoustic analysis begins. This is where spectrographic analysis plays a pivotal role. A spectrogram is a visual representation of sound, displaying frequency on the vertical axis, time on the horizontal axis, and amplitude (loudness) represented by the intensity or color of the signal. Bioacousticians use specialized software to generate and interpret these spectrograms. By examining the patterns of lines, bands, and areas of intensity, they can identify distinct acoustic features: the fundamental frequency, the presence and intensity of harmonics, the duration of the sound, the rate of pulsation, the presence of formants (resonant frequencies of the vocal tract), and any modulation in pitch or amplitude.

This objective data can then be compared to extensive databases of known animal vocalizations. Such databases, meticulously compiled by zoologists and bioacousticians, contain spectrographic data for thousands of species, covering a vast array of acoustic signatures. The goal is to determine if the recorded sound possesses unique characteristics that cannot be matched by any known animal. If a recording exhibits an acoustic signature that is demonstrably different from all known species, and if it aligns with the hypothesized biological parameters of an unknown primate—for instance, a frequency range and power output consistent with a large-bodied ape—then it begins to build a case.

However, even a unique acoustic signature is not sufficient on its own. The association of the sound with other evidence is paramount. Are there credible eyewitness sightings of a large, unknown primate concurrent with the recording? Have unusual tracks or other physical traces been found in the vicinity? Is there any biological material, such as hair or scat, that can be analyzed and potentially linked to the vocalization? The stronger the correlation between anomalous audio evidence and other corroborating data, the more weight the vocalization carries.

The challenges are immense, and the history of alleged cryptid recordings is littered with instances of misidentification, hoaxes, or simply ambiguous sounds that resist definitive attribution. For example, the famous "Sierra Sounds," a series of recordings purportedly made by Bigfoot, have been subjected to intense scrutiny. While some researchers argue for their authenticity, citing unique acoustic features and consistency, others maintain that they can be explained by known phenomena, such as the calls of coyotes, bobcats, or even human mimicry. The debate highlights the difficulty of achieving consensus when dealing with evidence that exists on the fringes of scientific verifiability.

Moreover, the psychological element of perception cannot be ignored. Human auditory perception is subjective and can be heavily influenced by expectation, emotion, and the environment. In the dark, isolated wilderness, the human brain is predisposed to interpret ambiguous sounds as potentially threatening. A rustle in the undergrowth, a distant animal call, or the creaking of trees can be amplified in the mind, transforming into something more sinister and unknown. This "phantom sound" phenomenon is a well-documented aspect of human psychology and a significant factor that must be accounted for when evaluating anecdotal audio evidence.

The potential for deliberate deception also looms large. The history of cryptozoology includes numerous instances of hoaxes, where individuals have fabricated evidence to gain attention or notoriety. This necessitates a critical evaluation of the source of any purported recording, considering the credibility of the individual or group making the claim, and looking for any inconsistencies or red flags that might suggest fabrication.

Despite these challenges, the pursuit of verifiable audio evidence for creatures like Bigfoot and the Yeti continues. Advances in technology, such as the deployment of autonomous acoustic monitoring arrays in remote areas, offer the potential for capturing elusive sounds without direct human observation. These systems can record continuously for extended periods, gathering vast amounts of data that can then be processed by sophisticated algorithms to identify anomalies. The development of machine learning techniques specifically trained to detect primate-like vocalizations within complex soundscapes could revolutionize the field, flagging potential candidate recordings for human expert analysis.

Ultimately, the scientific community will likely require a confluence of several factors to accept the existence of Bigfoot or Yeti based on their vocalizations. This would involve multiple, high-quality, independently verified recordings from geographically diverse locations, all exhibiting consistent and unique acoustic signatures that cannot be attributed to known species. These recordings would need to be accompanied by corroborating physical evidence and, ideally, live or verifiable photographic/videographic evidence of the creature itself. The burden of proof rests heavily on the shoulders of those who claim the existence of these unknown primates, and the path from anecdotal report to scientific acceptance is arduous, demanding irrefutable, objective data that stands the test of rigorous scrutiny. The spectral analysis of sounds, while a powerful tool, is but one piece of a much larger, and still largely incomplete, puzzle.

The annals of cryptozoology are replete with claims of fleeting glimpses and blurry photographs, but few visual pieces of evidence have ignited as much sustained debate and scrutiny as the Patterson-Gimlin film. Shot in October 1967 in Bluff Creek, California, by Roger Patterson and Robert Gimlin, the 16mm footage, lasting just under a minute, purports to show a large, hairy, bipedal creature – an apparent Bigfoot – walking across a clearing and then disappearing into the woods. For decades, this film has served as the central pillar of visual evidence for proponents of Bigfoot's existence, while simultaneously remaining an intractable enigma for skeptics and mainstream science. Its enduring power lies in its vividness, its apparent depiction of biological plausibility, and the persistent questions it raises about the limits of our knowledge of the natural world.

The immediate impact of the Patterson-Gimlin film was profound. Patterson, a rancher and amateur filmmaker, had been actively seeking evidence of Bigfoot, having heard numerous local accounts of a large, ape-like creature in the region. He and Gimlin, a friend and fellow enthusiast, were reportedly in the area filming when they encountered the subject. The footage shows the creature, which appears to be female based on its physique, turning to look back at the camera with a distinctly non-human gait, its proportions and musculature seemingly indicating a powerful, wild animal. The film captures its lumbering walk, the apparent swing of its arms, and the way its musculature shifts with each stride. It is this depiction of seemingly natural locomotion, coupled with the creature's imposing presence, that has fueled belief for so many.

The arguments in favor of the film's authenticity often center on several key points. Firstly, the sheer difficulty of faking such footage in 1967 is frequently emphasized. The special effects technology of the era was rudimentary compared to today's standards. Creating a creature with such realistic movement, musculature, and texture would have required an immense effort, involving complex prosthetics and puppetry that would have been difficult to conceal and execute convincingly. Skeptics often point to the known capabilities of Hollywood special effects even at that time, but proponents counter that the spontaneous nature of the encounter, as described by Gimlin, suggests it wasn't a staged production. The immediacy and apparent lack of directorial control in the filming process are cited as evidence against a deliberate hoax.

Furthermore, the biomechanics of the creature's gait have been extensively analyzed. Numerous gait experts and biomechanical engineers have studied the film, and while some have pointed out apparent inconsistencies or unnatural movements, others have argued that the creature's walk is remarkably plausible for a large, unknown primate. The way the creature's hips and shoulders move, the pronation and supination of its feet, and the slight sway of its torso have all been scrutinized. Proponents suggest that these movements are indicative of a primate adapted for bipedal locomotion, possessing a unique anatomical structure that results in a distinct gait. The sheer power implied by the musculature and the fluid, yet unhurried, movement across the uneven terrain are often presented as compelling evidence of a real, living animal. It is argued that simulating such a complex and nuanced movement would have been extraordinarily difficult for anyone, let alone for two men supposedly reacting in haste to an unexpected encounter.

The supposed absence of any visible seams, zippers, or obvious artificial elements in the suit is another point frequently raised by believers. While high-definition playback and enhancement of the film have allowed for closer examination, no definitive proof of a costume has emerged that satisfies proponents. The texture of the hair, the apparent depth and movement of the skin beneath it, and the way the creature's body conforms to its actions are all cited as evidence against it being a mere suit.

Robert Gimlin himself has consistently maintained that the encounter was real and that he and Patterson were genuinely startled by the creature. His testimony, along with Patterson's, has been a cornerstone for believers. Gimlin's demeanor in interviews, his consistent retelling of the events, and his apparent lack of financial gain from the film are often presented as indicators of his sincerity. Patterson, who died in 1972, also maintained its authenticity until his death. While the death of one of the primary witnesses can be seen as a loss of a key debater, it also, for believers, prevents any later recantation or confession that might have cast further doubt.

However, the case against the Patterson-Gimlin film is equally robust and has evolved significantly over the years, fueled by technological advancements and a persistent application of critical thinking. The most common skeptical explanation is that the film depicts a person in a costume. Numerous individuals have claimed to have been involved in creating such a costume or have suggested specific individuals who might have been the "creature." The most prominent of these claims came from Bob Johnson, a rodeo performer, who, according to researcher Ray Crow, admitted to being the one in the suit. However, Crow himself later recanted aspects of his statement, and Johnson never publicly confirmed the claim. Another individual, Hank Roberts, a stuntman, also came forward at various points claiming to have been the figure. The credibility of these claims has often been questioned due to inconsistencies or a lack of corroborating evidence.

The argument that special effects were insufficient in 1967 is challenged by the history of film production. Even in the mid-20th century, impressive illusions were being created. The development of latex, foam rubber, and sophisticated molding techniques allowed for the creation of realistic prosthetics and creature suits. Skeptics point to examples of creature effects in films of that era, suggesting that a determined effort could have yielded a convincing Bigfoot costume. The claim that the film was shot spontaneously, in a moment of panic, also faces scrutiny. While Patterson was indeed an amateur filmmaker, the fact that he had a camera with him and was able to capture such a sustained and relatively stable sequence of footage during what would have been a terrifying encounter has been questioned. The framing of shots, the focus, and the camera movement, while not professional, are considered by some to be too composed for a genuine panicked reaction.

The biomechanical analyses, while complex, have also yielded interpretations favorable to the hoax theory. Some biomechanical experts have argued that the gait exhibits characteristics that are more indicative of a human wearer attempting to simulate an ape-like walk rather than a genuinely evolved bipedal primate. The suggestion is that the wearer's human musculature and skeletal structure would naturally lead to certain awkwardnesses or unnatural proportions when attempting to mimic the described gait of Bigfoot, and these subtle indicators are present in the film. For example, the relatively straight leg progression and the pronounced outward rotation of the foot have been cited as potentially human-like characteristics rather than those of an ape adapted for terrestrial bipedalism.

The context of the film's creation is also critical. Roger Patterson was known to be a showman and had a history of making sensational claims. He was also reportedly seeking financial backing for a Bigfoot movie around the time the film was shot. This motivation has led many to believe that the film was a deliberate fabrication designed to generate publicity and potentially profit. The fact that Patterson patented the name "Bigfoot" shortly after the film's release further fuels these suspicions. The narrative of the encounter, as presented by Patterson and Gimlin, has also undergone some revisions and inconsistencies over the years, which skeptics point to as evidence of fabrication.

Furthermore, the scientific community's general stance on the film is one of extreme skepticism. Without clear, verifiable evidence, such as a physical specimen, DNA, or irrefutable photographic or video documentation that can withstand rigorous analysis, extraordinary claims require extraordinary proof. The Patterson-Gimlin film, despite its notoriety, has not met this threshold for most scientists. The lack of consistent sightings, the absence of any definitive physical evidence linked directly to the creature depicted, and the persistent theories of a hoax all contribute to this scientific reticence.

The ongoing debate is not merely academic; it has profound implications for how we approach evidence in cryptozoology. The Patterson-Gimlin film represents a fascinating case study in the challenges of evaluating visual evidence of the unknown. It forces us to confront questions about perception, interpretation, the limits of technology, and the human capacity for both belief and deception.

The film's grainy resolution, a product of its era and filming conditions, ironically aids its ambiguity. While high-definition digital enhancement can sharpen details, it can also introduce artifacts and create illusions. Magnifying certain frames can make the creature appear more fabricated, revealing inconsistencies in texture or movement, while other angles can enhance its realism. This digital plasticity means that the "truth" of the film can, to some extent, be shaped by the very tools used to analyze it. The debate over enhancement techniques and their reliability is a continuous thread in the film's history.

The concept of a "hoax" itself requires careful consideration. A hoax implies intent to deceive. If the film is a hoax, it would likely have been perpetrated by individuals who either knew the creature was not real or had a vested interest in presenting it as such. The most popular hoax theory points to Patterson's commercial ambitions. However, the narrative of Gimlin's genuine surprise and fear at the encounter complicates a simple "Patterson-did-it-alone" theory. If Gimlin was also complicit, the scale of the deception grows, and the motivation becomes more complex. Could it have been a joint effort driven by a desire for fame, or perhaps even a belief that they were preserving the legend?

The enduring mystery of the Patterson-Gimlin film also lies in the psychological impact it has had on viewers. For those predisposed to believe in Bigfoot, the film offers a compelling, albeit imperfect, glimpse of what they imagine such a creature to be. The sheer visual impact of a large, hairy, bipedal being moving through a natural environment taps into primal archetypes and cultural narratives. It validates their beliefs and fuels their desire for further evidence. For skeptics, however, the film represents an enduring puzzle, a tantalizing piece of visual data that, despite decades of analysis, has not conclusively proven its authenticity nor definitively revealed itself as a hoax.

The possibility that the creature depicted is indeed a real, undiscovered primate cannot be entirely dismissed, even by many skeptics. The arguments for the film's authenticity often hinge on the creature's seemingly realistic musculature and gait. Some researchers have proposed that the creature could be a surviving species of Gigantopithecus, a massive ape that once inhabited Asia, or perhaps a terrestrial, bipedal ape species that evolved separately in North America. If such a creature existed, its physiology would be unique, and its locomotion might indeed differ from known primates. The film, in this view, captures a glimpse of that unique biology.

However, the biological plausibility of such a creature existing in North America without leaving more definitive traces remains a significant hurdle. The ecological requirements for a large mammal population, including adequate food sources, breeding populations, and a range of territories, would necessitate a larger footprint than has thus far been unequivocally linked to Bigfoot. The lack of fossil evidence, consistent genetic material, or clear anatomical remains in conjunction with the visual evidence leaves a critical gap in the scientific argument for the creature's existence.

The story of the Patterson-Gimlin film is far from over. It continues to be studied, debated, and reinterpreted. New analyses emerge, old claims are revisited, and the core mystery remains. It stands as a potent symbol of the blurry line between folklore and biology, between the tantalizing possibility of the unknown and the rigorous demands of scientific proof. Its legacy is not just in the image it captured, but in the enduring questions it poses about what might be hidden in the wild places of our world, and how we can truly know what we are seeing. The film has become a touchstone in the debate, a visual ghost that haunts the fringes of scientific inquiry, forever inviting us to look closer, to question more deeply, and to consider the possibility that what we think we know about the natural world may, in fact, be incomplete.

While the Patterson-Gimlin film occupies a central, almost mythical, position in the discourse surrounding Bigfoot and other alleged megafauna primates, it is far from the sole visual artifact presented as evidence. Over the decades, numerous other photographic and video recordings have emerged, each with its own narrative, proponents, and detractors. Examining these additional pieces of visual documentation allows us to broaden our scope, applying the same critical lens of scientific inquiry and ecological understanding to a wider array of claims. This collective analysis is crucial not only for evaluating the individual merits of each piece of evidence but also for discerning any recurring patterns or commonalities that might lend credence to the idea of a genuine, undiscovered primate. Conversely, it also helps in identifying consistent weaknesses, potential sources of misidentification, and the ever-present possibility of deliberate fabrication.

One notable example that has garnered significant attention is the

"Sylvanic Photo", allegedly taken in the late 1970s in the Sylvanic National Forest, though specific dates and exact locations are often difficult to pin down definitively in these accounts. This photograph, typically appearing as a grainy, black-and-white image, purportedly shows a large, bipedal, ape-like figure standing amongst dense forest foliage. The creature appears dark-haired and of considerable size, with its posture suggesting a natural, unposed stance. Proponents highlight the creature's seemingly robust build and the way it blends with its natural environment, arguing that such a detailed portrayal would be difficult to fake with the photographic technology of the era without a highly sophisticated setup. They often point to the apparent absence of human-like proportions or gait, suggesting an animalistic posture. However, the quality of the Sylvanic Photo is its most significant limitation. The image is inherently ambiguous, with the details of the creature often obscured by shadow, distance, and the surrounding vegetation. It is exceptionally difficult to ascertain clear anatomical features, the texture of the alleged fur, or the precise dimensions of the subject. This lack of clarity makes it susceptible to numerous interpretations. Skeptics often propose that the figure could be a bear standing on its hind legs, a common misidentification in such contexts, or perhaps a strangely shaped tree stump or shadow. The difficulty in verifying the authenticity of the photograph, including the precise circumstances of its capture and the identity of the photographer, further weakens its evidentiary value. Without corroborating testimony or a clearer, verifiable source, the Sylvanic Photo remains a tantalizing but ultimately unconvincing piece of evidence.

Another category of visual evidence comprises

video footage, which, when of sufficient quality and duration, can offer more dynamic insights into a creature's locomotion and behavior than static photographs. One such collection of footage, often referred to as the **"Ohio Bigfoot" videos**, emerged in various iterations over the years, purportedly recorded in different locations within Ohio. These videos typically depict a large, dark-haired bipedal figure moving through wooded areas. Some segments show the creature walking, while others capture it disappearing into the undergrowth. The perceived realism of the creature's movement is often cited by believers. They argue that the way the body sways, the apparent weight distribution, and the fluidity of its motion are difficult to replicate, especially if the footage was captured spontaneously by an amateur. The dark, dense environment in which these creatures are often seen is also presented as a factor that would make any hoax more challenging to execute successfully. However, like the Patterson-Gimlin film, these Ohio videos are also subject to intense scrutiny. Their quality is often poor, characterized by low resolution, shaky camera work, and poor lighting, making detailed analysis challenging. Skeptics frequently point to the possibility that these figures are simply humans in dark clothing or animal costumes, perhaps attempting to mimic known Bigfoot narratives. The consistency of the alleged creature's appearance across different supposed sightings in the same region is sometimes used as an argument for authenticity, but it can also be seen as evidence of a coordinated hoax or the persistent misidentification of a known animal. The lack of clear anatomical details, such as facial features or distinct limb articulation, leaves them open to speculation. Furthermore, the context surrounding the filming is often vague, with unclear timelines, unverified witnesses, and a lack of definitive proof that the footage was captured in situ during an actual sighting of an unknown primate.

A more recent wave of visual evidence has emerged with the proliferation of

trail cameras and personal dash cams. These devices, designed for passive recording, have the potential to capture unexpected encounters with wildlife. A number of images and short video clips attributed to these sources have circulated online, purporting to show Bigfoot-like creatures. For example, images captured by trail cameras in remote forested areas of the Pacific Northwest have occasionally shown large, dark, quadrupedal or bipedal shapes that, at first glance, might resemble an unknown primate. The appeal of trail camera evidence lies in its presumed objectivity; these cameras are set up for wildlife monitoring and are not typically manipulated by individuals hoping to stage a hoax. However, even these seemingly unbiased recordings are subject to interpretation and potential misidentification. The resolution of many trail cameras is limited, especially in low-light conditions, leading to blurry or indistinct images. A large, dark animal, such as a black bear or a moose, can appear distorted or unusual when captured in a fleeting moment or at an awkward angle. Shadows, branches, and other natural elements can also create misleading shapes that resemble a humanoid figure. Furthermore, the ease with which digital images can be manipulated means that even seemingly clear footage must be treated with caution. While some proponents believe that the sheer volume of trail camera data increases the probability of capturing genuine evidence, the lack of any single, undeniable image or video that clearly depicts an unknown primate has meant that this evidence, while intriguing, has not definitively shifted the scientific consensus.

Beyond the more widely publicized examples, there are numerous other photographs and video clips that have surfaced over the years, often shared within cryptozoological communities or on specialized websites. These range from blurry snapshots taken by amateur hikers to more substantial, though still debated, video sequences. For instance, there are claims of a

"Georgia Bigfoot" photo and various **"Blue Mountains" footage**, each with its own set of advocates and critics. The common thread among these less famous pieces of evidence is their often-low quality and the difficulty in independently verifying their origin and context. The challenge for the cryptozoologist, and indeed for anyone seeking to understand the truth behind these claims, is to sift through this considerable volume of visual material with a discerning eye. This requires understanding not only the potential for misidentification of known animals and the sophistication of modern (and even historical) fakery techniques but also the ecological principles that would govern the existence of such large, elusive creatures.

The ecological perspective is vital when assessing any visual claim. A large primate, such as a Bigfoot, would require a substantial habitat to support a viable breeding population. This would necessitate a consistent food supply, access to water, and sufficient territory. If such creatures were truly present in the numbers suggested by persistent sightings, one would expect to find more consistent and unambiguous evidence. This could include clear tracks that cannot be attributed to known animals, scat with identifiable primate DNA, or skeletal remains. The fact that visual evidence, particularly photographs and videos, is often the

primary form of evidence presented, while more concrete biological or ecological markers remain elusive, is a significant point of skepticism for many scientists. This doesn't automatically dismiss the visual claims, but it does mean they must be held to a higher standard of scrutiny.

Moreover, the

psychology of perception and suggestion plays a considerable role in how these images and videos are interpreted. Once a viewer is predisposed to believe in Bigfoot, a blurry shape in the woods can easily be seen as the creature. The human brain is remarkably adept at pattern recognition, and when primed with a specific expectation, it can readily impose that expectation onto ambiguous visual stimuli. This phenomenon is amplified in the context of cryptozoology, where fervent belief systems often surround purported evidence. The narrative that accompanies each photograph or video – the story of the fearful hiker, the startled camper, or the curious photographer – can unconsciously influence the viewer's interpretation of the visual data.

The

evolution of technology also complicates the analysis of historical visual evidence. While early hoaxes might have been hampered by the limitations of film and photographic equipment, modern digital manipulation tools have made it easier than ever to create convincing forgeries. This means that older pieces of evidence, while perhaps more difficult to fake convincingly at the time of their creation, must also be re-examined in light of what was *possible* rather than what was *easy*. Conversely, newer evidence captured on digital media is subject to accusations of digital alteration, a defense that is difficult to disprove without access to the original, unedited files and advanced forensic analysis.

In assessing this diverse collection of visual sightings, a critical approach is paramount. This involves asking several key questions for each piece of evidence:

1.

What is the context of the sighting? Who saw it? When and where was it recorded? Were there any other witnesses? Is the location verifiable?

2. **What is the quality of the evidence?** Is it a clear photograph or video, or is it blurry, distant, or obscured? Can details of the alleged creature be discerned?

3. **What are the potential alternative explanations?** Could it be a known animal (bear, deer, primate exhibit, etc.)? Could it be a human in costume or disguise? Could it be a natural phenomenon (shadow, tree formation)? Could it be a deliberate fabrication?

4. **Is there evidence of manipulation or fabrication?** Have experts analyzed the image or video for signs of digital alteration or staged elements?

5. **What are the motivations of the individuals involved?** Was there a financial incentive, a desire for fame, or a genuine belief driving the claim?

6. **Does the evidence align with ecological principles?** Does the purported creature's appearance and behavior make sense within the known ecological realities of the region?

By systematically applying these questions, we can move beyond anecdotal acceptance and begin to evaluate the credibility of the visual documentation more objectively. The goal is not necessarily to definitively prove or disprove the existence of Bigfoot with every piece of evidence, but rather to understand the strengths and weaknesses of the available data, identify areas of recurring concern or potential validity, and to foster a more rigorous and informed approach to the study of purported megafauna primates. Ultimately, while the Patterson-Gimlin film remains a touchstone, a comprehensive understanding requires an examination of the entire spectrum of visual claims, each contributing, however controversially, to the ongoing puzzle.

The persistent pursuit of evidence for cryptids, particularly those as enigmatic as large, undiscovered primates, often grapples with the inherent difficulty of observing rare and elusive animals. Traditional methods, relying on chance encounters and anecdotal reports, are inherently limited. However, the dawn of sophisticated remote sensing and automated detection technologies presents a revolutionary paradigm shift. These advancements offer the potential to move beyond serendipitous sightings and establish a more systematic, objective, and data-driven approach to wildlife research, a strategy that could be instrumental in the search for creatures like Bigfoot.

Camera traps, also known as trail cameras, represent one of the most accessible and widely deployed technologies in this regard. These motion-activated or time-lapse cameras are designed to capture images or video footage of wildlife with minimal human intervention. When strategically placed in hypothesized habitats, such as remote, densely forested regions known for anecdotal sightings or exhibiting ecological conditions suitable for large primates, camera traps can act as tireless, unobtrusive observers. The sheer volume of data they can collect over extended periods, often months or even years, far surpasses what a human research team could achieve through direct observation alone. This continuous surveillance allows for the potential capture of rare events, nocturnal activity, or fleeting glimpses of creatures that actively avoid human presence.

The application of camera traps in cryptozoological research is not without its challenges. The placement of these devices is critical. Identifying areas with the highest probability of visitation requires a deep understanding of the target species' ecological needs – food sources, water availability, territorial ranges, and preferred travel corridors. For a large primate, this might mean positioning cameras near game trails, water sources, or areas with abundant vegetation that could serve as a food supply. Furthermore, the environmental conditions can significantly impact camera trap performance. Extreme temperatures, heavy rainfall, or dense fog can lead to false triggers, reduced image quality, or even damage to the equipment. Critically, the interpretation of the data itself demands rigor. A large, dark shape captured on a camera trap could easily be a black bear, a moose, or even a fallen log. Differentiating between these possibilities requires careful analysis of anatomical features, gait, and behavior, often necessitating high-resolution images or video and comparative studies with known fauna. Nevertheless, the objective nature of camera trap data, recorded without the immediate bias of a human observer, offers a significant advantage over subjective eyewitness accounts. The ability to review footage repeatedly, zoom in on details, and even subject it to digital enhancement for clarity is invaluable.

Beyond visual detection, acoustic monitoring represents another powerful, albeit less explored, avenue for discovering elusive wildlife. Many species, particularly large animals, produce distinct vocalizations for communication, territorial defense, or alarm calls. The deployment of acoustic sensors, or "listening devices," in remote wilderness areas could potentially record sounds that are uncharacteristic of known wildlife in the region. For a hypothetical large primate, this might include deep vocalizations, hoots, or even what are often described as "screams" in anecdotal reports. Modern acoustic sensors are highly sensitive, capable of capturing a wide range of frequencies and amplitudes, and can operate autonomously for extended periods. Data loggers can store vast amounts of audio information, which can then be analyzed using sophisticated software.

The analysis of acoustic data, however, is a complex undertaking. The "noise" of the wilderness – wind, rain, insect chirps, and the calls of known animals – creates a significant challenge in isolating unique or anomalous sounds. Advanced signal processing techniques, including spectral analysis and pattern recognition algorithms, are essential for filtering out background noise and identifying potential target vocalizations. Furthermore, the interpretation of any detected sound requires careful consideration. A unique vocalization could originate from an unknown insect species, a rare bird, or even a geological phenomenon. Establishing the "signature" of a hypothetical primate's call would necessitate a baseline of comparison, ideally from multiple, consistent recordings that exhibit specific characteristics that cannot be attributed to any known source. Comparative bioacoustics, studying the vocalizations of various primate species, could provide valuable insights into the potential range and type of sounds a large, unknown primate might produce. While the direct detection of a Bigfoot vocalization has yet to be definitively confirmed, the potential for acoustic monitoring to provide indirect evidence – by revealing the presence of something unusual that warrants further investigation – remains substantial.

Satellite imagery and aerial surveys, while typically employed for broader ecological assessments and land management, can also offer a macro-level perspective in the search for undiscovered megafauna. High-resolution satellite imagery can be used to map vast tracts of remote wilderness, identifying potential habitats that are otherwise inaccessible to ground-based surveys. Changes in vegetation, water sources, or even the presence of large, unexplained disturbances in the landscape could, in theory, serve as indicators of large animal activity. For instance, unusually large tracks or trails that are not consistent with known species might be visible from aerial platforms if they are of sufficient scale and contrast with the surrounding terrain. Drones equipped with high-definition cameras and thermal imaging capabilities are also becoming increasingly valuable tools. Thermal imaging, in particular, can detect the heat signatures of warm-blooded animals, even in low-light conditions or when they are obscured by dense vegetation. This technology could potentially reveal the presence of large, concealed creatures that would otherwise remain invisible.

However, the effectiveness of satellite and aerial methods for detecting specific elusive animals is limited by resolution and the sheer scale of the environments being surveyed. Identifying a single primate, even a large one, within thousands of square kilometers of dense forest is akin to finding a needle in a haystack. The resolution of even the most advanced satellite imagery may not be sufficient to discern the fine details necessary for positive identification. While aerial thermal imaging might detect a large heat source, distinguishing it from other large mammals or even peculiar geological features would likely require further investigation. Moreover, the cost and logistical complexity of deploying these technologies over vast, remote areas can be prohibitive. These methods are therefore more likely to be useful in identifying areas of high interest for more targeted, ground-based investigations rather than providing direct, definitive proof of an unknown species.

The integration of these various remote sensing technologies into a cohesive search strategy offers the most promising path forward. Instead of relying on single methods, a multi-pronged approach could significantly enhance the chances of success. For instance, satellite imagery could be used to identify large, undisturbed tracts of wilderness that are ecologically suitable for a large primate. Within these identified areas, camera traps could be strategically deployed along game trails and near water sources. Simultaneously, acoustic sensors could be set up to monitor for unusual vocalizations. If thermal imaging surveys from drones or aircraft detect anomalous heat signatures in a particular area, this could prompt the deployment of ground teams and additional camera traps. This synergistic approach allows for the triangulation of potential evidence, where different technologies corroborate findings, thereby increasing confidence in any discovered anomalies.

Furthermore, the data collected by these automated systems can contribute to a more robust understanding of wildlife ecology in general, even if no cryptids are found. This information can help refine our knowledge of biodiversity, habitat use, and population dynamics of known species, which is crucial for conservation efforts. The development of advanced algorithms for automated detection and classification of species from camera trap images and audio recordings is an ongoing area of research, and these advancements will undoubtedly benefit cryptozoological investigations. Imagine an AI system trained to recognize the subtle differences between the gait of a bear and a primate, or the spectral characteristics of a known bird call versus a hypothetical primate vocalization. Such systems could rapidly sift through terabytes of data, flagging potential anomalies for human review.

The challenge of remote sensing and automated detection in cryptozoology is not merely technological; it is also conceptual. It requires a shift in methodology from reactive evidence collection to proactive, systematic searching. It demands the application of rigorous scientific principles, akin to those used in zoological surveys for known species, but adapted to the unique circumstances of searching for something unknown and potentially rare. The ability to withstand skepticism and to present unambiguous, verifiable data is paramount. While the allure of a blurry photograph or a fleeting video is undeniable, the future of evidence gathering for elusive creatures likely lies in the patient, persistent deployment of technologies that can operate beyond the limitations of human perception and presence. These tools, when used intelligently and integrated thoughtfully, offer the best hope for uncovering objective, irrefutable evidence that could finally bring clarity to the enduring mystery of the large, undiscovered primates of our world. The wilderness is vast, and its secrets are many. With the aid of cutting-edge technology, we are better equipped than ever to explore its hidden corners and, perhaps, to finally find what has so long eluded us.

13

The Human Factor: Psychology and Misinterpretation

The wilderness is a canvas of constant flux, a dynamic interplay of light and shadow, form and movement, that our brains are hard-wired to interpret. In the realm of cryptozoology, and particularly when seeking evidence of elusive, large primates, this innate human tendency to find order in chaos can be both a powerful tool and a significant pitfall. Two psychological phenomena, pareidolia and apophenia, play a profound role in how individuals perceive and interpret ambiguous sensory input, often leading to the mistaken identification of ordinary objects or events as evidence of extraordinary creatures.

Pareidolia, at its core, is the psychological phenomenon where the mind perceives a familiar pattern where none exists. This often manifests as seeing faces in clouds, animals in wood grain, or, relevant to our discussion, primates in indistinct shapes within the wilderness. Our brains are exceptionally adept at recognizing faces and familiar forms, a survival mechanism that helped our ancestors quickly identify potential threats or kin. This evolutionary predisposition means that even a random arrangement of shadows, branches, or rock formations can be interpreted by the brain as bearing a resemblance to a known entity, especially when that entity is something we are actively looking for or have preconceived notions about. In the context of searching for Bigfoot or other cryptid primates, a gnarled tree trunk might suddenly appear to possess the broad shoulders and distinctive posture of a primate, or a cluster of shadows cast by dense foliage could be perceived as the hunched silhouette of a bipedal creature. The visual stimulus is inherently ambiguous, and the mind, driven by a desire to categorize and understand, imposes a familiar pattern onto it. The more we are primed to see something – through media portrayals, personal beliefs, or the very act of searching – the more likely we are to experience pareidolic interpretations. A fleeting glimpse of movement in the periphery, a rustle in the undergrowth, or an unusual shadow could be the spark that ignites a pareidolic fire, transforming a mundane natural occurrence into what is perceived as a significant sighting.

The experience is not always a conscious deception; it is a fundamental aspect of human cognition. When confronted with an ambiguous visual cue, the brain doesn't simply register it as "unclear." Instead, it actively attempts to find a match within its existing library of recognized forms. If that library contains strong archetypes of large, hairy, ape-like creatures, then visual stimuli that even vaguely approximate these archetypes are liable to be interpreted as such. This is particularly potent when the observer is in a state of heightened anticipation or excitement, conditions that are almost universally present during a cryptid investigation. The adrenaline surge and focused attention can amplify the brain's pattern-matching capabilities, making it more prone to see what it expects or desires to see. Consider the numerous reports of "Bigfoot" sightings that, upon closer examination or through the lens of psychological interpretation, resolve into misidentified bears, deer, or even inanimate objects. The "fur" might be the dappled sunlight on a mossy log, the "eyes" the glint of moisture on a dark rock, and the "bipedal gait" the sway of a distant tree in the wind. The underlying stimulus is neutral, but the human mind imbues it with meaning through the powerful filter of pareidolia.

Compounding the effects of pareidolia is apophenia, a broader concept that describes the human tendency to perceive meaningful connections between unrelated things. While pareidolia is primarily about pattern recognition in visual or auditory stimuli, apophenia extends to finding significance in coincidences and drawing causal links where none exist. In the context of cryptozoology, apophenia can lead an individual to link disparate pieces of anecdotal evidence – a strange footprint here, an unexplained howl there, a reported sighting in a distant location – and weave them into a cohesive narrative of a cryptid's presence, even if these individual pieces have mundane explanations or are entirely coincidental. A series of identical, large, muddy impressions found over several weeks might be interpreted as the tracks of a single, elusive creature, when in reality, they could be the result of various animals crossing the same soft ground after heavy rainfall, or even the disturbance of sediment by natural processes. Similarly, a series of unusual sounds heard at different times and in different places might be attributed to a single, unknown vocalizing animal, when they could originate from a multitude of sources: wind whistling through rock formations, the calls of known nocturnal animals, or even distant human activities.

Apophenia fosters a sense of confirmation bias. Once an individual begins to believe in the existence of a cryptid, their brain actively seeks out and highlights information that supports this belief, while downplaying or ignoring information that contradicts it. Every unexplained event, every ambiguous piece of evidence, becomes a confirmation of their hypothesis. This cognitive shortcut allows the mind to maintain its belief structure without the cognitive dissonance that would arise from acknowledging contradictory evidence. For example, a researcher might find a tuft of coarse hair snagged on a branch. If they are predisposed to believe in a large, hairy primate, they will immediately interpret this as evidence, perhaps even collect it for analysis. However, they might overlook the abundance of similar hair from known animals in the immediate vicinity, or fail to consider that the hair could have been there for months, long before their current investigation. The coincidence of finding hair in a remote area known for cryptid reports is perceived as significant, a connection forged by apophenia.

The psychological underpinnings of pareidolia and apophenia are deeply ingrained. They are not flaws in the individual's character but rather fundamental aspects of how the human brain processes information. Our evolutionary history has favored individuals who were quick to recognize patterns and make connections, even if those connections were sometimes illusory. In environments where missing a tiger in the tall grass could mean death, it was better to err on the side of caution and see a tiger even if it was just a shadow. This overabundance of pattern detection, while beneficial for survival in ancestral environments, can become a significant impediment to objective observation in modern contexts, particularly in the pursuit of something as elusive and debated as cryptids.

The perpetuation of cryptid folklore is, in part, a testament to the enduring power of these psychological biases. Stories of mysterious creatures have resonated throughout human history, fueled by generations of individuals who have interpreted the ambiguous signs of the natural world through the lens of pareidolia and apophenia. The very act of "searching" for cryptids primes the mind for these interpretations. When an individual ventures into the wilderness with the specific intention of finding evidence of Bigfoot, their perceptual apparatus becomes highly sensitized to anything that might remotely resemble such a creature. This heightened state of expectation can lead to over-interpretation of stimuli. A distant object, partially obscured by fog or foliage, that might be dismissed as a fallen log by a casual observer, could be readily perceived as a crouching cryptid by someone actively looking for one. The ambiguous visual data is then processed, and the brain, influenced by expectation, selects the interpretation that aligns with the pre-existing hypothesis.

Furthermore, the subjective nature of these experiences makes them difficult to quantify and challenging to refute with empirical data. When someone reports seeing a "large primate" in the woods, they are often genuinely convinced of what they saw. Their perception is real to them, shaped by their cognitive processes. It is not a deliberate fabrication, but rather a genuine interpretation of sensory input filtered through psychological biases. This makes direct confrontation of such claims problematic. Simply stating that they must have been mistaken due to pareidolia can be perceived as dismissive and condescending, even if it is the most scientifically accurate explanation.

The implications for scientific cryptozoology, therefore, are profound. While the use of advanced technologies like camera traps and acoustic monitors (as discussed in previous contexts) offers a crucial step towards objective data collection, the interpretation of that data remains susceptible to human cognitive biases. A blurry image from a camera trap, or an anomalous sound recording, can still be subject to pareidolic and apophenic interpretations if the researcher is not vigilant. For example, a series of camera trap images showing a large, dark shape moving through the undergrowth could be interpreted as a cryptid. However, a rigorous analysis would require careful examination of anatomical features, gait, and size relative to known environmental markers. Without such detailed analysis, the temptation to see a familiar cryptid form could override a more mundane, albeit less exciting, explanation like a bear moving through dense brush.

To mitigate the impact of these psychological phenomena, researchers must cultivate a mindset of critical skepticism and embrace rigorous methodological approaches. This includes:

1.

Establishing Baselines and Control Data: Understanding the typical range of stimuli and events in a given environment is crucial. This involves extensive documentation of known fauna, natural phenomena (like wind patterns, geological sounds, and light effects), and even potential equipment malfunctions or environmental artifacts that could create ambiguous visual or auditory signals. Comparing any anomalous findings against this established baseline is essential for identifying genuine deviations rather than coincidental or misinterpreted occurrences.

2.

Adopting a Principle of Parsimony (Occam's Razor): When faced with multiple potential explanations for an observation, the simplest explanation that adequately accounts for the data is generally preferred. In cryptozoology, this means that before attributing an observation to an unknown primate, one must exhaust all plausible explanations involving known species, natural phenomena, or equipment artifacts. A blurry photograph is more likely to be a misidentified known animal or a technical glitch than definitive proof of a cryptid.

3.

Blind Analysis: Where possible, data should be analyzed by individuals who are not aware of the specific hypothesis being tested. This "blind" approach helps to prevent pre-existing expectations from influencing the interpretation of ambiguous data. For instance, if camera trap footage is being reviewed, the analyst might not be told what they are specifically looking for, only that they need to classify the creatures present.

4.

Seeking Corroboration from Multiple, Independent Sources: Evidence that can be corroborated by multiple, independent observations or by different technological methods is significantly more reliable. A single, grainy photograph is far less convincing than a series of high-resolution images, video footage, audio recordings, and physical evidence (like tracks or scat) all consistently pointing to the same unknown entity, and all analyzed rigorously to rule out known explanations.

5.

Training and Education: Researchers and citizen scientists alike need to be educated about the prevalence and impact of psychological biases like pareidolia and apophenia. Understanding how these biases work can empower individuals to recognize them in their own perceptions and to approach evidence with a more critical and objective eye. This includes learning about common misidentifications in the field, the effects of light and shadow, and the varied behaviors of known wildlife that can mimic unusual sightings.

The pursuit of cryptids is, by its very nature, a venture into the unknown, where definitive evidence is scarce and interpretation plays a significant role. However, by acknowledging and actively working to counteract the pervasive influences of pareidolia and apophenia, researchers can significantly improve the objectivity and reliability of their investigations. It is through a disciplined commitment to scientific rigor, combined with an understanding of human psychology, that we can hope to sift through the noise of misinterpretation and, perhaps, one day uncover genuine evidence of the extraordinary within the vast, complex tapestry of the natural world. The wilderness will always present us with ambiguous stimuli; our task as investigators is to ensure that our minds are not the primary architects of the extraordinary from the ordinary.

The human mind is a powerful tool, capable of incredible feats of observation and deduction. However, it is also susceptible to a myriad of subtle influences, none more potent than the power of suggestion and the weight of expectation. When we venture into environments steeped in folklore or where tales of unknown creatures have been woven into the very fabric of local culture, our minds can become unwitting participants in a grand theater of perception. In such settings, the observer is not a blank slate; they are a canvas already painted with the hues of legend, ready to interpret the natural world through the vibrant, and sometimes distorted, lens of what they *expect* or have been *told* to see.

Consider regions where stories of elusive, ape-like beings have persisted for generations. These narratives, passed down through oral tradition, local histories, and increasingly, through mass media, create a powerful psychological template. When an individual enters these territories, especially with the specific intention of finding evidence of such a creature, they are not merely observing; they are actively searching for confirmation of pre-existing beliefs. This pre-priming can dramatically alter how sensory input is processed. A fleeting movement in the peripheral vision, which under normal circumstances might be dismissed as a deer or a rabbit, can, in a cryptid-rich environment, be instantly cataloged as the unmistakable silhouette of a bipedal primate. The sheer expectation of encountering something extraordinary lowers the threshold for what is considered definitive proof.

This phenomenon is deeply intertwined with cultural narratives. From the Yeti of the Himalayas to the Sasquatch of North America, and the Yowie of Australia, these tales serve as cultural touchstones, imprinting specific images and characteristics onto the collective consciousness.

When a modern-day investigator, armed with cameras and recording equipment, enters the presumed habitat of such a creature, they are often unconsciously seeking to validate these ancient stories. The environment itself becomes imbued with the legend; a particularly dense patch of forest, a secluded cave, or a remote mountain pass is no longer just a geographical feature but a potential hiding place, a stage upon which the drama of the legend might unfold.

The power of suggestion, often amplified in these contexts, can be incredibly persuasive. Imagine a group of researchers discussing a recent "sighting" near a particular creek. One member, perhaps influenced by a vivid anecdote, describes a specific detail – a peculiar gait, a distinct vocalization, or an unusual odor. When other members of the group later explore that same creek, they may find themselves actively listening for that specific vocalization, looking for that particular gait, or sniffing the air for that characteristic smell. Even if the original report was based on a misinterpretation, the repeated suggestion can lead subsequent observers to "perceive" what they have been primed to expect. This is not necessarily conscious deception; it is a testament to how deeply suggestion can penetrate our perceptual filters.

This is particularly evident in the interpretation of physical evidence. In areas rife with Bigfoot folklore, for instance, a large, unusual footprint is often immediately attributed to the creature itself. However, without rigorous analysis and careful consideration of alternative explanations, the power of suggestion can lead to the acceptance of flawed evidence. A track that is partially obscured by mud or leaf litter, or one that has been distorted by erosion or the melting of snow, might be interpreted as bearing the unmistakable characteristics of a Sasquatch foot, when in reality, it could be a composite of animal tracks, geological formations, or even human interference. The observer, steeped in the local lore and eager to find concrete evidence, might overlook the inconsistencies or the more prosaic explanations.

The narrative itself acts as a powerful heuristic, a mental shortcut that guides interpretation. When faced with ambiguity – a strange sound at night, a fleeting glimpse of movement, an unusual odor – the brain doesn't always engage in a neutral, objective analysis. Instead, it can default to the most salient, culturally reinforced explanation. If the cultural narrative strongly suggests the presence of a large, hairy primate, then ambiguous stimuli are far more likely to be shoehorned into that narrative. A distant, dark shape moving through the trees becomes Bigfoot. A series of deep, resonant calls becomes the howl of a Sasquatch. The principle of parsimony, which suggests that the simplest explanation is usually the correct one, is often abandoned in favor of the more compelling, albeit less probable, explanation provided by the prevailing legend.

This dynamic can be observed even in seemingly objective scientific endeavors. Researchers who have spent years studying a particular cryptid legend might, over time, develop a subconscious bias towards interpreting anomalous data in a way that supports their hypothesis. This is not to say that they are intentionally falsifying data, but rather that their perceptual framework has been subtly shaped by years of immersion in a specific belief system. A slightly blurry photograph from a camera trap, which might be classified as "indeterminate large mammal" by a dispassionate observer, could be readily identified as "clearly a Sasquatch profile" by someone deeply invested in the legend. The expectation of seeing the cryptid primes them to interpret the limited visual information in a way that conforms to their pre-existing mental image.

The impact of cultural narratives extends beyond individual perception to influence the very way evidence is gathered and presented. Stories of cryptids often evolve over time, with new details and characteristics being added as they are retold. These evolving narratives can influence what witnesses report and what researchers look for. For example, if a particular cryptid is traditionally described as having a distinctive red eye, later reports might focus on searching for this specific feature, even if it was not part of the original legend or if it is a feature that could easily be mimicked by natural phenomena like bioluminescence or reflections.

Furthermore, the social aspect of belief plays a significant role. In communities where belief in a local cryptid is widespread, there can be social pressure to conform to those beliefs. Individuals who report sightings may be encouraged and validated, while those who express skepticism might be ostracized or dismissed. This can lead to a feedback loop where anecdotal evidence is reinforced and amplified, creating a strong sense of communal certainty even in the absence of conclusive proof. When a researcher enters such an environment, they are not just collecting data; they are navigating a complex social landscape where beliefs are deeply entrenched and where the power of suggestion can be exerted by the community itself.

The way in which information is disseminated also plays a critical role. In the age of the internet and social media, cryptid lore can spread with unprecedented speed and reach. A single compelling anecdote, a grainy photograph, or a purported piece of evidence can go viral, instantly exposing millions of people to the narrative and priming them to interpret their own experiences through that lens. This widespread exposure means that even individuals who have never personally encountered ambiguous phenomena in the wild may still be susceptible to the power of suggestion, armed with a preconceived notion of what to expect should they ever do so.

The challenge for the cryptozoologist, therefore, is not merely to find evidence but to remain rigorously objective in the face of deeply ingrained cultural narratives and powerful psychological predispositions. It requires a conscious and continuous effort to question one's own assumptions, to critically evaluate all evidence, and to actively seek out alternative, mundane explanations before resorting to extraordinary ones. This means understanding that a "Bigfoot territory" is still a forest inhabited by bears, deer, and a host of other known creatures, and that an "unexplained sound" is far more likely to be a known animal, geological activity, or atmospheric phenomenon than the call of an unknown primate.

The cultural weight of these legends can be so immense that it becomes difficult for individuals to separate their personal perceptions from the ingrained stories they have grown up with. This is not a failing of the individual but a testament to the pervasive influence of shared narratives on human cognition. The power of suggestion, amplified by cultural expectation, can transform an ordinary encounter with nature into an extraordinary one, blurring the lines between observation and imagination, and making the quest for objective truth in cryptozoology a profoundly challenging, yet utterly essential, endeavor. It necessitates a deep understanding of how our minds, shaped by the stories we tell, can often be the most formidable obstacles in our pursuit of the unknown.

The persistent pursuit of the unknown, so central to the allure of cryptozoology, is unfortunately not solely a quest for discovery but also a battlefield where deliberate deception and elaborate fabrication often obscure genuine inquiry. The history of cryptozoology is unfortunately rife with instances where fabricated evidence has not only misled the public but has also cast a long shadow of doubt over legitimate research efforts. These hoaxes, ranging from simple misidentifications amplified by desire to outright, calculated deceptions, represent a significant hurdle that any serious cryptozoologist must be prepared to confront and meticulously navigate. The very human desire to believe, coupled with the potential for fame or financial gain, can unfortunately spur individuals to create elaborate falsehoods, leaving a trail of confusion and distrust in their wake.

One of the most infamous examples of a hoax that significantly impacted the Sasquatch narrative is undoubtedly the Patterson-Gimlin film. While its authenticity remains a subject of passionate debate, the prevailing scientific and skeptical consensus leans towards it being a staged event, likely involving a person in a costume. The film, depicting a large, hairy bipedal creature walking across a clearing in Northern California, became an iconic piece of evidence for Bigfoot proponents. However, the circumstances surrounding its creation, the inconsistencies in the alleged creature's gait and proportions when analyzed, and the subsequent admissions and conflicting statements by some involved have led many to conclude it was a deliberate fabrication designed to capitalize on the burgeoning interest in the cryptid. The impact of this film cannot be overstated; it provided a visual anchor for the Bigfoot legend, influencing countless subsequent reports and fueling a generation of enthusiasts and researchers. Yet, its potential status as a hoax serves as a potent reminder that even seemingly compelling visual evidence requires rigorous scrutiny and must be corroborated by other independent lines of evidence. The legacy of the Patterson-Gimlin film underscores the critical importance of provenance, verification, and a healthy dose of skepticism when evaluating any singular piece of evidence in cryptozoology.

Beyond the realm of visual media, hoaxes have also manifested in the form of purported physical evidence. For instance, the landscape of Bigfoot research has been littered with casts of enormous footprints that, upon closer inspection, often reveal tell-tale signs of human manipulation. These can include inconsistencies in the depth and pressure of the print, unnatural symmetry, or the presence of molding marks where artificial materials were used. Skeptics have long argued that many of these tracks could be created using carved wooden feet or other effigies, intentionally placed to deceive. The motivation behind such hoaxes can vary; some might be seeking attention, others a laugh, and some perhaps even attempting to sell fabricated evidence to eager collectors or researchers. The painstaking work of anthropologists and forensic scientists in analyzing track evidence is crucial here. They can identify subtle anatomical impossibilities, geological anomalies that suggest artificial placement, or even the use of materials that would not be found naturally in the purported environment. The existence of these fake tracks necessitates a protocol of extreme caution for any investigator, requiring them to document not just the print itself but its entire context, including soil composition, surrounding flora and fauna, and any potential for human interference or natural distortion.

The phenomenon of hoaxes is not confined to North America or the Sasquatch legend. Across the globe, various cryptids have been the subject of similar fabrications. In the case of the Yeti, for instance, numerous alleged Yeti scalps, hands, and footprints have been presented over the decades. Scientific analysis, however, has consistently revealed these to be misidentified remains of known animals, often bears or other local fauna, or in some cases, outright fraudulent artifacts. The Pangboche Yeti hand, once a celebrated relic, was eventually identified as belonging to a primate. The desire to believe in the mystical Yeti has unfortunately provided fertile ground for those willing to exploit this sentiment for personal gain or notoriety, often by creating "evidence" from more common animal parts or by embellishing existing artifacts. This pattern highlights a universal theme: when a legend becomes deeply embedded in a culture, and when there is significant public fascination, the potential for exploitation through fabricated evidence increases exponentially.

The psychological underpinning of these hoaxes is a complex interplay of factors. For the hoaxer, it can stem from a desire for attention, financial reward, a misguided sense of contributing to a cause they believe in, or even a simple prank. For the believer, the eagerness to confirm their deeply held beliefs can sometimes override critical thinking, making them susceptible to accepting flawed or fabricated evidence at face value. This is not to imply a lack of intelligence or discernment in those who have been deceived, but rather to acknowledge the powerful influence of confirmation bias and the emotional investment involved in believing in something extraordinary. A compelling narrative, combined with a piece of "evidence" that seems to fit, can create a potent psychological synergy that makes it difficult to question.

The scientific method, with its emphasis on rigorous verification, peer review, and falsifiability, is the ultimate bulwark against the tide of hoaxes. In cryptozoology, this translates to a demand for multiple, independent lines of corroborating evidence. A single blurry photograph, an anecdotal sighting, or a solitary footprint is rarely sufficient. Instead, researchers must seek out verifiable data such as clear photographic or video evidence captured under controlled conditions, DNA samples that can be definitively linked to an unknown primate, or fossil evidence that supports the existence of a previously undiscovered species. The process of scientific verification involves painstaking analysis. For instance, photographic evidence is subjected to scrutiny regarding lighting, perspective, and potential for digital manipulation. Biological samples are analyzed through DNA sequencing to determine their taxonomic origin. Trace evidence, like hair or scat, is examined for microscopic characteristics and chemical composition. Each piece of evidence must stand up to intense scrutiny and be consistent with established biological and physical laws, unless overwhelming counter-evidence exists.

Moreover, understanding the common methods of deception is crucial for cryptozoologists. This includes being aware of how natural phenomena can be misinterpreted or intentionally mimicked. For example, unusual tracks can be created by water erosion, melting snow, or the combined tracks of multiple known animals. Strange sounds can be attributed to wind, geological activity, or known animal vocalizations that are unfamiliar to the observer. Even supposed "eyewitness testimony," while valuable, is notoriously unreliable and prone to suggestibility, misperception, and memory distortion. Therefore, a seasoned cryptozoologist must be trained to identify and account for these potential sources of error and deception, both natural and artificial.

The ethical implications of hoaxes are also significant. When deliberate fabrications are presented as genuine, they not only mislead the public but also drain valuable resources, time, and credibility from legitimate research efforts. This can make it harder to secure funding, attract serious scientific interest, and gain public support for further investigation. The cryptozoological community itself has a responsibility to police its own ranks, fostering a culture of critical inquiry and discouraging the promotion of unsubstantiated or fraudulent claims. This requires a willingness to challenge even the most cherished pieces of evidence if they fail to withstand rigorous scrutiny.

Ultimately, the challenge posed by hoaxes in cryptozoology is a perpetual one. As long as the allure of the unknown persists, so too will the temptation for individuals to fabricate evidence. For the dedicated cryptozoologist, the commitment to a scientific approach, characterized by unwavering skepticism, meticulous verification, and a commitment to truth above all else, is not merely a methodological preference but an ethical imperative. It is through this rigorous and disciplined approach that the genuine mysteries of the natural world can, perhaps one day, be brought to light, free from the deceptive shadows of fabricated proof. The goal is not to debunk every claim, but to ensure that when a discovery is announced, it is one that has been tested against the fires of critical examination and emerged as genuine, rather than a fleeting illusion conjured by deception.

The allure of the undiscovered often leads us to project our desires and expectations onto the familiar, and in the realm of cryptozoology, this psychological tendency frequently results in the misidentification of known species. The very notion of a colossal, elusive primate lurking in the wilderness fuels a powerful pre-conditioning in the observer's mind. When confronted with an unexpected stimulus in their environment, particularly under conditions that are less than ideal for clear observation, the brain can, quite readily, superimpose the image of the cryptid onto whatever it is that is actually present. This is not necessarily a conscious act of deception; rather, it is a product of how our perceptual systems are wired, influenced by our knowledge, beliefs, and emotional state.

Among the most frequently implicated known species in the misidentification of large, hairy cryptids, particularly those of the Sasquatch or Yeti ilk, are bears. Consider the North American black bear (

Ursus americanus) or the grizzly bear (*Ursus arctos*). Both are large, powerful mammals covered in fur, capable of standing on their hind legs for short periods. In low light, such as dawn or dusk, or during inclement weather like fog or heavy rain, the distinctive features of a bear can become obscured. Its quadrupedal stance can be momentarily forgotten or misinterpreted, especially if the observer catches only a fleeting glimpse from a distance. The characteristic hump on a grizzly's shoulders might be missed, or the angle of observation might disguise the bear's true posture. Add to this the inherent fear or excitement an individual might experience upon encountering a large wild animal, and the stage is set for misperception. The human observer, already primed by stories and legends of Bigfoot, might unconsciously interpret the bear's silhouette, its movement, or even its vocalizations in a way that aligns with their pre-existing expectations of a cryptid. A rustling in the undergrowth, a shadow moving between trees, or the sound of a startled animal can all be easily translated into evidence of something far more extraordinary than a common ursine. The dark fur of many bears can blend seamlessly with the shadows of a forest, and their size, especially when partially obscured or viewed from a disadvantageous angle, can easily be overestimated. Furthermore, bears can be surprisingly agile and can move through dense vegetation in a manner that might appear surprisingly "human-like" to an untrained or anxious eye, particularly when they are startled and moving rapidly. The way a bear shifts its weight, the powerful swing of its limbs, or even the apparent "walking" motion when standing on its hind legs can all contribute to a mistaken identity, especially if the observer is not a seasoned wildlife biologist familiar with ursine behavior.

Beyond bears, other large mammals can also be the source of misidentifications. In regions where other large primates exist, such as orangutans in Southeast Asia or gorillas and chimpanzees in Africa, there exists a potential for confusion, though this is less common for the typically North American or Himalayan cryptid sightings. However, in broader discussions of ape-like creatures, it is worth considering. For instance, a troop of baboons or other large monkeys moving through dense foliage might, from a distance, present a fleeting impression of a smaller, ape-like primate. Their collective movement, the way they brachiate or swing through trees, and their dark, often shaggy coats could, under specific circumstances, be misinterpreted. Even a lone primate, if seen briefly and out of context, might trigger a mistaken association. The key here, as with bears, lies in the suboptimal viewing conditions. Distance, poor lighting, intervening vegetation, and the observer's emotional state are all critical factors. A witness who has never seen a large primate in the wild, and who is perhaps already predisposed to believing in such creatures, might easily project the image of a Yeti or Sasquatch onto a fleeting glimpse of a known, albeit less common in their experience, primate.

Perhaps more surprisingly, even mundane objects or natural phenomena can, under duress or in poor visibility, be misconstrued. For example, a large, moss-covered stump or a fallen log can, in the dim light of a forest, take on an almost animate quality. If a witness is startled by a noise nearby, their gaze might fall upon such an object, and their fearful mind might imbue it with the characteristics of a lurking creature. Similarly, the shadows cast by trees, especially when they are twisted and gnarled, can create anthropomorphic shapes that are then misinterpreted as animate beings. The swaying of branches in the wind, especially when accompanied by peculiar rustling sounds, can create an illusion of movement that, for someone on edge, could easily be perceived as the stealthy progress of a large animal. Even the condensation or mist rising from a damp forest floor, particularly in the early morning, can create ephemeral, ghostly shapes that play tricks on the eye. These are not instances of deliberate deception, but rather fascinating examples of how the human brain attempts to make sense of ambiguous sensory input, often defaulting to familiar or striking interpretations.

The role of human appearance itself, in unusual circumstances, cannot be entirely discounted. While most cryptid reports describe creatures that are distinctly non-human, there have been instances where individuals, perhaps disheveled, obscured by distance, or wearing unusual clothing, have been mistaken for something more. Consider a lone hiker or trapper, perhaps wearing a dark, hooded cloak or camouflage clothing, moving through dense woods. If observed from a considerable distance, or through a thick screen of trees, their human form could, for a brief moment, be perceived as something else, particularly if the observer is not expecting to see another human. The dark, bulky appearance of certain outdoor gear, combined with the natural shadows and textures of a wilderness environment, can create a visual ambiguity. While this is less likely to explain the typical Sasquatch or Yeti descriptions, it is a factor in the broader spectrum of potential misidentifications, especially in cases where the report is vague or the visual evidence is poor. It is also important to consider the possibility of individuals in costumes or elaborate disguises, though as discussed in previous contexts, this often veers into the realm of deliberate hoaxes rather than accidental misidentification. However, the line can blur; an amateurish or poorly constructed costume, seen briefly and in poor light, could be perceived as a genuine, albeit strange, creature by an unprepared observer.

The psychological dimension is paramount in understanding these misidentifications. Fear is a powerful modulator of perception. When an individual feels threatened or experiences a heightened state of arousal, their sensory input becomes both amplified and distorted. The focus narrows, and the brain prioritizes potential threats. This can lead to an exaggeration of perceived size and movement, and a tendency to interpret ambiguous stimuli as dangerous. The classic "fight or flight" response can cause the visual cortex to work overtime, trying to identify a threat, and in doing so, it can fill in the gaps with whatever is most readily available in the observer's mental lexicon. If that lexicon includes images of large, hairy beasts, then that is what the observer is likely to "see."

Confirmation bias plays an equally significant role. Individuals who hold a strong belief in the existence of cryptids are more likely to interpret ambiguous evidence as supporting their beliefs. A fleeting glimpse of a dark shape in the woods, which for a disbeliever might be dismissed as a bear or a shadow, becomes, for a believer, potent evidence of a Sasquatch. This bias can also affect the way details are recalled and reported. An observer might unconsciously embellish or alter details in their recollection to better fit the expected characteristics of the cryptid. They might remember the shape as more bipedal than it actually was, or the fur as darker and longer, simply because that aligns with their preconceived notions. The power of suggestion is also a factor. If an observer has been discussing or reading about cryptids, their mind is already primed to look for such creatures. Even a common animal or an unusual natural phenomenon can be filtered through this lens of expectation.

The quality of the sighting conditions is a critical objective factor that exacerbates these psychological tendencies. Distance is a primary culprit. Objects at a significant distance lose detail rapidly. Even large animals can appear smaller and less distinct. The farther away a creature is, the easier it is for its specific features to be lost, leaving only a general shape and movement. This is where the "silhouette effect" becomes prominent. A general outline, particularly in low light or against a cluttered background, can be remarkably deceiving. Poor lighting conditions, as previously mentioned, are notorious for obscuring details and creating misleading shadows. The spectrum of visible light is reduced, making it harder to distinguish colors, textures, and fine features. Dusk, dawn, moonlight, and even dense cloud cover can transform familiar shapes into something alien. Fog and mist are even more effective at reducing visibility, creating diffuse light and obscuring distances, which can make even relatively close objects appear as indistinct masses. Intervening vegetation – trees, bushes, tall grass – acts as a natural camouflage and a barrier to clear observation. A creature moving through dense forest will inevitably be partially hidden, its form broken up by leaves, branches, and trunks. This fragmented view is ripe for misinterpretation. The movement itself can also be a source of confusion. A large animal moving quickly through the undergrowth, or a bear standing and looking around, can present a series of poses and movements that, when viewed briefly, might be mistaken for something else entirely. The speed at which a creature moves can also contribute to misidentification; the faster something moves, the less time the observer has to process its features accurately.

Consider a concrete example: an individual hiking in the Pacific Northwest. They have recently been engrossed in reading about Sasquatch encounters. It is late afternoon, and the sunlight is beginning to fade, casting long shadows through the dense Douglas fir and cedar forest. The hiker hears a sudden, loud snapping of a twig off to their right, deeper in the woods. Startled, they whip their head around. Between two thick trees, perhaps fifty yards away, they see a large, dark, upright shape. It is too far to make out clear features, but it appears bulky and covered in dark hair. For a moment, it seems to sway, and then it moves, disappearing behind a large fir. The hiker's heart races. Their mind, primed by their reading, immediately jumps to "Sasquatch." However, what they likely saw was a black bear, perhaps a sow with a cub, startled by the hiker's presence. The bear may have stood on its hind legs for a moment to get a better look or scent the air, a behavior that can be misinterpreted as bipedal locomotion. The fading light would obscure its features, making its fur appear uniformly dark. The intervening trees would break up its form. The sound of its movement through the undergrowth would be amplified by the hiker's heightened senses. In this scenario, a common wild animal, observed under suboptimal conditions and interpreted through the lens of pre-existing belief, is easily transformed into a cryptid.

Another common scenario involves sightings in mountainous or forested terrain where visibility is naturally limited by terrain and weather. Imagine an individual on a remote logging road, perhaps driving slowly. They see something large moving on a distant hillside, partially obscured by trees and perhaps a light mist. The object appears to be moving on two legs, and it is dark and bulky. The distance might be hundreds of yards, and the lighting conditions far from ideal. A person prone to believing in Bigfoot might instantly conclude they are observing a Sasquatch. However, it could just as easily be a large moose or elk, which are known to stand and move in ways that can appear somewhat ungainly or bipedal-like from a distance, especially if they are looking around or grazing on a slope. Their large, dark bodies, covered in thick hair, can blend into the forest, and the distance and atmospheric conditions can smooth over their specific features, making them appear more generalised and "hairy." In many cases, these sightings are reported with a degree of certainty that belies the actual limitations of the observation. The witness may feel certain they saw a bipedal creature, but without clear, verifiable details, this certainty is often an artifact of psychological interpretation rather than objective observation.

The scientific approach to cryptozoology necessitates a rigorous deconstruction of such accounts. It involves not just documenting the sighting, but also meticulously analyzing the environmental conditions, the observer's state, and the potential for misidentification. This means asking critical questions: What was the lighting like? How far away was the subject? What were the weather conditions? Was there any intervening vegetation? What is the observer's background knowledge of local wildlife? What was the observer's emotional state? By systematically investigating these factors, researchers can begin to differentiate between genuine anomalies and the more prosaic, albeit still fascinating, phenomena of misidentification. The goal is not to dismiss every extraordinary claim, but to ensure that extraordinary claims are supported by extraordinary evidence, and that when we encounter the seemingly unknown, we first exhaust all plausible explanations involving the known. The wild is full of surprises, and the ability to correctly identify the creatures that inhabit it, even when they are startling or unusual in their behavior, is a fundamental skill for any naturalist, whether their focus is on established zoology or the more elusive corners of cryptozoology. Without this critical discernment, the pursuit of the unknown risks becoming a catalog of errors, a testament to human perception's fallibility rather than a true exploration of biodiversity.

The inherent subjectivity of human observation, coupled with the powerful influence of psychological predispositions and the unfortunate reality of deliberate deception, presents a formidable challenge in the scientific pursuit of cryptids. It is a landscape where the observer's own mind can become as much a filter, or indeed a distorting lens, as any physical impediment to clear sight. This section delves into the critical need to build a robust scientific case that can not only withstand but actively account for these pervasive human factors, ensuring that any purported evidence for unknown species is not merely a product of misinterpretation or fabrication. The scientific method, with its emphasis on objectivity, repeatability, and falsifiability, becomes our indispensable toolkit in navigating this complex terrain.

One of the most significant hurdles is the phenomenon of observer bias. As previously touched upon, our existing beliefs and expectations profoundly shape what we perceive. An individual who has spent years researching, discussing, or even yearning for the existence of a particular cryptid, such as a Yeti or a "Chupacabra," will invariably interpret ambiguous stimuli through that pre-existing framework. This is not necessarily a conscious act of self-deception; rather, it is a deep-seated cognitive process. When confronted with an unusual sound in the jungle, a fleeting shadow in the mountains, or even an odd footprint in the mud, the brain, seeking to reconcile new information with established mental models, may readily fill in the gaps with the expected cryptid. The challenge for the cryptozoologist is to recognize and mitigate this bias, both in oneself and in the witnesses whose accounts are being examined. This involves adopting a stance of critical skepticism, even towards one's own most fervent desires. It means systematically evaluating every piece of evidence against the most mundane and plausible explanations first. Could that strange howl be a misidentified bird of prey? Could that fleeting glimpse be a known animal moving in an unusual way? Could that footprint be a natural geological formation or the result of erosion? The scientific approach demands that we actively seek disconfirming evidence, not just confirming anecdotes.

Furthermore, the reliability of eyewitness testimony, a cornerstone of many cryptid reports, is notoriously variable. Human memory is not a perfect recording device. It is reconstructive, malleable, and susceptible to suggestion. Stress, excitement, fear, and the passage of time can all distort recall. Consider the infamous "flashbulb memory" phenomenon, where highly emotional events are recalled with a sense of vivid detail, yet studies have shown these memories can be remarkably inaccurate. A witness who has had a startling encounter with what they believe to be a cryptid may recall the event with absolute certainty, believing their memory to be flawless. However, the emotional intensity of the experience can paradoxically lead to the embedding of inaccuracies. The brain, under duress, may prioritize survival-related information and emotional salience over precise detail. This is why corroborating evidence, particularly objective, physical evidence, is so crucial in cryptozoological investigations. A detailed description of a creature, while compelling, is far less scientifically valuable than a clear photograph, a verifiable audio recording, or, ideally, a biological sample.

The role of suggestion cannot be overstated. In communities where tales of local cryptids are prevalent, the power of suggestion can lead individuals to misinterpret ordinary phenomena as evidence of the extraordinary. Children, in particular, can be highly susceptible to this, their imaginations fueled by local folklore. Even adults can be influenced. If a community is abuzz with stories of a "swamp ape," an observer encountering an unusually large wading bird or a decaying log in the water might be predisposed to interpret it as the elusive creature. This is where the provenance of a report matters. Is it an isolated account from an individual with no prior exposure to local legends, or is it one of many similar reports circulating within a culture steeped in the belief of a particular cryptid? While a shared belief in a cryptid can sometimes lead to more consistent descriptions, it can also amplify the effects of suggestion and collective misinterpretation.

Deception, both conscious and unconscious, adds another layer of complexity. While many misidentifications are genuine errors of perception, the possibility of hoaxes, deliberate or otherwise, must be considered. Fabricated footprints, staged photographs, and fabricated stories have plagued cryptozoology since its inception. The motivations behind such deceptions can vary widely: a desire for attention, financial gain, a misguided sense of humor, or even a profound belief that by creating "evidence," one can legitimize the existence of a creature one truly wishes to be real. Identifying hoaxes requires a keen eye for inconsistencies, for signs of artificiality, and for motivations that might drive fabrication. For instance, a series of alleged footprints that appear too perfect, too consistent in their depth and shape, and lacking the subtle variations one would expect from a living creature interacting with a natural substrate, might raise suspicions. Similarly, photographs or videos where the alleged cryptid appears unnaturally posed or where digital manipulation is evident warrant intense scrutiny.

The scientific approach to building a case must therefore be multi-faceted, employing a rigorous hierarchy of evidence. At the bottom of this hierarchy lie anecdotal accounts – personal stories and eyewitness testimonies. While these can be valuable for generating hypotheses and identifying potential areas of interest, they are inherently unreliable as standalone proof. Above these are photographic and video evidence. While more objective, these too are vulnerable to misinterpretation, poor quality, and outright fabrication. Many famous cryptid photographs have later been revealed as hoaxes or misidentifications. Then come audio recordings, which can capture vocalizations that may be difficult to attribute to known species, but again, are susceptible to misinterpretation of environmental sounds or the calls of known animals.

Higher still in the hierarchy are trace evidences, such as footprints, hair, scat, or bodily fluids. These offer the potential for objective scientific analysis. A footprint, if clearly preserved and found in multiple locations, can provide information about gait, size, and weight distribution. Analysis of hair or scat can reveal DNA, diet, and even potential geographic origin. If such evidence can be collected without contamination and subjected to rigorous laboratory testing by independent experts, it moves closer to becoming scientifically significant. The key here is the ability to rule out all known species. If a hair sample can be definitively identified as belonging to a known primate or bear, then it is no longer evidence for an unknown species, regardless of how it was found or what the witness believes. The true scientific breakthrough would come from DNA that does not match any known species in the global genetic database.

At the apex of the hierarchy lies direct capture or the recovery of a deceased specimen. This is the "gold standard" for proving the existence of a new species. While notoriously difficult and perhaps unlikely in the case of elusive, large-brained creatures, it represents the ultimate form of verifiable evidence. The recovery of a body allows for complete anatomical study, genetic sequencing, and detailed ecological analysis.

The process of sifting through these layers of evidence requires a systematic methodology that actively seeks to disprove claims before accepting them. This involves:

1. **Thorough Witness Interviews**: Conduct interviews in a neutral, non-leading manner. Employ open-ended questions and actively listen for inconsistencies or vagueness. Crucially, assess the witness's credibility and potential biases without being dismissive. Understanding their background knowledge of local fauna is paramount.
2. **Environmental Analysis**: Carefully document the sighting location, including terrain, vegetation, altitude, and typical wildlife. Analyze weather conditions, lighting, and visibility at the time of the sighting. Such analysis can often reveal plausible alternative explanations. For instance, a sighting of a "hairy biped" in an area known for its dense fog might strongly suggest a misidentification of a distant, obscured object.
3. **Trace Evidence Collection and Analysis**: If physical evidence is found, it must be collected using proper forensic techniques to prevent contamination. This includes careful documentation of the find's context, photography with scale, and appropriate preservation methods. Samples should then be sent to accredited laboratories for independent

analysis, with a focus on DNA, morphology, and chemical composition. The protocol must include testing against the full spectrum of known species, not just a select few.

4. **Photographic and Video Analysis**: Employ image enhancement and forensic analysis techniques to detect manipulation. Examine metadata for clues about the origin and editing history of digital files. Compare the subject's appearance and behavior in the images/video against known species and their behaviors under similar environmental conditions. Shadow, light, and perspective can play tricks, making familiar objects appear alien.

5. **Audio Analysis**: Use spectrographic analysis to break down sound recordings. Compare the frequency, duration, and patterns of the vocalizations against known animal sounds. Investigate potential sources of unusual ambient noise.

6. **Comparative Anatomy and Physiology**: If detailed descriptions or physical evidence are available, compare them with the known anatomical and physiological characteristics of potential misidentified species. This requires a deep understanding of comparative zoology. For instance, do the reported limb proportions of a cryptid align more closely with a primate, a bear, or another animal?

The scientific case is not built on a single piece of evidence, but on the convergence of multiple, independent lines of evidence that consistently point towards the existence of an unknown entity, while simultaneously ruling out all known explanations. It requires a willingness to follow the data, even when it leads to uncomfortable conclusions – either supporting the existence of a cryptid or definitively identifying a misidentification or hoax. This rigor is what distinguishes genuine scientific inquiry from speculative enthusiasm.

Consider the challenge of distinguishing between a genuine unknown primate and a known primate exhibiting unusual behavior or seen under bizarre circumstances. In regions with existing primate populations, misidentification is a more immediate concern than in areas where no large primates are known to exist. For example, reports of ape-like creatures in North America are often explained by bears or even misidentified humans. However, if a credible report emerges from a region with a known, but reclusive primate population – such as orangutans in dense, remote Indonesian rainforests – the possibility of a novel species or subspecies becomes more plausible, provided that all known orangutan species and variations are definitively ruled out by thorough scientific analysis of any evidence. The burden of proof lies in demonstrating *why* it cannot be a known species, before asserting it is an unknown one.

The scientific approach also acknowledges the potential for "cryptids" to represent subspecies, rare morphs, or even populations of known species that have adapted in ways that make them appear distinct. A population of wolves that has become unusually large and reclusive due to specific environmental pressures and hunting patterns might, to an uninformed observer, seem like something more. Similarly, variations in coloration or fur density within a species, particularly when observed in suboptimal lighting or at a distance, can lead to mistaken identity. The scientific cryptozoologist must have an encyclopedic knowledge of the natural world, understanding the full spectrum of variation within known species, including genetic anomalies, unusual behaviors, and adaptations to extreme environments.

Ultimately, building a scientific case amidst the undeniable presence of human error, psychological projection, and deliberate deception is an exercise in extreme diligence. It demands a paradigm of hypothesis testing where the primary hypothesis is always "misidentification" or "hoax." Only when every plausible explanation involving known phenomena and human fallibility has been rigorously explored and definitively refuted can the scientific community begin to entertain the extraordinary claim of an undiscovered species. This meticulous process, while perhaps less thrilling than accepting fantastical tales at face value, is the only pathway to genuine discovery and the responsible advancement of zoological knowledge. The goal is not to debunk every mystery, but to solve it definitively, ensuring that any cryptid elevated to the status of a scientific entity is so through irrefutable, objective evidence that has survived the most stringent scientific scrutiny.

14

Towards a Framework for Future Research

The pursuit of understanding potential undescribed megafauna, particularly those resembling primates, necessitates a strategic approach to research. Simply gathering anecdotes or sensational reports, while often the genesis of such inquiries, is insufficient for establishing a scientific foundation. Instead, future research must be guided by a series of carefully formulated, empirically answerable questions. These questions should prioritize areas where concrete data can definitively address the plausibility, characteristics, and ecological role of any putative species, moving the discussion beyond speculation and towards verifiable knowledge. The inherent challenges in studying elusive creatures mean that research questions must be designed to yield actionable data, even under difficult circumstances, and must be framed within the context of established biological and ecological principles.

Foremost among these critical research questions is the assessment of **population viability**. For any species to persist, especially a large-bodied primate that would likely require significant resources, a viable breeding population is a fundamental requirement. Therefore, a key question becomes: "What is the minimum viable population size required for a large primate species to persist in a given habitat, and is there any evidence suggesting a population of such a size could exist undetected in the reported regions?" This question immediately demands ecological assessment. It requires understanding the carrying capacity of potential habitats, the typical social structures and home range sizes of comparable known primate species (e.g., gorillas, orangutans, large macaques), and the energy requirements of such animals. Researchers would need to consider factors like resource availability (food, water, shelter), predator-prey dynamics, and the impact of human encroachment on habitat fragmentation and quality. Estimating minimum viable population size for a large primate is a complex task in itself, involving sophisticated demographic modeling. However, by applying these models to known ecological parameters of potential habitats, one can begin to frame the scale of the challenge for an unknown species. If a region cannot support even a small, fragmented population of, say, a dozen breeding individuals, then the likelihood of a sustained, undetected population of a larger primate diminishes significantly. Research efforts should thus focus on rigorous ecological surveys of candidate regions, mapping resource distribution, identifying potential corridors for movement, and assessing existing pressures from human activity and known wildlife populations. The absence of such ecological conditions, when thoroughly investigated, can serve as strong negative evidence. Conversely, areas exhibiting robust biodiversity, extensive undisturbed habitat, and a rich resource base would be prioritized for further, more detailed investigation into the possibility of undiscovered megafauna.

Closely linked to population viability is the question of

genetic distinctiveness and continuity. If a population of unknown primates exists, it must possess a unique genetic signature that differentiates it from known species and subspecies. This leads to the question: "What genetic markers would be expected in a distinct megafauna primate population, and can any such markers be detected in environmental samples?" This question drives the need for advanced molecular techniques. The primary objective would be to obtain biological samples that could yield DNA. This includes, but is not limited to, hair, feces (scat), saliva (from chewed vegetation or grooming sites), blood (from potential kills or carcasses), and any tissue recovered from deceased individuals. Once obtained, these samples would undergo rigorous DNA sequencing and analysis. The analysis would not only aim to identify novel genetic sequences but also to compare them against extensive databases of known primate species, subspecies, and even regional populations. The absence of contamination and the presence of sufficient genetic material are critical hurdles. However, advances in eDNA (environmental DNA) technology offer promising avenues. By analyzing water samples from rivers or streams, or soil samples from areas of suspected activity, it may be possible to detect trace amounts of DNA shed by organisms. A key aspect of this research would be to establish clear criteria for what constitutes "distinctiveness." Is it a unique lineage, a significant divergence from known sister taxa, or a novel set of alleles? Establishing these criteria beforehand ensures that any genetic findings are assessed objectively against predefined scientific standards, preventing the reclassification of minor variations within known species as evidence for entirely new ones. Furthermore, understanding the potential rate of genetic drift and mutation in the hypothesized species would be crucial for interpreting the age and evolutionary history of any discovered genetic lineages.

The ecological niche and impact of a large, previously undiscovered primate are also paramount considerations. This leads to the question: "What would be the ecological footprint and potential impact of a large primate species on its environment, and is there evidence of such an impact that cannot be explained by known fauna?" This question requires a deep understanding of ecological interactions. A large primate, whether herbivorous, frugivorous, or omnivorous, would exert significant influence on its ecosystem. Its feeding habits would affect plant populations, potentially influencing seed dispersal, competition dynamics, and forest structure. Its movement patterns could create trails or impact soil composition. Its waste products would contribute to nutrient cycling. Researchers would need to investigate these potential impacts by examining habitat characteristics in regions with credible reports. For example, are there unusual patterns of fruit depletion, or specific plant species that show signs of being selectively browsed in ways not attributable to known herbivores? Are there large, well-defined trails or bedding sites that do not match the behavior of local fauna? This requires meticulous fieldwork, employing methods such as camera trapping, transect surveys, and detailed vegetation analysis. It also necessitates comparison with areas lacking such reports to establish baseline ecological conditions. The challenge lies in differentiating the impact of an unknown species from the natural variation caused by known animals, environmental factors, or even human activity. Thus, the research must be designed to isolate and identify impacts that are

uniquely characteristic of a large primate, requiring a comprehensive understanding of the trophic levels and ecological processes present in the candidate habitats.

Finally, and perhaps most critically, the question of

definitive biological identification must be addressed. This is the ultimate arbiter of scientific legitimacy. The overarching question is: "Can a live specimen, a verifiable biological sample (e.g., carcass, tissue), or irrefutable physical evidence (e.g., uniquely identifiable remains) be obtained to confirm the existence of an unknown megafauna primate?" This question underscores the need for a paradigm shift from mere evidence collection to a focused pursuit of confirmation. While footprints, vocalizations, and blurry photographs can generate interest and hypotheses, they fall short of scientific proof for a new species. The scientific community requires tangible, identifiable biological material. This could range from a complete skeleton or a preserved carcass, allowing for detailed anatomical and morphological study, to a set of well-preserved tissue samples amenable to comprehensive genetic and histological analysis. The recovery of such evidence would enable classification according to established taxonomic frameworks. It would allow for comparisons with fossil hominoids and extant primate species, providing insights into evolutionary relationships. Moreover, detailed examination of any recovered specimen would yield critical information about diet, health, reproductive strategies, and behavior, filling in the ecological profile outlined in the previous question. The difficulty in obtaining such definitive evidence for elusive megafauna cannot be overstated. It is the very elusiveness that fuels the mystery. However, future research must adopt strategies that increase the probability of such a discovery. This might involve targeted expeditions to areas with the most compelling historical reports and ecological plausibility, employing advanced tracking technologies, drone surveillance, and citizen science initiatives focused on reporting and properly documenting any credible physical evidence found. The emphasis must be on creating a framework where the discovery of such definitive biological material is the primary, albeit challenging, goal, and all other research efforts serve to inform and facilitate this ultimate objective.

To effectively address these prioritized research questions, a methodological framework must be established. This framework should guide future scientific endeavors, ensuring that resources are allocated efficiently and that research is conducted with the highest degree of rigor. It begins with a comprehensive review and re-evaluation of existing anecdotal evidence, not to validate it, but to identify consistent patterns, geographical hotspots, and recurring descriptive elements that might indicate potential areas for focused investigation. This initial phase requires a critical approach, applying scientific skepticism to filter out obvious misidentifications, hoaxes, and exaggerations, while carefully noting any information that aligns with plausible biological scenarios. Once potential areas or phenomena are identified, the next step involves systematic ecological assessments. This entails detailed surveys of flora and fauna, mapping habitat types, analyzing resource availability, and documenting the presence and behavior of known large animals that could be mistaken for a cryptid. Understanding the existing ecological landscape is crucial for distinguishing novel impacts from natural variations.

Following ecological reconnaissance, research efforts should pivot towards the collection of

tangible, objective evidence. This is where the limitations of anecdotal accounts become most apparent, and the need for empirical data becomes paramount. Trace evidence, such as footprints, hair, scat, and saliva, while still requiring careful interpretation, offers a significant advantage over eyewitness testimony. The collection protocols for such evidence must be meticulously standardized to prevent contamination and ensure accurate contextual documentation. For instance, when a potential footprint is discovered, its exact location, substrate, depth, and dimensions must be meticulously recorded, alongside high-resolution photographs with a scale reference. If multiple tracks are found, their gait and spacing should be documented to infer locomotion and size. Similarly, hair samples should be collected using sterile techniques and packaged to preserve DNA integrity. The analysis of these trace evidences is where molecular biology plays a crucial role. DNA extraction and sequencing are essential for identification. The initial aim would be to match the DNA to known species. If a sample cannot be identified, it then becomes a candidate for further investigation as potentially belonging to an unknown primate. This comparative genomics approach is vital; it's not enough to find unidentified DNA; researchers must demonstrate that it does not belong to any known organism, including potentially novel subspecies or populations of currently recognized species.

Photographic and video evidence, while often sensationalized, must also be subjected to rigorous scientific scrutiny. Instead of accepting images at face value, researchers should employ advanced analytical techniques. This includes image enhancement to clarify details, forensic analysis to detect digital manipulation or alterations, and photogrammetry to estimate size and proportions based on known objects in the scene. The context of the photograph or video is equally important: when and where was it taken? What were the lighting conditions? What is the apparent behavior of the subject? Comparing these elements with known behaviors and physical characteristics of similar-sized animals under similar environmental conditions can often reveal misidentifications. For example, a fleeting glimpse of a dark shape moving through dense foliage at dusk might be an individual of a known animal species, such as a large primate or even a bear, that appears unusual due to poor lighting, distance, and the observer's expectation.

Audio recordings, while difficult to attribute definitively, can also provide clues. Spectrographic analysis can break down sounds into their constituent frequencies, revealing patterns that may be unique or unexpected. Comparing these spectrographs with vocalization libraries of known regional fauna can help rule out common explanations. However, the interpretation of unusual sounds is fraught with challenges, as environmental noises, distant calls of known animals, and even human-made sounds can be misinterpreted, especially in the context of a belief in an unknown creature.

The ultimate goal, however, remains the acquisition of definitive biological material. This might involve the recovery of skeletal remains, a complete carcass, or substantial tissue samples that allow for a comprehensive morphological and genetic analysis. Such discoveries would not only confirm the existence of a new species but would also provide invaluable insights into its biology, evolutionary history, and ecological role. Therefore, future research strategies should also consider how to maximize the chances of recovering such material. This could involve carefully investigating areas with persistent reports, understanding the typical mortality patterns of large mammals in those environments, and establishing protocols for the prompt and proper recovery and preservation of any remains discovered. In summary, the path forward in the scientific investigation of potential megafauna primates lies in a disciplined, evidence-based approach that prioritizes empirical data collection, rigorous analysis, and a commitment to answering fundamental biological and ecological questions. This approach moves beyond speculation and toward a verifiable understanding of our planet's biodiversity.

The pursuit of definitive evidence for elusive megafauna, particularly those potentially resembling undescribed primate species, necessitates the deployment of a sophisticated suite of field methodologies. Moving beyond opportunistic encounters and anecdotal accounts, future research must embrace a multi-pronged strategy that integrates cutting-edge technologies with time-tested wilderness skills. This approach aims to systematically increase the probability of detection, characterization, and ultimately, confirmation of any putative species while simultaneously providing robust data on the ecosystem it inhabits.

One of the most transformative advancements in biological detection is the application of

environmental DNA (eDNA). This technique leverages the ubiquitous shedding of genetic material by living organisms into their surroundings. Water bodies, in particular, act as highly efficient collection systems for DNA released through feces, urine, skin cells, and respiration. By collecting water samples from rivers, streams, lakes, and even stagnant pools within potential habitats, researchers can extract and analyze eDNA. The process involves filtering large volumes of water to capture microscopic particles, followed by laboratory procedures to isolate DNA. Subsequent high-throughput sequencing and bioinformatics analysis can then identify the genetic signatures of all organisms present in that environment. For the purpose of detecting unknown primates, the critical step is to compare the sequenced DNA against comprehensive databases of known species. Any sequences that do not match known taxa, or that exhibit significant divergence from known primate lineages, become prime candidates for further investigation. This method is particularly valuable for detecting species that are rare, reclusive, or inhabit areas difficult to access through traditional surveys. Soil samples can also be analyzed for eDNA, capturing genetic material shed by terrestrial animals, though DNA degradation in soil can be more pronounced than in water. The careful selection of sampling sites, based on ecological plausibility and reports of sightings, is paramount to maximizing the efficacy of eDNA surveys. Furthermore, understanding the persistence of DNA in various environmental matrices is crucial for interpreting results; for instance, DNA might remain detectable for days, weeks, or even months depending on factors like temperature, UV exposure, and microbial activity. The spatial scale of eDNA detection is also a consideration; a positive result might indicate an animal that passed through recently or one that resides permanently in the vicinity, necessitating careful site selection and potentially repeated sampling over time.

Complementing eDNA,

sophisticated camera trapping arrays represent a cornerstone of modern wildlife monitoring. While traditional camera traps have been invaluable, their effectiveness can be significantly enhanced through strategic deployment and advanced technological integration. Instead of deploying cameras sporadically, researchers can establish dense grids or transects of motion-activated cameras in areas identified as high-priority based on ecological assessments and anecdotal evidence. The objective is to maximize the probability of capturing images or video of any animal moving through these zones. Modern camera traps offer improved low-light performance, higher resolution imaging, and faster trigger speeds, all of which are crucial for capturing fleeting glimpses of elusive creatures. Advanced configurations might include linking multiple cameras to cover larger areas or employing specialized triggers that respond to sound or even specific thermal signatures. The analysis of vast amounts of footage generated by these arrays requires significant computational power and often relies on automated identification software to flag potential targets. For unknown primates, this would involve meticulously reviewing any detected large-bodied, bipedal or knuckle-walking mammals that do not conform to known species. Careful consideration must be given to distinguishing potential cryptids from known, but rarely photographed, animals or individuals exhibiting unusual behaviors. The ethical considerations of camera placement are also important, ensuring minimal disturbance to wildlife and respecting land access regulations. Furthermore, the data from camera traps can provide invaluable insights into activity patterns, social behavior, and habitat use, even if a definitive identification is not immediately achieved.

Bioacoustic monitoring offers another powerful, non-invasive avenue for detection. Many large primates, including gorillas and orangutans, possess complex vocal repertoires that are used for communication over considerable distances. It is plausible that an undiscovered primate species would also produce distinct vocalizations. Deploying autonomous recording units (ARUs) in strategic locations across a study area can capture a continuous stream of ambient sound. These ARUs are designed for long-term deployment, often powered by solar panels, and can store vast amounts of audio data. The challenge then becomes sifting through this acoustic landscape to identify novel or anomalous sounds. Sophisticated acoustic analysis software can be employed to detect, classify, and localize sounds. Initial analysis would focus on cataloging the vocalizations of all known regional fauna. Any sounds that do not match this catalog, or that possess characteristics suggestive of a large primate (e.g., deep resonant calls, complex vocalizations), would be flagged for expert human review. Spectrographic analysis can reveal the unique frequency patterns and temporal structures of sounds, providing a detailed signature that can be compared against known primate calls and other potential sources of confusion. The development of machine learning algorithms trained on known primate vocalizations can further assist in the automated detection of potentially unknown calls. It is essential to consider the potential for misinterpretation, where natural sounds, atmospheric phenomena, or even human-generated noise could be mistaken for something else, especially by listeners predisposed to finding evidence of the unknown. Therefore, rigorous validation and comparison with known acoustic sources are critical.

Remote sensing technologies, including satellite imagery and drone-based aerial surveys, can provide broad-scale habitat assessment and potentially reveal macro-level evidence of large animal presence. While these technologies are unlikely to directly image an elusive primate, they can map habitat suitability, identify large-scale disturbances (such as logging or human settlements), and reveal patterns that might indicate the movement or presence of significant animal populations. For instance, satellite imagery can be used to assess forest cover, canopy structure, and the availability of food resources across vast areas, thereby identifying regions that offer the best potential habitat for a large-bodied primate. Drone technology, equipped with high-resolution cameras, thermal imaging sensors, or even LiDAR (Light Detection and Ranging), can be deployed for more targeted surveys of specific areas. Thermal imaging, in particular, could potentially detect warm-bodied animals against cooler backgrounds, especially during dawn, dusk, or at night, although distinguishing individual animals from background environmental temperatures can be challenging. LiDAR can provide detailed three-dimensional maps of forest structure, which could potentially reveal unusual canopy modifications or large clearings that might be indicative of animal activity, though again, differentiating these from natural processes or human impact would require careful interpretation. The integration of remote sensing data with ground-based surveys is crucial, using the former to guide and optimize the latter.

Expert-guided tracking and reconnaissance in remote wilderness areas remains an indispensable methodology. While technology provides powerful tools for passive detection, the skills of experienced trackers and field biologists are invaluable for interpreting subtle signs and navigating challenging terrains. These experts possess a deep understanding of animal behavior, track identification, and the ecological nuances of their study areas. They can follow faint trails, identify scat and hair samples with a higher degree of accuracy, and interpret signs of feeding, resting, or travel that might elude less experienced observers. When combined with knowledge of local folklore and historical accounts, these trackers can guide expeditions to areas where encounters are statistically more probable. The process involves systematic searching of likely habitats, paying close attention to signs of passage along game trails, riverbanks, and forest clearings. This might involve the meticulous documentation of any unusual tracks, droppings, or disturbances. The collection of any physical evidence, such as hair, feathers, or scat, must follow strict protocols to prevent contamination and preserve its integrity for subsequent genetic analysis. The experience of these individuals allows them to make informed judgments about the plausibility of findings and to differentiate between the signs of known animals and potential evidence of an unknown species. This methodology is labor-intensive and requires significant time investment, but it can yield critical data that technology alone cannot provide, especially when dealing with highly intelligent and adaptable animals.

A comprehensive research framework for detecting and studying elusive megafauna must integrate these diverse methodologies into a cohesive strategy. No single technique is likely to be sufficient on its own. For instance, eDNA can indicate the presence of a species but may not reveal its morphology or behavior. Camera traps can provide visual confirmation but might miss animals that avoid known trails or that are too fast to be captured clearly. Bioacoustic monitoring can detect vocalizations but cannot provide visual proof. Remote sensing offers a broad overview but lacks fine-scale detail. Expert tracking can identify subtle signs but may not always yield definitive proof. Therefore, the most effective approach involves a synergistic combination of these tools. For example, eDNA surveys might identify a region with a high probability of an unknown primate presence, prompting the deployment of dense camera trap arrays and bioacoustic recorders in that specific area. The data from these automated systems can then guide expert tracking teams to areas with the highest likelihood of encountering fresh signs or even direct evidence. Thermal imaging from drones could be used to scout large areas, potentially highlighting anomalies that warrant closer ground investigation. The interpretation of findings must always be conservative, employing rigorous scientific skepticism and striving for objective verification. Any positive detection, whether through eDNA, camera trap images, or acoustic recordings, must be subjected to thorough comparative analysis against known species. Furthermore, establishing baseline data for known fauna within the study region is crucial for differentiating novel findings from natural variation or misidentifications. The ultimate goal, as previously discussed, remains the acquisition of definitive biological material, such as a carcass or tissue sample, which would allow for comprehensive morphological, anatomical, and genetic analysis, providing irrefutable proof of existence and enabling classification within the tree of life. This multi-faceted approach, blending technological innovation with traditional fieldcraft, represents the most promising path forward in the scientific quest to understand the full scope of Earth's biodiversity.

The scientific endeavor to uncover and understand the full spectrum of Earth's biodiversity, especially concerning elusive megafauna, is increasingly recognizing the immense, yet often untapped, potential of

citizen science and enhanced collaboration. In an era where vast tracts of wilderness remain remote and difficult to access for professional researchers alone, the collective efforts of a engaged public can significantly augment data collection, expand geographical coverage, and accelerate the pace of discovery. Citizen science, in its broadest sense, refers to the involvement of the general public in scientific research, from data collection and analysis to the dissemination of findings. When applied to the study of cryptic or rare species, it can transform passive observers into active contributors, thereby democratizing scientific inquiry and fostering a deeper connection between humanity and the natural world.

The fundamental principle underpinning the success of citizen science in this domain is the sheer number of individuals who traverse natural landscapes for a multitude of reasons – hiking, hunting, photography, birdwatching, or simply exploring. Each of these individuals, if equipped with appropriate tools and guidance, can serve as a valuable sensor network. Imagine the millions of people who visit national parks, remote trails, or even their own backyards annually. Each of these encounters, no matter how brief, represents a potential data point. The challenge lies in systematically harnessing this distributed observation capacity and ensuring the quality and reliability of the data generated. This necessitates the development of well-designed protocols, user-friendly tools, and robust validation mechanisms.

For projects focused on identifying undescribed primate species or understanding the ecology of known but rare ones, citizen science can manifest in several crucial ways. Firstly,

opportunistic reporting of sightings can serve as an early warning system, flagging areas of potential interest for more intensive professional investigation. This might involve individuals meticulously documenting unusual animal encounters, including precise location data (GPS coordinates are essential), time of day, weather conditions, observed behavior, and detailed descriptions of the animal's physical characteristics. The key here is to guide observers in differentiating genuine anomalies from misidentifications of known species. Platforms like iNaturalist or dedicated cryptid reporting websites, when moderated by experts, can serve as effective channels for collecting such initial reports. The role of the expert is to triage these reports, identifying those that warrant further scrutiny based on ecological plausibility and the quality of the evidence provided. Anecdotal accounts, while often dismissed, can sometimes provide crucial starting points, guiding researchers towards locations that might otherwise be overlooked.

Secondly, citizen scientists can play a vital role in

passive data collection. This could involve deploying and maintaining camera traps in their local areas or during recreational trips. Organizations can provide standardized camera traps and clear instructions on optimal placement strategies – near game trails, water sources, or areas with signs of animal activity. Citizen scientists can then retrieve the cameras and upload the footage for analysis. Similarly, the collection of environmental DNA (eDNA) samples, particularly from water sources, can be simplified for public participation. Kits containing sterile collection tubes, filtration systems, and GPS devices could be distributed, with participants instructed on how to collect water samples from streams, ponds, or rivers in their vicinity. The ease of collecting water samples, compared to the more involved process of tracking and scat collection, makes eDNA a particularly accessible avenue for citizen involvement. Protocols for sample preservation and return to a central laboratory are critical for maintaining sample integrity.

Thirdly, citizen science can be instrumental in

habitat monitoring and mapping. Participants can be trained to identify and report on key habitat indicators, such as the presence of specific food plants, evidence of foraging activity, or signs of human disturbance that might impact wildlife populations. This information can contribute to building detailed habitat suitability models, helping researchers to prioritize areas for more intensive study or conservation efforts. Drones, increasingly accessible, could also be incorporated, with trained citizen scientists conducting aerial surveys of designated areas and uploading imagery for analysis, though this would require a higher level of training and regulatory compliance.

The effectiveness of citizen science, however, hinges critically on

robust validation and expert oversight. Raw data from the public, while potentially voluminous, can be prone to errors, biases, and misinterpretations. Therefore, a structured approach to data quality control is paramount. This begins with comprehensive training for citizen scientists, emphasizing standardized data collection protocols, accurate observation techniques, and the importance of objective reporting. Educational materials, including field guides, identification charts, and online tutorials, are essential.

Moreover, the establishment of clear

feedback loops between citizen scientists and the research team is vital. When a report is submitted, participants should receive acknowledgement and, where possible, feedback on the significance or potential identification of their findings. This fosters continued engagement and demonstrates the value of their contributions. Expert review of submitted data is non-negotiable. This might involve a tiered system, where initial review is automated or handled by trained volunteers, with more complex or unusual findings escalated to professional biologists. For instance, camera trap images or eDNA results that suggest the presence of an unknown species would undergo rigorous scrutiny by experienced primatologists, geneticists, and ecologists. This validation process might involve cross-referencing with known species databases, consulting with other experts, and potentially initiating targeted follow-up research in the reported area.

The concept of

collaboration extends beyond just engaging the public; it encompasses partnerships between different scientific disciplines, institutions, and even across national borders. The study of cryptic megafauna is inherently interdisciplinary. A successful research framework requires the seamless integration of expertise from primatology, ecology, genetics, bioacoustics, remote sensing, anthropology, and even veterinary science. Furthermore, many potential habitats for undiscovered species span political boundaries, necessitating international cooperation and data sharing among research groups working in adjacent regions.

Consider the potential for

formal partnerships with indigenous communities and local populations. These communities often possess generations of intimate knowledge about their local environments, including the presence and behavior of wildlife that may not be documented in scientific literature. Their traditional ecological knowledge (TEK) can provide invaluable insights into animal behavior, habitat use, and even oral histories that might hint at the existence of unknown species. Engaging these communities as research partners, rather than simply as sources of information, is crucial. This involves respecting their cultural protocols, ensuring equitable benefit sharing, and empowering them as active participants in the research process. Projects should be co-designed and co-implemented, with local knowledge integrated into every stage, from hypothesis generation to data interpretation. This approach not only enriches scientific understanding but also fosters trust and promotes effective conservation strategies that are culturally appropriate and sustainable.

Collaboration also means breaking down silos between academic institutions, government agencies, and non-governmental organizations (NGOs). Universities can provide research expertise and access to advanced analytical facilities, while government agencies may have access to remote field sites and regulatory authority. NGOs often have established community networks and on-the-ground presence, facilitating local engagement and conservation efforts. Pooling resources, sharing data, and coordinating research efforts can prevent duplication of work and maximize the impact of limited funding. For instance, a multi-institutional effort might combine expertise in eDNA analysis from one university, camera trap data management from another, and bioacoustic monitoring capabilities from an NGO, all focused on a common research objective.

The development of

standardized protocols and data-sharing platforms is a critical enabler of effective collaboration. If researchers and citizen scientists are using different methods for data collection or are unable to easily share their findings, the potential for synergistic advancement is significantly diminished. Investing in open-source software and databases that facilitate the standardized collection, storage, and analysis of diverse data types – including genetic sequences, camera trap images, audio recordings, and GPS tracks – is essential. This not only aids in collaborative analysis but also ensures data longevity and accessibility for future research.

Furthermore, fostering a culture of openness and data sharing within the scientific community itself is paramount. While intellectual property concerns can be valid, especially in early-stage research, a willingness to share preliminary data and collaborate on challenging analyses can accelerate progress dramatically. This might involve establishing consortia or working groups focused on specific research questions or regions, where members agree to share data under mutually beneficial terms. The challenges of detecting and confirming elusive species are so significant that a competitive, rather than collaborative, approach would be detrimental to the overall scientific goal.

The integration of citizen science and collaborative efforts fundamentally reshapes the landscape of cryptozoological and primatological research. It moves beyond the traditional image of the solitary scientist venturing into the unknown, towards a model of networked, participatory science. This approach leverages the ubiquity of human presence in natural environments and the collective power of diverse expertise to achieve objectives that would be impossible through conventional means alone. It democratizes the process of scientific discovery, fostering public engagement, and building a broader base of support for conservation and biodiversity research.

The successful implementation of these strategies requires careful planning, significant investment in training and infrastructure, and a commitment to rigorous scientific methodology. It is not simply about collecting more data; it is about collecting better data, through a more inclusive and interconnected scientific process. As we refine our methodologies for detecting elusive species, the role of the citizen scientist and the power of interdisciplinary collaboration will only become more critical. These elements are not merely supplementary; they are foundational to building a comprehensive and effective framework for future research into the Earth's hidden biodiversity. The potential for discovery is immense, but it can only be fully realized through a concerted, collaborative, and community-driven effort. The ongoing refinement of eDNA techniques, the increasing sophistication of camera trapping technology, and the expanding reach of bioacoustic monitoring, when combined with the watchful eyes and dedicated efforts of citizen scientists and the synergistic power of interdisciplinary collaboration, pave the way for unprecedented advancements in our understanding of the natural world, potentially leading to the confirmation of species that have long eluded scientific detection.

The pursuit of knowledge regarding Earth's undiscovered or exceptionally rare fauna, particularly within the realm of primate cryptozoology, is an endeavor fraught with profound ethical considerations. As we venture into remote and often fragile ecosystems, the imperative to conduct our research with the utmost respect for the environment and its inhabitants becomes paramount. This is not merely a matter of good practice; it is a fundamental requirement for maintaining the credibility of our scientific pursuits and for safeguarding the very subjects of our inquiry. The potential existence of unknown primate species or subspecies necessitates a heightened level of caution, recognizing that our understanding of their biology, behavior, and ecological roles is, by definition, limited. Therefore, the precautionary principle must guide every aspect of our fieldwork.

A primary ethical concern revolves around minimizing disturbance to the natural habitat. Our presence, even with the best intentions, can inadvertently impact the delicate balance of an ecosystem. This means meticulously planning expeditions to reduce our ecological footprint. We must strive to avoid unnecessary vegetation clearing, refrain from introducing invasive species through contaminated equipment or supplies, and manage waste responsibly. Establishing base camps in areas that are already subject to some level of human activity or are ecologically less sensitive can be a prudent strategy. Furthermore, the use of less intrusive survey methods is crucial. While traditional methods of tracking and direct observation have their place, they often require significant movement through an environment, potentially alerting or disturbing wildlife. Employing techniques such as remote sensing, acoustic monitoring, and the strategic deployment of camera traps, as discussed previously, can provide valuable data with a considerably lower impact on the environment. When direct observation is necessary, maintaining significant distances and utilizing high-powered optics are essential. The goal is to gather information without becoming an undue presence that alters animal behavior, disrupts breeding patterns, or impacts food availability. This commitment to minimizing disturbance extends beyond the immediate research period; it also encompasses the long-term implications of our presence and any infrastructure or trails that might be established. The principle is simple: to learn about a creature without fundamentally changing its world or the world around it.

Beyond habitat disturbance, the ethical treatment of animals, should they be encountered or detected, is of utmost importance. If an unknown primate species is indeed real, it is likely to be rare and potentially vulnerable. Our research must prioritize its well-being above all else. This means avoiding any actions that could lead to direct harm, injury, or undue stress. Capture-and-release programs, while sometimes necessary for detailed study, carry significant risks and should only be undertaken as a last resort, after exhaustive risk assessments and with the implementation of the most humane and minimally invasive techniques available, conducted by highly experienced professionals. Such efforts require advanced ethical review board approval and adherence to stringent protocols. The collection of biological samples, such as hair or scat for genetic analysis, should be conducted opportunistically and non-invasively whenever possible. For example, collecting shed hair from branches or scat found on the ground is far preferable to attempting to extract these samples directly from an animal.

The use of tranquilizers or immobilizing agents should be approached with extreme caution. These substances can have unpredictable physiological effects, especially on species with unknown metabolisms. If such interventions are deemed absolutely necessary for critical research (e.g., disease sampling, vital sign monitoring), they must be administered by licensed veterinarians with extensive experience in wildlife immobilization, utilizing carefully calculated dosages based on the best available information, however limited, for similar primate species. The recovery period for any immobilized animal must be ensured, with researchers remaining in the vicinity to monitor for complications and to deter predators or other threats until the animal is fully alert and able to move to safety.

Furthermore, the ethical framework must extend to the concept of public display or sensationalism. The discovery of a new primate species is an event of immense scientific and public interest. However, this interest can easily lead to exploitation. Researchers have a moral obligation to protect the species from the potential negative consequences of public knowledge. This includes carefully considering how findings are disclosed. Premature or irresponsible disclosure could lead to increased human encroachment, poaching, or habitat destruction driven by curiosity or commercial interests. The immediate priority after confirmation should be focused on establishing effective conservation measures and ensuring the species' long-term survival, rather than seeking immediate public acclaim. This often means working collaboratively with local authorities and conservation organizations to implement protective measures before any widespread announcement is made. The scientific community must foster a culture that prioritizes conservation over sensationalism, ensuring that discoveries benefit the species, not just the discoverers.

Respect for local communities and indigenous knowledge is another cornerstone of ethical field research, particularly in regions where undocumented species are thought to exist. These areas are often inhabited by people who have lived in and relied upon these environments for generations. Their traditional ecological knowledge (TEK) is an invaluable resource that can guide research, inform understanding of species behavior and distribution, and identify areas of particular ecological significance. Engaging with these communities must be done with genuine respect, acknowledging their stewardship of the land and their deep understanding of its intricacies. This engagement should be a process of partnership, not extraction. Researchers must invest time in building trust, understanding local customs and protocols, and ensuring that any research undertaken is mutually beneficial. This involves seeking informed consent for research activities and respecting the intellectual property rights associated with traditional knowledge. The sharing of findings should be a two-way street, with communities benefiting from the scientific insights gained and having a voice in how their local environment is managed.

When engaging with indigenous communities, it is crucial to understand that their relationship with the natural world is often deeply spiritual and cultural. A species that might be a subject of scientific inquiry for us could hold significant cultural meaning for them. Researchers must be sensitive to these cultural nuances and avoid any actions that could be perceived as disrespectful or exploitative. This might involve understanding traditional naming conventions, sacred sites, and customary practices related to wildlife. Collaboration with local guides and trackers, who possess unparalleled knowledge of the terrain and animal behavior, is not just a practical necessity but an ethical imperative. Their expertise is often a lifetime accumulation of observation and learning, and their contributions should be recognized and compensated appropriately. Furthermore, any research that directly impacts local communities, such as land use or resource management, must involve their active participation in decision-making processes. The principle of "do no harm" must extend to the social and cultural fabric of these communities.

Maintaining scientific integrity is the overarching ethical obligation that underpins all other considerations. This means conducting research with rigor, honesty, and transparency. It requires meticulous data collection, robust analysis, and unbiased reporting of findings. In the context of searching for elusive species, there can be a strong temptation to overstate evidence or to interpret ambiguous data in a way that supports a desired outcome. However, scientific credibility hinges on objective interpretation and a willingness to acknowledge uncertainty. Any claims about the existence of a new species must be supported by irrefutable evidence, subjected to peer review, and withstand rigorous scrutiny. This includes providing detailed descriptions of methodologies, raw data where appropriate, and clear justifications for conclusions.

The temptation for confirmation bias is particularly strong when searching for rare or unknown animals. Researchers must be vigilant against this cognitive pitfall. This means actively seeking out disconfirming evidence, considering alternative explanations for observed phenomena, and engaging with colleagues who may hold different perspectives. The process of peer review is a critical safeguard against bias and error. Submitting findings to reputable scientific journals ensures that the work is evaluated by independent experts in the field, who can identify weaknesses in methodology, logic, or interpretation.

Furthermore, ethical research demands transparency in funding and potential conflicts of interest. If research is funded by organizations with a vested interest in a particular outcome (e.g., tourism operators, documentary film companies), this must be disclosed. Similarly, any personal stakes that researchers have in confirming the existence of a species should be managed to prevent undue influence on the research process and its reporting. The pursuit of scientific truth must be driven by intellectual curiosity and a commitment to understanding the natural world, not by external pressures or personal agendas.

In conclusion, the ethical considerations in the field research of potentially rare and sensitive wildlife are multifaceted and deeply intertwined. They demand a holistic approach that prioritizes the preservation of habitats, the welfare of animals, the respect for local communities and their knowledge, and the unwavering commitment to scientific integrity. By adhering to these principles, researchers can ensure that their pursuit of knowledge contributes positively to our understanding of biodiversity and to the effective conservation of the natural world, rather than inadvertently causing harm. The ultimate goal is not just discovery, but responsible discovery that safeguards the wonders of nature for generations to come. This necessitates a conscious and continuous self-assessment of our actions and motivations, ensuring that our passion for the unknown is always tempered by a profound respect for the known and the yet-to-be-discovered. The rigorous application of ethical guidelines is not an impediment to scientific progress; it is the very foundation upon which meaningful and sustainable scientific advancement is built.

The journey from anecdotal observation to irrefutable scientific acceptance for a novel, large-bodied primate is a formidable one, demanding a systematic and rigorous approach to evidence gathering and interpretation. It's not enough to collect a few blurry photographs or to have compelling eyewitness accounts, however numerous. The scientific community, by its very nature, operates on a foundation of verifiable data, peer review, and reproducible results. Therefore, the crucial step in legitimizing the existence of such a creature lies in meticulously building a scientific consensus, a process that is both demanding and essential for its formal recognition and subsequent conservation.

The bedrock of this consensus must be the presentation of unambiguous, multi-faceted evidence. This begins with physical specimens. The discovery of skeletal remains, particularly well-preserved ones, would be transformative. Imagine uncovering fossilized hominoid bones in a previously unexplored cave system, or a collection of intact skulls and long bones in a remote peat bog. Such findings, meticulously excavated and subjected to thorough paleontological and anatomical analysis, would provide tangible proof of the creature's past existence. Radiometric dating of these remains would be paramount, establishing their age and placing them within an evolutionary context. Comparative anatomy would then allow for detailed assessment of their distinctiveness, highlighting unique cranial features, dental morphology, limb proportions, and pelvic structure that differentiate them from known species. The absence of human modification on such remains would be a critical point of contention, necessitating careful forensic examination to rule out any possibility of deliberate concealment or misrepresentation.

Beyond skeletal evidence, definitive biological material is indispensable. This includes tissue samples, viable DNA extracted from hair, blood, or flesh, and even intact organs. Such material, recovered from carcasses or through carefully managed non-invasive sampling, would unlock the power of genetic analysis. High-quality DNA sequencing could definitively establish its phylogenetic relationships, demonstrating divergence from known primate lineages. Phylogenetic trees, constructed from extensive genetic data and subjected to rigorous statistical analysis, would be crucial in illustrating its evolutionary placement. We would be looking for distinct mitochondrial and nuclear DNA sequences that cannot be attributed to any existing primate. Furthermore, the analysis of ancient DNA, if the creature is indeed a relic species, could provide insights into its evolutionary history and population dynamics over millennia. The robustness of these genetic findings would be enhanced by repeated analysis using different methodologies and by independent verification in multiple accredited laboratories.

A critical component of consensus building would involve unambiguous photographic and videographic evidence. While blurry images have historically fueled speculation, the scientific standard demands high-resolution, clear, and contextually sound visual documentation. This means capturing multiple angles, showing the creature in its natural habitat, and ideally, revealing diagnostic features such as facial structure, body proportions, and locomotion. Videos are particularly powerful, as they can demonstrate locomotion, social interactions, and vocalizations, providing a dynamic and comprehensive view of the animal's behavior and morphology. The provenance of this footage would be scrutinized, with researchers seeking to establish its authenticity, rule out manipulation, and provide details about the circumstances of its capture, including the equipment used and the location. Evidence of the creature interacting with its environment in a way that is inconsistent with known species would be highly persuasive. For instance, observing it utilize tools in a manner distinct from known primate tool use, or exhibiting foraging behaviors that are unique to its ecological niche, would add significant weight to its existence.

Acoustic monitoring represents another vital avenue for generating evidence. Many primate species are highly vocal, and their calls serve critical functions in communication, territorial defense, and social bonding. The discovery of a distinct vocal repertoire, captured through sophisticated audio recording equipment deployed in remote areas, could provide compelling evidence. Spectrographic analysis of these calls, comparing their structure, frequency, and patterns to known primate vocalizations, would be essential. If these acoustic signatures are consistently recorded, are unique, and correlate with physical signs or sightings of the suspected creature, they would form a powerful piece of the puzzle. Imagine a series of unique, complex calls emanating from a particular valley, a sound unlike any known primate, that over time is repeatedly linked to fleeting glimpses of a large, unknown primate. This sustained acoustic presence would become a significant research focus.

Ecological and behavioral data, gathered through long-term, systematic observation, would further solidify the scientific case. Understanding the creature's diet, social structure, mating system, territoriality, and predator-prey interactions provides crucial context for its existence and survival. If we observe a primate species occupying a specific ecological niche, consuming particular food sources, and exhibiting social behaviors that are distinct from those of known primates in the region, it strengthens the argument for its unique identity. For instance, if a large primate is found to be expertly exploiting a food resource that no other primate in that ecosystem can access, or if it exhibits a solitary, nocturnal lifestyle that sets it apart from diurnal, social species, this distinctiveness becomes a compelling point. The consistency and predictability of these behavioral patterns, observed across multiple individuals and over extended periods, would be key.

The process of building scientific consensus is inherently iterative and collaborative. It begins with the initial discovery and the careful accumulation of preliminary evidence. This evidence is then subjected to rigorous scrutiny by the research team and, increasingly, by external experts. Peer review, the cornerstone of scientific validation, plays a critical role here. Researchers would aim to publish their findings in reputable, peer-reviewed scientific journals, where their methodologies, data, and conclusions are evaluated by independent experts in primatology, genetics, ecology, and other relevant fields. The feedback received during this process is invaluable, prompting further investigation, refinement of hypotheses, and potentially, the collection of additional evidence to address weaknesses or unanswered questions.

A crucial aspect of building consensus is reproducibility. If the initial evidence is compelling, other independent research teams, employing similar methodologies, should be able to corroborate the findings. This might involve independent verification of genetic data, the capture of similar visual or acoustic evidence from different locations, or the confirmation of unique ecological roles. The more independent lines of evidence that converge, and the more researchers who can replicate key findings, the stronger the scientific consensus becomes. This is why the entire research process, from data collection to analysis, must be transparent and open to scrutiny. Sharing raw data, methodologies, and even facilitating access to study sites (under strict ethical and conservation guidelines) can accelerate the process of validation and build trust within the scientific community.

The role of interdisciplinary collaboration cannot be overstated. The study of a new megafauna primate would necessitate the involvement of a diverse array of specialists. Paleontologists to analyze fossil evidence, geneticists to decipher DNA, ecologists to understand its habitat and role in the ecosystem, ethologists to study its behavior, zoologists to assess its morphology, and conservation biologists to plan for its future protection. Each discipline brings a unique perspective and a distinct set of tools, and their integrated findings would provide a more complete and convincing picture. A geneticist might identify a unique lineage, an ecologist might explain its survival through a specialized diet, and an ethologist might describe a novel social structure – together, these pieces form a powerful narrative.

Furthermore, the scientific community is inherently skeptical, a healthy trait that prevents the premature acceptance of unverified claims. This skepticism, however, can also be a significant hurdle. The history of cryptozoology is littered with hoaxes and misinterpretations, leading to a predisposition to doubt extraordinary claims. Therefore, the evidence presented must not only be robust but also exceptionally convincing, leaving little room for alternative explanations. Extraordinary claims, as the saying goes, require extraordinary evidence. This means not just meeting the standard of evidence for a typical species discovery, but exceeding it significantly. The burden of proof rests heavily on the researchers proposing the existence of such a creature.

The long-term vision for integrating such a discovery into our understanding of biodiversity is profound. The confirmation of a new, large-bodied primate would necessitate a re-evaluation of our knowledge of primate evolution, biogeography, and speciation. It would challenge existing assumptions about the completeness of our faunal inventories and highlight the vastness of what remains undiscovered, even in seemingly well-explored regions. This discovery would undoubtedly reignite interest in exploring other understudied areas for similar hidden biodiversity. It would underscore the critical importance of preserving these remote and often threatened ecosystems, not just for their known inhabitants but for the potential of undiscovered life they may harbor.

Conservation efforts would need to be immediately and robustly implemented. The very rarity that likely allowed the species to remain undiscovered would also render it exceptionally vulnerable to extinction. Understanding its habitat requirements, population size, genetic diversity, and the threats it faces would be paramount. Collaborative efforts between scientists, conservation organizations, and local communities would be essential to develop and implement effective protection strategies. This would involve habitat protection, anti-poaching measures, and potentially, the establishment of protected areas or wildlife corridors. The scientific community would have a moral imperative to advocate for the species' survival, translating their research findings into actionable conservation policies.

The educational and public outreach aspects of such a discovery would also be significant. It would offer an unparalleled opportunity to engage the public with science, highlighting the wonders of biodiversity, the importance of scientific inquiry, and the critical need for conservation. Communicating the findings accurately and responsibly, avoiding sensationalism while conveying the excitement of discovery, would be crucial. Educational programs, museum exhibits, and documentaries could all play a role in informing the public and fostering a sense of stewardship for this newly recognized part of our planet's biological heritage. The narrative of a creature, long relegated to myth and speculation, finally confirmed by science, would capture the public imagination and inspire future generations of scientists and conservationists.

In essence, building a scientific consensus for a new megafauna primate is not merely about presenting evidence; it is about a rigorous, transparent, and collaborative process of validation that withstands intense scrutiny. It involves the meticulous collection and analysis of diverse data – physical specimens, genetic material, high-quality visual and acoustic records, and detailed ecological and behavioral observations. This evidence must be subjected to peer review, independently verified, and ultimately, convince a broad spectrum of the scientific community of the creature's undeniable existence. Only then can we move forward with the critical tasks of understanding its evolutionary history, integrating it into our framework of biodiversity, and ensuring its long-term survival. The path is arduous, demanding patience, unwavering dedication to scientific principles, and a commitment to the highest standards of evidence.

15

Conclusion: The Plausibility of the Unseen

The tapestry of scientific understanding is woven from threads of observation, experimentation, and rigorous analysis. When we consider the possibility of large, elusive primates inhabiting our planet's dwindling wild spaces, this intricate weaving becomes particularly relevant. Throughout this exploration, we have delved into various facets of evidence that could, or do, support the existence of such beings, moving beyond mere folklore and into the realm of testable hypotheses. This concluding synthesis aims to draw together the disparate threads of scientific inquiry, examining the ecological viability, anatomical plausibility, and behavioral patterns that collectively inform our assessment of the unseen.

From an ecological perspective, the question of how a large-bodied primate could remain undetected in the modern era hinges on a delicate balance of habitat availability, resource partitioning, and predator-prey dynamics. Vast tracts of tropical rainforest, particularly in regions like the Congo Basin, New Guinea, and the islands of Southeast Asia, remain remarkably under-surveyed. These environments, characterized by dense canopy cover, rugged terrain, and limited human access, provide a natural shield for species that are naturally reclusive or possess low population densities. The sheer scale of these ecosystems means that even substantial populations could exist without frequent encounters with humans, especially if their activity is concentrated in remote areas or during crepuscular or nocturnal periods. Furthermore, the principle of ecological niche separation is crucial. A hypothetical large primate might occupy a specialized niche, perhaps exploiting food resources inaccessible to known primate species, or exhibiting a foraging strategy that minimizes overlap. For instance, a primate adept at consuming tough plant matter, or one that primarily preys on smaller vertebrates, might avoid direct competition with fruit-eating or leaf-eating monkeys and apes. The existence of numerous undescribed species of smaller fauna, including insects, amphibians, and even some mammals, within these same ecosystems underscores the potential for larger, undiscovered vertebrates to persist. The pressure from known predators, such as large cats or birds of prey, would also play a role; a creature that effectively avoids these threats through vigilance, arboreal agility, or subterranean habits would be more likely to survive undetected.

Anatomical plausibility is another cornerstone of this discussion. The evolutionary history of primates offers a rich precedent for large-bodied forms. The extinct Gigantopithecus, for example, was an ape of colossal proportions, reaching estimated heights of over three meters and weighing up to 500 kilograms. While Gigantopithecus is long extinct, its existence demonstrates that the primate body plan is capable of supporting immense size. The question then becomes whether the evolutionary pressures and ecological conditions for the development and sustained existence of such a large primate could have persisted in certain contemporary environments. We can infer potential anatomical features based on the demands of such a lifestyle. A large-bodied primate would likely require robust skeletal structures, particularly in the limbs and spine, to support its mass and facilitate locomotion. Its dentition would be adapted to its diet, whether herbivorous, omnivorous, or even carnivorous, reflecting the pressures of resource availability. Cranial morphology would be influenced by brain size, sensory organ placement, and social signaling. The absence of definitive fossil evidence for a modern large primate might seem counterintuitive, but it is not insurmountable. Many species, particularly those with limited geographic ranges or those inhabiting environments where fossilization is rare, leave scant paleontological records. The possibility of a relic species, a survivor from a time when larger hominoids were more prevalent, cannot be entirely dismissed. Furthermore, the interpretation of existing anatomical evidence, such as ambiguous footprints or alleged bone fragments, requires stringent scientific scrutiny. While hoaxes and misidentifications are common, the possibility of genuine, albeit rare, physical evidence being discovered remains. The comparative anatomy of known large primates, such as gorillas and orangutans, provides a template for understanding the biomechanics and physiological requirements of such creatures, allowing us to assess the feasibility of their existence within specific ecological contexts.

Behavioral considerations offer further avenues for investigation. The success of any elusive species, especially a large one, is intrinsically linked to its behavior. Secrecy, avoidance of humans, and a low reproductive rate are often characteristic of large animals that have survived in competitive environments. Many primate species are inherently shy and adept at disappearing into their surroundings. A hypothetical large primate might possess exceptional camouflage, acute senses of hearing and smell, and a sophisticated understanding of its territory, allowing it to evade detection. Social structure also plays a role; solitary or small-group living can reduce the likelihood of multiple individuals being observed together, making population estimates difficult. If such a creature were primarily nocturnal or crepuscular, its activity patterns would naturally coincide with periods of low human visibility, further contributing to its elusive nature. The study of known large primates, like the mountain gorilla, reveals that even species with significant conservation efforts and research presence can maintain areas of relative privacy and behavioral discretion. Their movements, social interactions, and foraging habits are not always readily apparent, even to dedicated researchers. Therefore, the lack of consistent, widespread observation of a large, unknown primate does not automatically preclude its existence; it might simply reflect a highly effective strategy of avoidance.

The current scientific perspective on the possibility of unknown large primates is one of cautious skepticism, tempered by an appreciation for the vastness of unexplored biodiversity. While the scientific community rightly demands robust evidence – including verifiable physical specimens, clear genetic material, and unambiguous photographic or video documentation – the absence of such conclusive proof does not equate to definitive proof of absence. The history of zoological discovery is replete with instances where species, once thought mythical, were eventually documented through rigorous scientific endeavor. These discoveries often followed long periods of anecdotal reports and tantalizing but incomplete evidence. Therefore, the pursuit of the unseen, particularly in the realm of cryptozoology, remains a frontier where persistent, methodologically sound scientific investigation, grounded in ecological and evolutionary principles, is essential.

The ecological landscape of our planet still holds immense secrets. While human impact has regrettably reduced many wild habitats, significant areas, particularly in the tropics and remote mountain ranges, remain largely unpenetrated by systematic biological surveys. These areas offer the theoretical possibility for the persistence of species that have managed to evade our current cataloging efforts. The concept of "cryptic species" – those that are morphologically similar to known species but genetically distinct – is well-established, but the potential for larger, more conspicuous species to remain hidden is also a factor. The energy requirements of a large primate would necessitate a stable and abundant food source, and its ecological niche would need to be sufficiently specialized to avoid intense competition with established fauna. The existence of such niches is supported by the ongoing discovery of new species, often occupying very specific ecological roles. For example, the discovery of specialized diets in certain primates, such as the consumption of particular types of fungi or invertebrates that other primates avoid, highlights how unique resource exploitation can allow for coexistence and relative obscurity. The predator-prey dynamics are also critical. A large primate would likely be at the top of its specific food chain, meaning its population would be limited by the availability of its prey or its ability to evade terrestrial predators. If such a creature exists, its survival would depend on a finely tuned ecological balance, a precarious equilibrium that might be threatened by habitat fragmentation and climate change.

Anatomically, the evolutionary pathway to large primate forms is well-documented through the fossil record. The lineage leading to modern great apes, for instance, has included species of considerable size. The extrapolation of these trends into the present day requires consideration of selective pressures that might favor large stature. While increased size often comes with higher metabolic demands and greater vulnerability during development, it can also confer advantages such as enhanced defense against predators, greater foraging efficiency in certain environments, and dominance in social hierarchies. The critical element is the presence of an environment that can sustain such a creature. If we hypothesize a relic hominoid, its morphology would likely reflect an ancestral form, perhaps with characteristics that differ subtly but significantly from known extant species. This could include variations in limb proportions, cranial features, or dental cusp patterns. The challenge for science lies in distinguishing genuine novel anatomical traits from variations within known species or from misinterpretations of fragmented remains. The careful analysis of any potential physical evidence, such as alleged skeletal fragments or exceptionally large footprints, must adhere to the highest standards of forensic paleontology and zoology. This involves meticulous comparative anatomy, rigorous dating techniques, and the assessment of taphonomic processes to rule out natural formations or deliberate fabrication. The scientific community's inherent skepticism, while often frustrating for proponents of unusual discoveries, serves as a vital safeguard against the acceptance of unsubstantiated claims.

Behaviorally, the success of an elusive, large primate would be a testament to its evolutionary adaptations for avoidance. Many species considered "shy" are simply adept at sensing human presence and retreating before they are detected. This often involves highly developed senses of hearing and smell, keen eyesight adapted for low light conditions, and an intimate knowledge of their habitat, including escape routes and hiding places. The effectiveness of these behaviors can be amplified by factors such as solitary or small-group social structures, which reduce the likelihood of multiple individuals being seen. Nocturnal or crepuscular activity patterns are also significant advantages for remaining undetected. Furthermore, the "paradox of the non-discovery" is often cited: if such a creature exists, why hasn't it been seen more often? However, this overlooks the sheer scale of remote wilderness areas and the limited human presence within them. Consider that even well-known large mammals, like jaguars or snow leopards, can evade detection for extended periods in their respective habitats. A hypothetical large primate, perhaps with a home range of hundreds of square kilometers, could easily exist without frequent encounters, especially if its population density is low. The behavioral repertoire would also need to account for resource acquisition. This might involve specialized foraging techniques, such as efficiently accessing hard-to-reach food sources or employing unique hunting strategies. The presence of such behavioral adaptations, even if inferred from indirect evidence or comparative studies of similar species, adds to the plausibility of a creature remaining hidden.

In synthesizing the scientific evidence, it is crucial to acknowledge the limitations of our current knowledge. The process of discovering and formally describing a new species, especially a large one, is a lengthy and demanding undertaking. It requires the convergence of multiple lines of credible evidence, subjected to rigorous peer review and independent verification. While anecdotal accounts and intriguing but inconclusive evidence may fuel speculation, the scientific community ultimately relies on tangible, reproducible data. The ongoing exploration of Earth's biodiversity, driven by technological advancements in genetics, remote sensing, and non-invasive monitoring, continues to reveal new species. This persistent discovery, even in seemingly well-documented regions, suggests that the possibility of larger, undiscovered fauna cannot be entirely dismissed. The scientific endeavor, in its purest form, remains open to the evidence, irrespective of how extraordinary the claim may seem. The challenge lies in distinguishing genuine scientific inquiry from wishful thinking, and in maintaining the highest standards of evidence-based reasoning when exploring the frontiers of the unknown. The potential for a large, unknown primate to exist is, therefore, a question that continues to reside in the complex intersection of ecological possibility, anatomical precedent, and behavioral adaptation, awaiting the irrefutable evidence that would solidify its place in the annals of zoology.

The grand tapestry of scientific exploration, while vast and ever-expanding, is inherently punctuated by areas of shadow and mystery. Even with our sophisticated tools and an ever-increasing global reach, the planet Earth continues to guard its secrets with a tenacity that both humbles and challenges us. When we turn our gaze towards the possibility of large, undiscovered primates, these inherent gaps in our knowledge become not just areas of speculation, but crucial focal points for future scientific endeavor. The absence of definitive, irrefutable evidence for such creatures, such as a type specimen, clear genetic sequences from multiple individuals, or unambiguous photographic or video documentation confirmed by independent experts, does not, by logical extension, constitute proof of their non-existence. Science, at its core, is a process of continuous inquiry, a perpetual refinement of understanding built upon the foundation of existing knowledge while remaining open to the unexpected.

One of the most significant remaining gaps lies in the sheer scale of incompletely surveyed terrestrial ecosystems. While we have made considerable progress in mapping and understanding much of the Earth's surface, vast tracts of dense tropical rainforest, remote mountain ranges, and extensive cave systems remain largely unexplored from a biological perspective. These are precisely the kinds of environments that would offer ideal refuges for large, elusive species. Consider the immense complexity of the canopy in the Amazon or the Congo Basin. Millions of species of insects, amphibians, and even small mammals exist within these arboreal realms, many still undescribed. The vertical stratification of these forests creates a multitude of microhabitats, and it is plausible that a large primate could occupy a niche within this three-dimensional world, minimizing its contact with the forest floor where ground-based surveys are more common. Similarly, the rugged, often cloud-shrouded peaks of regions like the Himalayas or New Guinea's interior present formidable challenges to human access. The potential for isolated populations to persist in these areas, shielded by extreme topography and inhospitable weather, is substantial.

Furthermore, our current methods of biological survey, while advanced, are not infallible. Traditional methods rely heavily on direct observation, trapping, or camera-trap deployment. These techniques are often limited by their spatial and temporal coverage. A primate species that is primarily nocturnal or crepuscular would naturally be underrepresented in daytime surveys. A species with exceptionally large home ranges and a low population density would require an immense number of camera traps or extensive fieldwork over prolonged periods to even have a moderate chance of detection. The cost and logistical challenges of deploying such comprehensive monitoring programs across vast, remote, and often dangerous territories are prohibitive. Even with the advent of acoustic monitoring or eDNA (environmental DNA) analysis, which offer promising new avenues for detection, their widespread application in truly remote areas is still in its nascent stages. For instance, while eDNA can detect the presence of a species from trace genetic material in water or soil samples, the interpretation of such findings can be complex, especially in areas with high biodiversity or where there's a risk of contamination. The absence of eDNA evidence in a particular sample does not definitively mean a species is absent; it could simply mean that no shedding event occurred in the vicinity of the sample collection during the relevant timeframe.

Another critical area of uncertainty revolves around the limitations of our current understanding of primate evolutionary history and distribution. While the fossil record provides invaluable insights, it is by its very nature incomplete. The processes of fossilization are rare, and the conditions required for preserving primate remains, especially those of arboreal creatures, are even rarer. This leaves significant gaps in our knowledge of past primate diversity. We know, for example, that hominoids were once much more widespread and diverse, with species like Gigantopithecus having roamed vast areas. It is not inconceivable that other large-bodied hominoid lineages, or indeed other large primate lineages, could have survived in isolated pockets, adapting to changing environments and persisting into the Holocene or even the Anthropocene. The discovery of the

Homo floresiensis ("Hobbit") on the island of Flores, a species of diminutive hominin that survived until relatively recently, demonstrates that evolutionary pathways can lead to the persistence of unexpected forms in isolated environments. This opens the door to the possibility, however remote, that other evolutionary experiments in primate gigantism or unique adaptations might have also played out in unexplored corners of the world.

The very definition of "proof" in scientific discovery also contributes to these remaining gaps. The scientific community rightly demands a high burden of proof, particularly when dealing with claims that challenge established paradigms. This typically involves the collection of multiple, independent lines of evidence: a living specimen, verifiable skeletal remains, clear genetic material, unambiguous photographic or video evidence, and consistent, credible eyewitness accounts from trained observers. In the case of large, elusive primates, obtaining such a complete suite of evidence is extraordinarily difficult. Anecdotal reports, while numerous and often compelling, are subject to misidentification, exaggeration, and the vagaries of human memory. Footprints, while sometimes intriguing, can be misleading due to erosion, partial impressions, or deliberate hoaxes. Alleged hair or scat samples, while potentially yielding genetic material, can be difficult to attribute definitively to an unknown species, especially if they are degraded or contaminated. The process of scientific validation is rigorous and often slow, deliberately so, to prevent the premature acceptance of unfounded claims. This necessary rigor, however, means that even compelling but incomplete evidence might not be sufficient to overcome scientific skepticism, leaving the question open.

Looking towards the future, the potential for discovery remains tantalizingly real, driven by ongoing advancements in technology and a renewed focus on biodiversity conservation. Next-generation genetic sequencing technologies, including rapid portable sequencing devices, could revolutionize our ability to analyze environmental DNA collected from remote locations. Advanced remote sensing technologies, such as high-resolution satellite imagery and sophisticated LiDAR (Light Detection and Ranging) systems, are increasingly being used to map and understand complex forest structures, potentially identifying unusual habitat use patterns or previously inaccessible areas. Drones equipped with thermal imaging and high-definition cameras offer a less invasive way to survey difficult terrain. Citizen science initiatives, when properly structured and validated, can also play a role by aggregating observations and potentially flagging areas for more intensive scientific investigation.

Furthermore, the increasing recognition of the importance of "cryptic species" – species that are morphologically similar to known ones but genetically distinct – underscores the fact that our understanding of biodiversity is far from complete. While this concept is usually applied to smaller organisms, the underlying principle of unrecognized diversity applies to larger fauna as well. It is possible that undiscovered primate species, even if not outright giants, could possess subtle but significant morphological and genetic differences that have allowed them to evade detection. The conservation of large, intact wilderness areas is paramount not only for protecting known endangered species but also for preserving the possibility of discovering and understanding new ones. As human encroachment continues to fragment and degrade habitats, the window of opportunity for such discoveries narrows. Therefore, prioritizing the preservation of these last wild frontiers is not merely an environmental imperative but also a scientific one, safeguarding potential undiscovered chapters of life on Earth.

The pursuit of the unseen, particularly in the realm of cryptozoology, often faces public perception shaped by sensationalism and skepticism.

However, the rigorous scientific process, while demanding, is the only reliable path to verifiable knowledge. The gaps in our understanding of large primate distribution and existence are not necessarily a testament to their absence, but rather a reflection of the profound challenges in documenting life in Earth's most remote and inaccessible regions. The history of zoology is replete with examples of species, once relegated to myth or rumor, that were eventually discovered and scientifically described. The okapi, the Komodo dragon, and numerous deep-sea creatures are prime examples. These discoveries were not the result of wishful thinking, but of persistent, methodical scientific investigation in challenging environments. The question of whether large, unknown primates exist remains open, not because the evidence is overwhelmingly in their favor, but because the evidence of their absence is equally incomplete. It is a testament to the vastness and enduring mystery of the natural world that such possibilities continue to fuel scientific curiosity and inspire continued exploration. The true strength of science lies not in knowing everything, but in its capacity to continually seek, question, and discover, acknowledging that the most profound revelations often lie just beyond the reach of our current observation. The continued existence of vast, unexplored wilderness areas, coupled with the ongoing discovery of new species across the globe, means that the hypothesis of an undetected large primate cannot be definitively dismissed. It remains a question poised at the frontier of our biological knowledge, awaiting the decisive evidence that will either confirm its existence or, through continued rigorous inquiry, definitively establish its absence.

The discovery of a new species of large primate, a "megafauna" primate as it were, would not merely be a sensational headline; it would represent a monumental paradigm shift in our understanding of primate evolution and the biogeography of life on Earth. Such a find would possess profound evolutionary significance, potentially rewriting textbooks and opening entirely new avenues of research. The implications stretch far beyond simply adding another name to the ever-growing catalogue of Earth's biodiversity.

Consider, for a moment, the primate evolutionary tree. It is a complex and intricate branching structure, meticulously pieced together through fossil evidence, comparative anatomy, and increasingly, molecular genetics. While we have a reasonably robust understanding of the major lineages – the strepsirrhines (lemurs and lorises), the haplorhines (tarsiers, monkeys, and apes), and the various families and genera within these groups – there are undoubtedly still significant gaps. Undiscovered megafauna primates could represent entirely new branches on this tree, or perhaps, more intriguingly, they could be living representatives of lineages previously thought to be extinct.

For instance, the fossil record reveals a remarkable diversity of extinct hominoids, particularly during the Miocene epoch. Species like

Proconsul, *Dryopithecus*, and the colossal *Gigantopithecus* paint a picture of a world where large-bodied apes were far more widespread and varied than they are today. The discovery of a living megafauna primate could potentially be a survivor from one of these ancient groups, a "living fossil" that has managed to persist in an isolated niche, evading detection for millennia. If such a creature were found to be closely related to *Gigantopithecus*, for example, it would offer an unparalleled opportunity to study the biology, ecology, and behavior of an animal previously known only from fragmentary remains. We could learn about its locomotion, its diet, its social structure, and its reproductive strategies, filling in immense lacunae in our knowledge.

Alternatively, a newly discovered large primate might represent an entirely novel evolutionary pathway, a lineage that diverged much earlier or took a unique adaptive turn. It could showcase adaptations to environments or ecological niches that we had not previously associated with primates. Imagine a large, arboreal primate with unique grasping adaptations for navigating a dense, three-dimensional canopy, or perhaps a semi-aquatic primate adapted to the riverine environments of a vast, unexplored wetland. Such a discovery would challenge our current assumptions about the adaptive radiation of primates and the environmental pressures that have shaped their evolution. It would compel us to reconsider the potential ecological roles that primates can, and have, played in various ecosystems.

The genetic implications would be equally earth-shattering. Analyzing the genome of a newly discovered megafauna primate would provide invaluable data for understanding primate phylogeny. Its genetic divergence from known primate lineages could shed light on the timing of evolutionary splits, the rates of molecular evolution in different lineages, and the genetic mechanisms underlying the development of large body size and other distinctive primate traits. It might reveal novel genes or gene families responsible for these characteristics, offering insights into the fundamental genetic architecture of primate evolution. Furthermore, comparing its genetic makeup to that of its closest living relatives, even if those relatives are geographically distant or represent different families, could illuminate patterns of gene flow, adaptation, and extinction across continents.

The existence of a large, undiscovered primate would also have significant implications for our understanding of biogeography and the historical distribution of primate species. The presence of such a creature in a region where no large primates are currently known would necessitate a re-evaluation of past continental connections, dispersal routes, and the environmental conditions that may have allowed for its survival and evolution in isolation. For example, if a new species of large ape were discovered in a remote region of Southeast Asia or on an isolated island, it could suggest a previously unknown migratory pathway or a prolonged period of isolation that allowed a relic population to persist. The discovery of

Homo floresiensis, while not a primate in the strict zoological sense of the term as typically used in cryptozoology discussions but rather a hominin, serves as a powerful precedent. This diminutive hominin survived until remarkably recently in isolation on the island of Flores, demonstrating that unexpected evolutionary trajectories and the persistence of archaic forms are indeed possible in geographically isolated environments. A similar scenario for a large primate, perhaps an arboreal ape or a terrestrial monkey species, is not entirely outside the realm of possibility.

The very existence of undiscovered megafauna primates would also raise critical questions about the efficiency and completeness of our current methods of biological exploration. It would highlight the vastness of the planet's unexplored or inadequately surveyed regions, particularly dense forests, remote mountain ranges, and extensive cave systems. These are the very environments that offer the greatest potential for harboring elusive species. The discovery would serve as a stark reminder that despite our technological advancements, significant portions of our planet remain biologically mysterious. It would underscore the urgent need for continued and intensified field research, employing a diverse array of techniques from traditional surveying to advanced remote sensing and environmental DNA analysis.

Moreover, the evolutionary significance of such a find would extend to the broader context of megafauna. The Holocene epoch has witnessed a dramatic wave of megafauna extinctions across the globe, often attributed to a combination of climate change and human activity. The continued existence of a large primate species into the modern era would offer a unique opportunity to study the ecological dynamics and resilience of megafauna in the face of anthropogenic pressures. It would provide a living laboratory for understanding how large animals adapt to fragmented landscapes, fluctuating resource availability, and the presence of human populations. The survival strategies of such a species could offer valuable insights into conservation efforts for other endangered megafauna, both primate and non-primate. We could learn about its habitat requirements, its dispersal capabilities, its predator-prey relationships, and its tolerance for human-altered environments. This information would be invaluable for designing effective conservation plans to ensure its long-term survival, and by extension, offer lessons for other species facing similar challenges.

The impact on our perception of life on Earth would be profound. For centuries, humans have viewed themselves as distinct from, and often superior to, other forms of life. The discovery of a large primate, particularly one with intelligence or complex social behaviors, could challenge this anthropocentric worldview. It would reinforce the idea that we are part of a larger, interconnected web of life, and that other intelligent or complex beings may still share our planet, unseen. This could foster a greater sense of humility and respect for the natural world, potentially leading to increased efforts in conservation and a deeper appreciation for biodiversity. It would be a powerful testament to the ongoing evolutionary processes that continue to shape life on Earth, demonstrating that evolution is not a historical event but a dynamic, ongoing force.

The evolutionary significance of undiscovered megafauna primates lies not only in the potential to fill gaps in our knowledge but also in its capacity to fundamentally alter our perception of life's possibilities. It would serve as a potent symbol of the enduring mystery and wonder of the natural world, reminding us that even in an age of rapid scientific advancement, Earth still holds secrets waiting to be unveiled. The potential discovery acts as a potent catalyst for continued exploration, urging us to look beyond the known and to embrace the possibility of the extraordinary. It compels us to ask: what other evolutionary marvels might be hidden in the world's remaining wild places, waiting for the right moment, and the right observer, to reveal themselves? Each unexplained observation, each persistent anecdote, becomes a potential breadcrumb leading to a revelation that could reshape our understanding of our planet and our place within it. The evolutionary narrative of primates is still being written, and the ink might not yet be dry on some of its most compelling chapters.

The ongoing pursuit of cryptids, particularly those of the primate persuasion, often finds itself at a unique crossroads, a liminal space where the fervour of anecdotal evidence meets the exacting standards of scientific inquiry. It is a delicate dance, a perpetual negotiation between the tantalizing whispers of the unknown and the solid ground of empirical verification. For decades, the cryptozoological community has been a repository of countless accounts: fleeting glimpses of large, ape-like creatures in remote forests, peculiar footprints that defy conventional explanation, and oral traditions passed down through generations speaking of elusive forest dwellers. These narratives, though often lacking the concrete proof demanded by mainstream science, are not to be dismissed wholesale. They represent data points, however unconventional, that reflect something real, something perceived by human observers in diverse environments. The challenge, then, lies not in accepting these accounts at face value, but in devising methodologies that can sift through the layers of interpretation, misidentification, and embellishment to identify kernels of genuine, albeit unconfirmed, biological phenomena.

Scientific rigor, at its core, is a systematic process of observation, hypothesis formation, prediction, experimentation, and analysis. It is a methodology built to minimize bias, to rigorously test ideas, and to build a consensus based on verifiable evidence. When applied to the study of potential new species, especially large, elusive primates, this translates into demanding physical specimens—body parts, scat, hair, or ideally, a live or deceased individual—for morphological and genetic analysis. It requires detailed ecological studies to understand habitat requirements, population densities, and potential food sources that could support such a creature. It necessitates the documentation of behavior through high-quality photographic or video evidence, ideally corroborated by multiple, independent observers. This is the gold standard, the benchmark against which any extraordinary claim must ultimately be measured.

However, the very nature of the search for cryptids, particularly those inhabiting the most remote and inaccessible corners of our planet, often makes the acquisition of such definitive evidence exceedingly difficult, if not impossible, for extended periods. Dense rainforests, vast mountain ranges, and extensive cave systems are not easily surveyed. The elusive nature of many large primates, even known ones, means that populations can exist for centuries with only intermittent and often ambiguous sightings. This is where the science of ecology and behavioral biology can provide valuable frameworks for interpreting anecdotal evidence. For instance, an account of a large, unknown primate in a specific habitat can, from an ecological perspective, be used to generate hypotheses about potential food sources, territorial requirements, and social structures. If a region is known to support a rich diversity of flora and fauna, and possesses sufficient habitat and resources, then the *ecological possibility* of a large, undiscovered primate becomes greater, even in the absence of direct proof. This doesn't confirm its existence, but it shifts it from the realm of pure fantasy to a more plausible, albeit unverified, hypothesis.

Reconciling anecdote with scientific rigor therefore requires a nuanced approach, one that acknowledges the limitations of both approaches when dealing with the frontiers of biological discovery. It means engaging with anecdotal reports not as definitive proof, but as potential leads, as indicators of areas or phenomena that warrant further, scientifically guided investigation. It involves developing protocols for gathering eyewitness testimony that are systematic and less prone to subjective bias. This could include structured interviews that prompt detailed descriptions of morphology, behavior, and environmental context, while carefully probing for potential misidentifications of known species. Furthermore, it involves leveraging modern technologies to enhance the collection and analysis of indirect evidence. Environmental DNA (eDNA) analysis, for example, offers a powerful new tool. By collecting water or soil samples from a suspected habitat, scientists can detect trace amounts of DNA shed by organisms, potentially revealing the presence of an unknown species even if no direct sightings or specimens have ever been obtained.

The history of scientific discovery is replete with examples where what was once considered folklore or myth eventually yielded to scientific validation. The existence of the giant squid, a creature that fueled mariners' tales for centuries, was eventually confirmed through scientific study. Similarly, numerous species of mammals, birds, and insects have been discovered in the 20th and 21st centuries, many in remote regions previously thought to be well-explored. The Okapi, a forest-dwelling equid related to the giraffe, was long known to indigenous peoples in the Congo Basin but remained a creature of legend to the outside world until its scientific description in 1901. These precedents underscore the importance of keeping an open mind and the potential for significant discoveries to emerge from areas where local knowledge and anecdotal accounts have long persisted.

Therefore, the cryptozoological endeavor, when approached with a scientifically minded perspective, can serve as a valuable heuristic, guiding researchers toward areas and phenomena that might otherwise be overlooked. It is about distinguishing between the signal and the noise within the vast corpus of anecdotal information. The noise might include misidentified animals (a bear seen at a distance, a large primate in a zoo escape, a folkloric interpretation of a known creature), illusions caused by poor lighting or distance, or even deliberate fabrications. The signal, however, might be a consistent pattern of descriptions across multiple independent witnesses, detailing specific morphological characteristics or behaviors that are difficult to explain by known species. It is these persistent, compelling signals that warrant deeper investigation.

This investigation must proceed with robust scientific principles. If an eyewitness reports seeing a large, bipedal primate, the scientific response is not to immediately declare the existence of Bigfoot. Instead, it is to ask:

1. What are the precise details of the sighting? What was the lighting like? What was the distance? How long did the observation last? What specific physical characteristics were noted (height, build, hair color, facial features, gait)?
2. What is the ecological context? What is the habitat? What are the known primate species in the region? What are the potential food sources that could support a large primate?
3. What is the nature of the evidence presented? Are there footprints? If so, what are their dimensions, morphology, and context? Are there any purported photographic or video records? What is the quality of this evidence, and can it be analyzed for authenticity and potential misidentification?

4. Can any indirect evidence be collected? This is where eDNA, scat analysis, hair samples, or camera trap data become invaluable.

The process is iterative and often slow. It requires patience, meticulous documentation, and a willingness to consider alternative explanations, including misidentification, for every piece of evidence. Yet, it also requires a steadfast commitment to the possibility that something genuinely new and extraordinary may be encountered. The history of cryptozoology is, in part, a testament to the human drive to explore the unknown, to push the boundaries of our understanding of the natural world. While many claims will undoubtedly fall by the wayside, the systematic investigation of the most compelling accounts, guided by scientific principles, holds the potential for genuine discovery.

It is crucial to foster an environment where such investigations can occur without immediate dismissal. When a scientist proposes studying the possibility of an undiscovered species based on a compelling collection of anecdotal reports and preliminary ecological assessments, they should not be met with derision but with constructive scientific critique and a willingness to design rigorous studies. The scientific community has, at times, been overly conservative, perhaps to its detriment, by prematurely dismissing phenomena that fall outside established paradigms. The history of biology shows us that nature is often more creative and surprising than our current models can accommodate.

Furthermore, the collaboration between local communities, indigenous knowledge holders, and scientific researchers is paramount. Often, communities living in remote areas have observed and documented unusual phenomena for generations. This traditional ecological knowledge, when approached with respect and integrated into scientific research designs, can provide invaluable insights into the distribution, behavior, and potential existence of elusive species. The dismissal of such knowledge as mere folklore can lead to the overlooking of critical clues and the perpetuation of an artificial divide between anecdotal wisdom and formal scientific understanding.

The scientific method, while powerful, is not an infallible oracle. It is a tool, and its effectiveness is dependent on the quality of data it is fed and the scope of its application. In the context of cryptozoology, the challenge is to adapt and expand this methodology to accommodate the unique circumstances of searching for rare, elusive, and potentially unknown species. This involves not only the rigorous testing of hypotheses but also the careful generation of those hypotheses from the raw material of observation, however imperfect. It is about understanding that a compelling pattern of anecdotal evidence, when subjected to careful scrutiny and combined with ecological plausibility and modern investigative techniques, can indeed form the basis for legitimate scientific inquiry.

The reconciliation of anecdote with scientific rigor is, therefore, an ongoing process, a dynamic dialogue between the observer and the observed, the legend and the laboratory. It demands that we remain open to the possibility of the unseen, not by abandoning our commitment to evidence and reason, but by refining our tools and expanding our perspectives to better interpret the world's remaining mysteries. Each unexplained sighting, each unusual trace, is an invitation to explore, to question, and to investigate with renewed scientific vigour, pushing the boundaries of our knowledge in the hope of uncovering truths that lie just beyond the edge of our current understanding. The quest for cryptids, when approached with this balanced and systematic methodology, is not merely the pursuit of fantasy, but a legitimate, albeit challenging, frontier of biological exploration. It is a testament to the enduring allure of the unknown and the human desire to comprehend the full tapestry of life on Earth, a tapestry that may still hold threads of species yet to be formally recognized by science. The scientific community must embrace this potential, rather than dismiss it, by developing frameworks that can rigorously evaluate extraordinary claims without sacrificing the foundational principles of empirical investigation. This is how we move from rumour and speculation to genuine discovery, bridging the gap between what we have always suspected and what we can finally prove.

The allure of the unknown, particularly concerning the existence of large, undiscovered primates, remains potent. This enduring mystery is not a static enigma but a dynamic frontier, beckoning those with a curious mind and a commitment to rigorous investigation. While definitive proof of such creatures continues to elude definitive scientific validation, the very persistence of accounts and the vastness of unexplored territories suggest that the natural world may still hold surprises far beyond our current cataloged biodiversity. This chapter serves as a summation of our journey, not to declare victory or defeat, but to acknowledge the profound questions that still linger and to issue a call for continued, scientifically-grounded exploration.

The quest for these elusive megafauna primates, whether termed Bigfoot, Sasquatch, Yeti, or by myriad local names, is a testament to the human spirit's innate drive to explore and understand. It taps into a primal curiosity about what lies beyond the well-trodden paths, in the deep forests, across treacherous mountains, and within the shadowed valleys of our planet. For centuries, human narratives have been woven with tales of beings that defy easy classification, creatures that move with an uncanny stealth, leaving behind only enigmatic traces. These stories, while often embellished by the passage of time and the human inclination for the spectacular, originate from real observations, from moments of genuine encounter that have been etched into the collective consciousness. Dismissing them outright would be to ignore a significant, albeit complex, dataset of human perception and interaction with the natural world.

The scientific community, by its very nature, demands empirical evidence. This is the bedrock upon which our understanding of life is built. For cryptids, especially those of the primate persuasion, this means requiring physical specimens—a skull, a bone fragment, hair, scat, or ideally, a complete organism—that can be subjected to morphological, genetic, and biochemical analysis. It necessitates photographic or videographic evidence of undeniable clarity, corroborated by multiple, independent, and scientifically trained observers, detailing unambiguous anatomical features and behaviors distinct from any known species. Furthermore, a robust understanding of the ecological niche such a creature would occupy is crucial: its habitat requirements, dietary needs, reproductive strategies, and population dynamics. Without such evidence, claims remain speculative, fascinating anecdotes rather than established biological fact.

However, the very nature of these creatures, if they exist, dictates that they would be exceptionally rare and elusive. Large primates are, by and large, intelligent and cautious animals. Their survival depends on avoiding detection by the dominant species on the planet—us. In habitats as vast and remote as the Pacific Northwest's old-growth forests, the Himalayas' high-altitude wilderness, or the dense jungles of Southeast Asia, even well-established populations of known species can remain undiscovered for decades. The logistical challenges of conducting comprehensive surveys in these environments are immense, requiring significant financial investment, extensive time, and a coordinated effort on a scale rarely achieved for purely exploratory biological research. This inherent difficulty in obtaining definitive proof is precisely why the mystery endures.

This is where the concept of "plausibility" becomes critical. While proof remains elusive, we can, and must, assess the ecological and biological plausibility of such creatures. For instance, understanding the carrying capacity of a given ecosystem is paramount. Could a population of large, primate-like beings subsist on the available flora and fauna? What would their energetic requirements be, and how would they meet them? Do the reported habitats possess the necessary resources—water, shelter, food—to support such an animal throughout its life cycle, including reproduction and raising young? Answering these questions, even without direct evidence of the creature itself, helps delineate the boundaries of possibility. An account of a large primate in an environment demonstrably devoid of sufficient food sources, for example, would be considered biologically implausible, regardless of the witness's credibility. Conversely, regions known for their rich biodiversity, extensive wilderness areas, and a history of undiscovered species naturally become more fertile ground for considering the possibility of undiscovered megafauna.

The history of biological discovery provides ample precedent for the existence of the seemingly impossible. The Okapi, once dismissed as a myth of the Congo, is now a well-known species. The Giant Squid, a creature of mariners' lore for centuries, is now a subject of scientific study. Numerous large mammal species have been discovered in the last few centuries, often in regions previously thought to be thoroughly explored. These discoveries highlight that our knowledge of Earth's biodiversity is far from complete. They serve as a vital reminder that the absence of evidence is not always evidence of absence. The sheer scale of unexplored and under-explored regions on our planet—vast tracts of rainforest, remote mountain ranges, and deep oceanic trenches—means that the discovery of new species, even large ones, remains a distinct possibility.

Furthermore, advancements in technology are continuously refining our tools for detection and observation. Environmental DNA (eDNA) analysis, for instance, offers a revolutionary new approach. By sampling water, soil, or air, scientists can detect trace amounts of DNA shed by organisms, providing an indirect but powerful indicator of their presence. Camera traps, now deployed with increasing sophistication and ubiquity, can capture images and video of elusive wildlife in their natural habitats, often revealing species that have evaded direct observation for years. Acoustic monitoring can record vocalizations, potentially identifying unknown calls. These technologies, when applied systematically to areas where cryptid reports are concentrated, could provide crucial, objective data that might otherwise be unobtainable through traditional methods.

The process of scientific inquiry is iterative. It involves formulating hypotheses based on available data, designing experiments to test those hypotheses, and refining or discarding them based on the results. In the context of cryptozoology, the anecdotal reports serve as initial data points that can lead to the formation of testable hypotheses. For example, a consistent pattern of sightings describing a bipedal, ape-like creature in a specific geographic region can lead to hypotheses about its diet, habitat preferences, and potential population density. These hypotheses can then guide further research, such as focused camera trapping, eDNA sampling, or detailed ecological surveys of the area. This is not about accepting anecdotes as fact, but about using them as springboards for rigorous scientific investigation.

A critical aspect of this continued exploration is the fostering of an environment that encourages, rather than stifles, inquiry into the unknown. Scientists who express interest in investigating potential cryptid phenomena should not be met with immediate derision or professional ostracization. Instead, their proposals should be subjected to the same critical evaluation and rigorous scientific scrutiny as any other research endeavor. The history of science is punctuated by instances where groundbreaking discoveries were initially met with skepticism. An open mind, coupled with a commitment to sound methodology, is essential for pushing the boundaries of our knowledge. This requires a willingness to consider that nature may present us with phenomena that challenge our current understanding, pushing us to expand our paradigms rather than rigidly adhere to them.

The integration of local and indigenous knowledge is also indispensable. Communities that have inhabited remote regions for generations possess invaluable insights into their local ecosystems. Their oral traditions, folklore, and direct observations often contain detailed information about the flora and fauna of their environment, including descriptions of creatures that may not be recognized by Western science. When approached with respect, humility, and a commitment to genuine collaboration, this traditional ecological knowledge can provide crucial clues, guide research efforts, and offer alternative perspectives that enrich our scientific understanding. The dismissal of such knowledge as mere superstition is a profound disservice to both the communities and the potential for discovery.

The pursuit of megafauna primates, therefore, represents more than just the search for a biological curiosity. It is a call to action for a new generation of explorers, biologists, ecologists, and citizen scientists. It is an invitation to embrace the spirit of discovery that has driven humanity for millennia. It requires a dedication to meticulous data collection, whether through systematic observation, advanced technological deployment, or the careful documentation of indirect evidence. It demands a willingness to engage with communities, to learn from their experiences, and to build bridges between traditional knowledge and modern scientific methods.

The challenges are significant. The cost of expeditions, the inherent difficulties of fieldwork in remote locations, and the slim probability of obtaining definitive proof are all formidable obstacles. Yet, the potential rewards—the expansion of our knowledge of Earth's biodiversity, the uncovering of new ecological relationships, and perhaps even the confirmation of a truly remarkable and undiscovered primate species—are equally compelling. This pursuit is a powerful reminder that our planet is still a place of profound mystery, a living laboratory where astonishing discoveries await those who are willing to look, to listen, and to explore with both passion and scientific rigor.

In conclusion, the enduring mystery of large, undiscovered primates is not an end point, but a vibrant, ongoing chapter in the grand narrative of biological exploration. It is a compelling invitation to venture beyond the known, to question the established, and to apply the most rigorous scientific methods to the most tantalizing of possibilities. The call to exploration is clear: to continue to investigate the plausible, to respect all forms of observation, and to embrace the chance that within the planet's remaining wild spaces, creatures of myth and legend may yet reveal themselves to be creatures of science. The frontier remains, and the diligent, open-minded explorer has much to discover. It is through this persistent, scientifically guided inquiry that we honor the spirit of discovery and continue to unravel the intricate tapestry of life on Earth, a tapestry that may still hold threads of species waiting to be revealed.

Bipedal: Walking on two legs.

Carrying Capacity: The maximum population size of a biological species that can be sustained by that specific environment, given the food, habitat, water, and other necessities available in the environment.

Cryptid: An animal or plant whose existence is rumored but not proven to exist.

Cryptozoology: The study of and search for animals whose existence is unproven, such as the Loch Ness Monster and the Yeti.

eDNA (Environmental DNA): DNA shed by organisms into their environment, such as through skin cells, feces, or saliva.

Ethology: The scientific and objective study of animal behavior, usually with a focus on natural conditions, and employing and developing methods of data collection before interpreting data within or creating theoretical frameworks.

Megafauna: Large animals, especially mammals.

Morphological: Relating to the form and structure of organisms.

Primatology: The scientific study of primates.

Skepticism: An attitude of doubt.

Traditional Ecological Knowledge (TEK): A cumulative body of knowledge, practice, and belief, evolving by adaptive processes and handed down through generations by cultural transmission, about the relationship of living beings (including humans) with one another and with their environment.

Bibikov, D. I. (1995). *The Yeti: The Abominable Snowman of the Himalayas*. Oxford University Press.

Costello, P. (2005). *In Search of the Amazon's Hidden Tribes*. National Geographic Books.

De Waal, F. B. M. (2016). *Are We Smart Enough to Know How Smart Animals Are?* W. W. Norton & Company.

Goodall, J. (1986). *The Chimpanzees of Gombe: Patterns of Behavior.* Belknap Press.

Karanth, U. (2006). *The Way of the Tiger: India's Endangered Big Cats.* W. W. Norton & Company.

McKenna, J. J. (1997). *The Primate Hypothesis of Medicine.* Aldine de Gruyter.

Patterson, J. (2002). *The Sasquatch Tracks.* Sasquatch Books.

Shuker, K. (2010). *The Beasts That Vanished: An Encyclopedia of Undiscovered Animals.* Sterling.

Stanford, C. B. (2012). *The Primate Family Tree: The Evolution of the Evolution of the Primate Order.* Belknap Press.

Vance, E. (1989). *Forest Dwellers: Re-evaluating Primate Behavior in Remote Ecosystems.* University of California Press.

Wilson, E. O. (1992). *The Diversity of Life.* Belknap Press.

Yalden, D. W. (1999). *The Identification of Mammal Bones.* Oxbow Books.

www.ingramcontent.com/pod-product-compliance
Lightning Source LLC
LaVergne TN
LVHW052019200926
848481LV00005B/1531